YEW

A HISTORY

FRED HAGENEDER

Forewords by David Bellamy
and Robert Hardy

With photographs by Andy McGeeney, Christian Wolf, Chris Worrall, Tim Hills, Edward Parker, Archie Miles, Christopher Cornwell, Paul Greenwood, the author and others

SUTTON PUBLISHING

Morning sun in the yew grove that covers Hambledon Hill, the site of an Iron Age hillfort in Dorset.

Page 2 *Female ancient yew at Tandridge, Surrey, girth 1,077cm just above the bulging roots (1999).*

First published in 2007 by
Sutton Publishing Limited · Phoenix Mill
Thrupp · Stroud · Gloucestershire · GL5 2BU

British Library Cataloguing in Publication Data
A catalogue record for this book is available from the British Library.

Hardback ISBN 978-0-7509-4597-4
Paperback ISBN 978-0-7509-4598-1

Picture editor: Fred Hageneder
Typeset in Optima
Typesetting and origination by
Dragon Design UK Ltd
Printed and bound in England

While every effort has been made to ensure the accuracy of information in this book, in no way is any material presented here offered as a substitute for professional medical care or attention by a qualified medical practitioner.

Note on age estimates of (old) yew trees

For the reasons discussed in Chapter 20, this book does not give age estimates for individual yews. Instead, they are simply classified as young, mature, old and ancient.

• ***Young*** refers to trees of small and medium size, responding to life stage 2 (see pp. 75–6).

• ***Mature*** refers to fully grown trees with compact trunks (life stage 3).

• ***Old*** refers to hollowing trees (life stage 4), in Britain usually from about 15ft (*c.* 4.50m) girth and above.

• ***Ancient*** refers to life stages 5 to 7, and in most cases to ages certainly with four-digit numbers.

For Carran
with very best wishes!
[illegible]
23 Jan 2010

Remember a saying of Homer's, and cherish it –
'A good messenger,' he said, 'heightens
The honour of any errand.'
Even the Muse's stature
Is more, if she be well reported.

Pindar, *Pythian Ode*, IV, XIII

He who cannot be far-sighted,
Nor three thousand years assay,
Inexperienced stays benighted,
Let him live from day to day.

Johann Wolfgang von Goethe,
West-eastern Divan

CONTENTS

A HISTORY

FOREWORD *by* DAVID BELLAMY

This is a book to take pride of place in your library, allowing you and your descendants to glean the wisdom of its words and savour their meaning. A landmark work of great scholarship, it is the best monograph on a single plant species I have ever read and I will of course browse it again and again. Like all great monographs, it is a working manual of research, posing as many questions as it answers, while willing you to join the quest for understanding that concerns the living mantle of Mother Earth of which we are but a transient part.

Fred does both the Yew and 'You' the courtesy of not calling his amazing book a natural history, for both are inextricably entwined with the roots of scholarship of *Homo sapiens*. Neither does he fall into the trap of concerning himself with the age of individual trees, for their remarkable powers of regeneration give them the stamp of immortality.

As one who has dabbled in the botany of the yew for over half a century, I am awestruck to be back on the trail of perhaps the real Golden Bough. My first ports of call are the ancient yew forests of Turkey, and I wish I could take Fred, Frazer and Goethe with me.

David Bellamy
Bedburn, 2007

One of the oldest trees of continental Europe: the yew at Barondillo o Valhondillo in the Sierra Guadarrama.

FOREWORD *by* ROBERT HARDY

This book is a milestone in the long journey of archaeology and history. Not only is it an example of what research and scholarship can achieve, it is also majestic in its breadth and scope and unique as a complete study of a subject. Much has been written through the years about the strange ancient family of yew trees, but no work has previously so totally correlated examination of every aspect of the yew's history. From pagan worship, attributions of magic, extraordinary millennial longevity, survival in adverse conditions, the history of its absorption by the Christian religion, often in the very places where it was previously part of pagan ritual, is meticulously traced and astonishingly illuminated.

Its various uses are investigated, perhaps the most striking of which is its adoption as the raw material for the manufacture of millions of longbows, as hunting weapons from thousands of years before the Christian calendar to its adaptation as a weapon of war from the year one to the end of the sixteenth century. Fred Hageneder's study traces the gradual development of this potent, increasingly powerful weapon until its great climax as a battle winner, all but a war winner, in the long, fearsome conflicts that England and Wales fought against France and Scotland. As an historian of those conflicts, what has astonished me in this study is the research – not yet complete says the author – into the major cause of the decline of the yew longbow, long before its power, its accuracy, its range and speed of shooting were overtaken by the military handgun. Hageneder reveals the extent to which the endless military demands for bow timber, by the Anglo-Welsh particularly, so depleted the yew stands of Europe that prices became prohibitive, felling was severely restricted, export rigorously licensed and the longbow, deprived of its vital component, gave early way to the gun and the pistol.

But that is only a small part of the fascinating story of a natural history and man's religious, military and decorative use of a unique tree that this brilliant piece of work unfolds.

Robert Hardy CBE, FSA
Charlbury, 2007

Yew seedling at Tomiyoshi, Japan.

Part I Nature

CHAPTER 1
BACCATA – 'BERRY-BEARING'

A conifer that bears scarlet 'berries' with sweet juicy pulp instead of cones? A tree species that grows in rainy Edinburgh as well as in hot Istanbul, in Canada and Scandinavia but also in Mexico, north-west Africa and Sumatra? A tree whose scope of altitude ranges from the coastal areas of the British Isles and North America to the mountains of Japan, and even higher in the Himalayas? What kind of tree is this that is highly poisonous in all its parts except the red fruit pulp, and yet intensely browsed by wild and domesticated animals alike? A tree that is included in the lists of threatened species of a number of nations, and yet is not protected in just as many countries?

One thing is for sure: the yew has always stimulated the human psyche and triggered many questions. We are still far from being able to give all the answers, and, indeed, often an answer will only bring new questions. The yew continues to surprise.

The challenges that the yew poses to modern science begin with its position in the classification of plants and with the definition of its status as a species.

***1.1** The red aril that surrounds the seed of* Taxus *is not a berry.*

A CONIFER?

The yew's position within the class of the Coniferophytes (plants with forked or needle-shaped leaves) is controversial. Since Stewart (1983), the Taxales ('yew-like') are an independent order *next to* the Coniferales ('cone-bearing'). They comprise the genera *Taxus, Austrotaxus, Pseudotaxus, Torreya* and *Amenotaxus*.

On the other hand, the characteristic of single ovules (see Appendix I) and of the special form of the fleshy aril (*arillus*) connect the family of the Taxaceae ('yew plants') with the Podocarpaceae (a large family of mainly southern hemisphere conifers) and the Cephalotaxaceae (a small grouping of conifers which, apart from two species of *Torreya* in the southern USA, is restricted to east Asia) and hence they could be combined with these in the sub-order Taxineae (next to the sub-order Pineae) in the order Pinales.

1.2 Taxus *in a botanical book from 1888.*

Standard botanical works of more recent date[1] have seized upon the idea of separating the family group Taxidae from the Pinidae, the true conifers. The Podocarpaceae and the Cephalotaxaceae also produce a fleshy aril. This aril originates from the stalk or the base of the ovule and *not* from the integument (see diagram p. 38), hence these plants clearly belong to the gymnosperms. In *Podocarpus* and *Cephalotaxus* (plum yew), the flower develops from a seed scale located in the axil of a cover scale. The close similarity between *Taxus* and *Cephalotaxus* suggests that *Taxus*, too, must once have had the same such fusion of seed scale and cover scale, which then disappeared in the course of evolution. Hence the new classification joins the Taxaceae to the other families of the Pinidae (conifers, cone-bearing), which indeed have cones.

Finally, the yew has now become a conifer, not because of new discoveries about the tree itself, but because the definition of conifer has been extended.

ONE OR MANY?

There is no agreement among botanists as to whether or not the various members of the genus *Taxus* are species, subspecies, or mere varieties of *Taxus baccata* L., the Common or English or European yew.[2]

The viewpoint of a single species is inviting. The various (sub)species are incredibly similar to each other, while on the other hand an unimaginable spectrum of morphological plasticity can be found just within *Taxus baccata*. Also, when different 'species' grow near each other, hybrids frequently occur.[3] The fact that over seventy garden cultivars[4] have been bred from *T. baccata* gives further confirmation of the rich potential and adaptability of its genetics. However, a taxonomic revision of the genus has been proposed for some time now by a long-time collector of plant samples for the US National Cancer Institute's (NCI) natural products screening programme,[5] Richard W. Spjut (Bakersfield, CA). In August 2000, Spjut presented a pictorial key to twenty-four species and fifty-five varieties at the Botany 2000 conference held in Portland, Oregon. His taxonomy is based on morphological characteristics (see Appendix II).[6]

1.3 *Distribution zones of the* Taxus *genus worldwide. (After Ferguson 1978, de Laubenfels 1988, modified)*

CHAPTER 2

EVOLUTION AND CLIMATE HISTORY

FOSSIL RECORDS

The earliest conifers (Coniferales) date back to the late Carboniferous (360–286 million years ago) and Permian (286–245 million years ago) geological periods. The Taxads, which include the Taxaceae, are believed to have originated from early cone-bearing plants of the Voltziaceae family in the Early Triassic period (beginning 248 million years ago). *Palaeotaxus rediviva*, the predecessor of the genus *Taxus* found in strata dating back 200 million years,[1] was widely distributed across the global landmass before its separation into the continents we know today. *Marskea jurassica* from the Upper Jurassic is about 140 million years old and contains many characteristics of the modern genus *Taxus*.[2] Since the Cenozoic era (66.4 million years–present) the genus *Taxus* has become restricted to the northern hemisphere, because of the continental drift. More recent fossils include *Taxus grandis, T. engelhardtii* and *T. inopinata* from the Middle Oligocene *c.* 32 million years ago. *Taxus baccata* itself appears in the Upper Miocene *c.* 15 million years ago.[3]

Various fossil specimens from the Yorkshire Lower Deltaic, formerly known as *Taxus jurassica*, were classified as *Marskea jurassica* in 1958. *Marskea* combines characteristics from among the diverse genera of the *Taxaceae*, but it also differs from each other genus therein in at least one important aspect. Its microscopic differences from *Taxus* include, among others, *Marskea's* monocyclic stomata, the sometimes wavy shapes of their cell walls, and its ovules being borne singly in leaf axils and having smooth stalks while those of *Taxus* have tiny scales. On the other hand, fluctuations throughout both genera are so great that another contemporary, *Taxus harisii*, is considered but a form of *Marskea jurassica*.[4] This difficulty in clearly defining the yew is to be found in almost all scientific disciplines engaging with *Taxus*.[5]

THE ICE AGES

Taxus pollen grains are a tricky object of study in vegetation history. In pollen samples from peat or lacustrine deposits, they can be easily overlooked by inexperienced researchers because of their small size or their resemblance to the pollen of poplar, oak, and grasses of the sedge family (*Populus* sp., *Quercus* sp., Cyperaceae), among others. Nevertheless, pollen records show that the yew, to varying degrees, was a constitutive element of the European mixed forest during various interglacial periods. Safe, but by no means the earliest, evidence comes from the Cromerian Interglacial period (700,000–450,000 years ago), but the greatest yew density occurred in the warm, oceanic climate of the Hoxnian (400,000–367,000 years ago). In north-western Europe, yew was associated with ash (*Fraxinus*) and alder (*Alnus*) fen woodland with wet soil conditions.[6]

During the last interglacial period, the Eemian (128,000–115,000 years ago), *Taxus* pollen reached significant values, comprising up to 20 per cent of all tree pollen precipitation. For 2–3,000 years the yew became an important species in the mixed (pine-)oak-hazel-woodland.[7] In the Northern Alps, local values even reach 65 per cent (Mondsee lake, Salzkammergut, Austria) and 80 per cent (eastern Upper Bavaria), which indicates that *Taxus* constituted about half of the woodland trees.[8] Ultimately, however, a steady decline followed, as the climate changed towards the next glacial period.

POST-GLACIAL

During the last Ice Age (*c*. 115000–11000 BCE), the yew, like the other woodland species, was once again pushed to the southern fringes of Europe (Spain, Italy and Greece). In Asia Minor, *Taxus* spent its glacial exile in the Amanus and Taurus Mountains (southern Turkey bordering north-west Syria), and from there spread north when temperatures began to rise. It probably took about two millennia to cross Anatolia and reach the Caucasus Mountains in the east and the Black Sea shores in the west.[9]

In the Western Mediterranean, *Taxus* began to migrate northwards as well. It reappeared in Germany between 7800 and 7200 BCE. It spread continuously and reached a climax during the pine-oak(-beech) mixed forest of the late warm period *c*. 3800 to 900 BCE.[10] In Britain, its post-glacial return occurred at the time of the transition from pine (*Pinus*) woodland to mixed deciduous woodland just after 5000 BCE. In the thousand years that followed, yew was widespread on calcareous peat, as in the Somerset levels in fen-carr (a type of bog with rather alkaline soil, partly covered with water and dominated by grasslike plants) with alder (*Alnus*), birch (*Betula*) and oak (*Quercus*), and in East Anglian fen woods with ash (*Fraxinus*) as well.

But the continuing warming of the climate and increasing human influence since the Neolithic kept changing the landscape and vegetation. For example, by *c*. 4600 BCE *Taxus* had largely disappeared from the eastern Swiss Plateau.[11] North of the Alps, the decline of yew on peat lands was partly compensated for by its spreading into drier woodlands with the elm decline *c*. 3800 BCE, and by finding new habitats in coppice and other areas of extensive forest management. Thus pollen records for Britain, for example, show another rise for *Taxus* around 2000 BCE. Later, however, a general decline set in, partly the result of wetter conditions but mainly because of the rise in human populations, agriculture and pasture. In the eastern German lowlands, for example, human pressure and forest destruction escalated from *c*. 1150 CE until the pine plantations that began in 1750.[12]

2.1 *(opposite page)* Palaeotaxus rediviva *Nathorst from Skromberga, Bjuv, Scania, Sweden, Late Triassic period.*
2.2 Marskea jurassica, *Yorkshire, Upper Jurassic period.*

CHAPTER 3

THE 'TREE ARCHETYPE'

Originally, the European yew occurred in the woodlands of western, southern and central Europe, the Baltic, the Atlas Mountains (north-west Africa), Asia Minor and northern Iran. The biggest stands, however, are located at the Black Sea border of Turkey, and particularly in the Caucasus Mountains, where over 130 sites are known.[1] The yew is an evergreen, non-resinous tree that grows extremely slowly, a 20–30cm annual height increment is normal in the open, and less in woodland. It rarely exceeds 15m in height, particularly in the cool maritime climate where (with much light in the open) it is more likely to extend horizontally. In the Killarney woodlands (south Ireland), for example, the canopy is 6–14m high.[2] Many of the monumental yews of northern Turkey, however, reach 20m and more, and the

Table 1 Constitutional characteristics of yew (*Taxus baccata* L.)
in comparison with Scots pine (*Pinus sylvestris* L.) as a typical pioneer species, and common beech (*Fagus sylvatica* L.) as a climax species (*after Leuthold 1998*)[3]

Ecological parameter (factors independent of location)	*pioneer constitution*	←						→	*climax constitution*
relation to climate									
general characteristics	extreme/ unstable	▲					●	■		balanced / predictable
light demand	high	▲						■	●	low
sensitivity to frost	low	▲					●	■		high
sensitivity to late frost	low	▲		●				■		high
sensitivity to drought	low	▲			●			■		high
relation to soil										
grade of soil development	low	▲		●				■		mature
water household	extreme	▲		●				■		balanced
nutrition	poor	▲				●		■		good
relation to relief	often extreme, e.g. cliff faces	▲	●					■		flat ground to moderate slopes
relation to vegetation										
inter- and intraspecific competition ability	low	▲			●			■		high
physical height	(low)	●			▲			■		tall
speed of growth	high	▲					■		●	low
relation to time										
phylogenetical age	old	▲		●				■		young
biological ageing	fast		▲				■		●	slow

▲ pine ● yew ■ beech

tallest of the old yews in the mixed forests of the Caucasus is reported to be *c*. 32m in height.

The *Taxus* genus dates back to the Upper Jurassic (about 140 million years ago), and the species *baccata* is 15 million years old. This makes the yew the *oldest tree species in Europe*. (In Asia, the oldest tree is *Ginkgo biloba*, which is 160 million years old.) But it is not only its great age in natural history that makes this tree so exceptional. The fact that it still lives today proves its remarkable adaptability. The yew occurs as a single-stemmed or multi-stemmed tree, as a bush or, in high altitudes, even as a creeper. Its extraordinary capabilities of vegetative reproduction include root suckers, the ability to regrow from cuttings, adventitious growth (buds appearing from the stem anywhere in the tree, most often at the base of the trunk), layering (branches touching the ground taking root), the growth of vertical branches (out of other branches or out of a fallen trunk), and the encasing of a decaying trunk with secondary wood growth (during the regeneration to a new trunk which is created by layering, e.g. an 'interior root', see Chapter 19). All these aspects express an almost unique ecological strategy (see box p. 23) that differs considerably from nearly all other European forest trees. Furthermore, *Taxus baccata* has a notably vast climatical and geographical range (see world map p. 13).

An investigation of the yew's ecological and phytosociological position reveals *Taxus* as occupying a rare position as a 'triple intermediary' (Leuthold, 1998):[4] ecologically, in an intermediate position between pioneer and climax species (Table 1); structurally, as a typical co-dominant tree of the second strata, mediating between the top of the canopy and the ground region; and morphologically-physiologically, as a vivid interim form between deciduous leaf-bearing trees and evergreen conifers (Table 2). Thus the yew is phylogenetically the *oldest* tree species of Europe, and at the same time, because of its incomparable vitality and its open constitution, the most *juvenile*. All these factors determine it as a 'tree archetype of Europe' (Leuthold, 1998).[5]

3.1 *A towering monumental yew at Alapli, Turkey.*
3.2 *Low-growing yew tree in Powys, Wales.*
3.3 *Creeper at Gait Barrows, Cumbria.*

Table 2 Characteristics of the yew as similar to conifers and broadleaved trees *(after Leuthold 1998, modified)*

characteristics like deciduous broadleaf trees, mesomorph *i.e. in its eco-physiological architecture and local behaviour adapted to moderate conditions, like linden or elm*		***characteristics like evergreen needle-bearing trees, scleromorph*** *i.e. in its eco-physiological architecture and local behaviour adapted to dry conditions, like Scots pine*
	shape	
after sexual maturity, tendency to a shrub-like multi-stemmed habit	highly variable and malleable, great plasticity (also among the various taxa of *Taxus*), e.g. used in topiary	in youth, coniferous habit (clear main axis)
	branches	
branches tend to the horizontal plane (leaf-like)		the single twig shows the axial-dominated, geometrical strictness and the structure typical for conifers
	foliage	
• internal architecture (anatomy) rather leaf-like • energy-efficient 'needle' composition • stomata able to rapid reaction (similar to the seasonal leaf)	dorsiventral, relatively broad, soft (among the Coniferophytina, the only needle without sclerenchyma)	• needle-shaped • evergreen, continuously able to assimilate • needles are long-lived
	seed	
single 'fruit' with coloured, juicy, sweet fleshy covering (arillus), as typical of deciduous trees (e.g. cherry, rowan, elder)	fleshy red aril	• dioeciousness as a phylogenetically old feature in trees • the seed sometimes stands upright on the twig (similar to cones)
	wood	
wood with short fibres, as typical for deciduous hardwood trees	wood with extremely narrow thracheids and spiral thickenings	great elasticity of the wood, usually a feature of long-fibred coniferous woods
	root system	
root type like in many deciduous trees: very efficient root and water transport system which ensures quick provision even in stress conditions	adapted to ground conditions: a richly branching heart-shaped root system, or wide and flat	(the root systems of conifers are generally of a simpler composition and less efficient)
	vitality	
ability of the yew to create root suckers, layering, and propagation by cuttings, as typical for deciduous trees	high ability to regenerate, extremely long-lived	The oldest tree specimen of the world (e.g. Great Basin bristlecone pine, *Pinus longaeva*, redwood, *Sequoia giganteum*) are not found among the deciduous trees (angiosperms) but among the phylogenetically older needle trees (gymnosperms). The age of yews too can reach four-digit numbers.

CHAPTER 4

CLIMATE AND ALTITUDE

The yew grows best in the moderate temperatures of the mild oceanic climate. Favourable are mild winters, cool summers, abundant rainfall and high humidity, for example, frequent mist. Severe winter cold, late frosts, or strong, cold and drying winds in exposed positions restrict yew growth. The ecological factors limiting its distribution are low temperatures in the north, severe continental climate (east of Poland, in central North America, and inland in north-east China and eastern Siberia), long droughts (e.g. Anatolia), and drought and high temperature in north-west Africa. Near these extremes, the yew becomes restricted to moist niches such as areas near bogs and marshes, in rock crevices, or in the understorey of a protecting forest. In the Mediterranean, yew is mostly found in the cool and moist higher elevations.

WATER

Taxus is often found in those areas with the highest rainfall of a region, for example, the Pacific rainforest of North America (*T. brevifolia*), the Reenadinna rainforest in south-west Ireland, and the western Taurus Mountains in southern Turkey. Particularly important is the rainfall in July and August when the leaf buds are formed, and from March to May when the leaf buds swell and open.[12] The annual wood production of the trunk is increased by abundant rain during the seasonal growth period.[13]

Table 3 Annual rainfall at various important yew stands

location	*altitude a.s.l.*	*average precipitation*
Reenadinna, Ireland[1]	20–30m	1,585mm
South Downs, Britain[2]	50–200m	800–over 1,000mm
Paterzell, Germany[3]	600–750m	1,050mm
Bakony Mountains, Hungary[4]	300–510m	795mm
Carpathians, Ukraine[5]	–	650–1,080mm
southern Crimea, Ukraine[6]	–	500–1,000mm
western Caucasus[7]	–	400–2,500mm
Hyrcanian forest, north Iran[8]	800–1,800m	580–1,850mm
Amanus Mountains, Turkey[9]	100–518m	785–1,173mm
western Taurus, Turkey[10]	20m	1,288mm
western Taurus, cedar-forests[11]	1,000–2,200m	1,500–2,000mm

Taxus, however, is moderately tolerant of drought. This is because of the swift reaction of the stomata as well as the wood structure since the narrow diameters of the xylem cells put certain limits on water transport, and hence, also on water loss. Furthermore, *Taxus* continuously invests into the root system,[14] where resources are being stored. If drought does cause damage, it can be seen in the needles older than two years (particularly those in the top of the crown) turning yellow from their base upwards before falling,[15] and in the adventitious growth withering and dying.

TEMPERATURE

The temperature range for photosynthesis in yew is extraordinarily large, that of all other European tree species is narrower and falls within this broad band of *Taxus*. Among other things, this means for forest yews that they can also assimilate in winter when they receive more light (compare 'Photosynthesis') because the leaf trees of the upper tier are bare. Storing the resources (carbohydrates, mainly as sugar compounds) produced in this season enables *Taxus* to balance its rather low photosynthetic effectivity during the summer half of the year. Because such weather (cool, but not too cold) occurs more often in oceanic climates, *Taxus* favours these regions.[16]

Generally, yew responds much less sensitively to climatic fluctuations than, for example, beech. This is partly owing to its occurrence in the lower tier, where it is not as directly exposed to the weather conditions as the trees above. A low sensitivity is made possible by, among other things, a high degree of adaption to periods of shortage, the ability to store resources effectively for such periods, and a low consumption of resources in times of abundance.[17]

Yew is intolerant of severe and prolonged frost and icy winds. Recorded frost damage occurred, for example, in western Scotland during the winter of 1837/8. But tolerance levels vary from region to region. In southern Sweden, for example, needles showed a maximum winter resistance of –33 to –35°C, and the male flower buds were damaged below –21 to –23°C. In the Austrian Alps, a temperature of –23°C for three hours was shown to damage 100 per cent of the needles.[18] Frost tolerance changes with the seasons as well, its maximum occurring in winter (January), and declining rapidly in early spring when the tissues become more vulnerable.

Table 4 Temperature range at various yew stands (in °C)

Location	*average annual*	*summer daytime*	*January*
Reenadinna, Ireland[19]	10.5	18 max. (Aug.)	4.2 min.
South Downs, Britain[20]	10.5	20.5 max. (Aug.)	6.7 min.
Paterzell, Germany[21]	7	15 average (July/Aug.)	0 average
Bakony Mountains, Hungary[22]	9.8	20.2 max. (July)	-1.6
Carpathians, Ukraine[23]	5–9.3	–	–
southern Crimea, Ukraine[24]	10–16	–	–
western Caucasus[25]	3.5–14-5	–	–
Hyrcanian forest, north Iran[26]	–	30	1.5
Amanus Mountains, Turkey[27]	18.5	28.0 average (July/Aug.)	8.9 average
western Taurus, Turkey[28]	18.2	27.6 average (July/Aug.)	10.5 average
western Himalaya[29]	–	19.2 average (June)	5.7 average

Taxus baccata also exhibits different frost tolerance in different regions. In Britain, damage can start at –13.4°C in midwinter, and then by March has risen to –9.6°C in the hardiest area (south England) and to –1.9°C in the most susceptible (north-east England).[30] In the Fisht Mountains of the Caucasus, the yew grows to an altitude of 2,000m a.s.l. and survives winters with 5–7m of snow. The Japanese yew *(T. cuspidata)* bears severe frost down to –40°C before needle damage occurs.[31]

Heat resistance, on the other hand, does not vary significantly in the course of the year. However, in cool moist summers the tolerance is lower than in hot dry ones, an indicator that trees in hot countries probably have adapted at least to some extent. Half an hour of 48–50°C will damage needles. Summer maxima can reach *c.* 52°C and winter ones are still surprisingly high reaching 49°C. The low spring resistance of *c.* 44°C is probably a result of the sensitive buds and young needles.[32]

Its thin bark means that *Taxus* is intolerant of fire, unlike redwood trees *(Sequoia)*, for example. Because of the absence of resin, however, its combustibility is much lower than that of other

4.1 In the dry summer in La Tejeda de Carazo, Burgos, Spain. ***4.2*** *In the short maritime winter in southern England.* ***4.3*** *In the temperate rainforest at Reenadinna, Killarney, Ireland.*

conifers, being more in league with the values for deciduous trees.[33] Another help against forest fires is the actual positioning of yews: most 'forest' fires in the Mediterranean are actually savannah fires, and *Taxus* generally does not grow among flammable grasses, bracken or undershrubs but rather in mixed forests or clusters of its own, and in higher elevations. Shading trees often account for unburnt islands, and many fires (even of pine plantations) stop at the edges of old forest (e.g. live oak, *Q. rotundifolia*, in Spain).[34] However, shortly after the year 2000, two ancient yews were lost to a meadow fire in western Sardinia.

ALTITUDE

Taxus baccata extends northward to *c.* 62°30′N (Norway) and southward to *c.* 33°N (Algeria); mainly to meet its moisture needs, the altitudes of yew stands increase from north to south.[35]

In mountainous areas, yew tends to grow on north-western to north-eastern slopes, but for different reasons: in southern countries (e.g. Turkey) it simply avoids the full sun exposure and dry heat of the southern slopes,[36] while in the vicinity of a more northern, continental climate (e.g. the Alps) the low spring sun on southern slopes would tempt it to bud early and then become susceptible to late frost damage.[37] Conversely, in the low altitudes of the British South Downs (90–250m above sea level) yew woods are primarily linked with south, east, and (to a lesser extent) west slopes, avoiding the cold northern exposure and the prevailing west winds. Another limiting factor in coastal lowlands (or sea cliffs as at the Great Orme, Wales) are the salt winds from the sea. *Taxus* is sensitive to airborne salt spray, which can turn its foliage brown-red.[38]

Table 5 Altitude of yew stands from north to south[39]
(in metres above sea level)

0–*c.* 470	British Isles
660–1,200	Slovakia
1,100–1,400	in the Alps
1,400–1,650	in the Pyrenees
up to 1,660	in the Carpathians
1,600–1,950	southern Spain
up to 1,700	Sardinia
up to 1,800	Macedonia
up to 2,000	central Greece
up to 2,500	north-west Africa
0–2,000	Caucasus
up to 3,333	Guatemala
2,800–3,570	Yünnan (China)

CHAPTER 5

PLANT COMMUNITIES

MIXED WOODLANDS

The European yew generally appears as scattered individuals or in groups or clusters, usually in the understorey, rarely breaking through into the canopy of the upper tier. The type of woodland is most often mixed oak *(Quercus)*, pure beech *(Fagus)*, or beech mixed with conifers. Other dominant trees in such woods include ash *(Fraxinus)*, sycamore *(Acer pseudoplatanus)*, fir *(Abies)*, spruce *(Picea)*, hornbeam *(Carpinus)*, lime *(Tilia)* and elm *(Ulmus)*. In the Mediterranean, yew also occurs with holm oak *(Quercus ilex)*, other oak species, and plane *(Platanus)*, with an understorey of myrtle *(Myrtus)*, laurel *(Laurocerasus)*, paradise plant *(Daphne)*, and others. The yew's most frequent companions in the understorey are holly *(Ilex)*, box *(Buxus)*, hazel *(Corylus)* and hawthorn *(Crataegus)*, occasionally also whitebeam *(Sorbus aria)*, blackthorn *(Prunus spinosa)* and elder *(Sambucus nigra)*.[1]

5.1 This ancient yew in Ortachis, Sardinia, provides a habitat for many small plant species.

The beech is often described as a major and more efficient 'competitor' to the yew, but this is a matter of interpretation. Pollen diagrams for various areas do attest to yew declines paralleling beech invasions, which has for a long time supported the opinion that yew is a 'weak competitor' and easily driven away by beech. Today we know that in various regions yew was very capable of establishing itself alongside those tree species that are still dominant today.[2] It can be noted that the spreading of beech in many regions of Central Europe occurred *exactly* within the period of colonisation by prehistoric farmers. Hence we can assume that the human effect on the forest composition and structure was unfavourable for yew, as for other species, and its decline paralleling the spread of beech would have occurred by chance only.[3] Additionally, even in prehistory yew wood was highly sought after, a fact that probably led to medium-term overuse in some regions. Still today, *Taxus* occurs in natural stands across the entire physiological range of its habitats.[4]

In any case, the establishment of woodland species is a matter of centuries (not of single trees competing). In general, a species is dominant in places for which it is better equipped. In this case, beech is dominant on deeper soils while yew does better on soils that are too wet or too poor for beech, or in exposed conditions. At locations with favourable climate and soil, for example at Paterzell in Bavaria, yews also grow well under large beech trees.

5.2 Young beech and yew trees on the chalk rocks in the park of Prunn Palace, Germany.

Some technical terms in ecology

Adaptability and adaptedness

'In genetics', says Dr U. Pietzarka from the Forest Botanical Garden at Tharandt, Germany, 'adaptedness is the condition of a population to survive perpetually under the given environmental conditions. Adaptability is its ability to adapt again to a changed environment by modifying its genetic structure. In the precise sense of the word, we cannot talk of adaptability in a single plant because it has no means of altering its genetic code.'[5] He adds, however, that the individual, too, 'does have means, within the confines of its genetic frame, to respond to a changed environment'.

Strategy 

All too often the term 'strategy' is associated with the human condition. It presupposes a conscious reflection upon one's actions that generally is not accredited to plants, and yet this term is widely used in ecological literature. Nevertheless, it can be understood as a dynamic event involving active participation of the organism in question. The ecological strategy of a species encompasses the entirety of its genetically fixed characteristics, which ensures adaptedness and adaptability and therefore supports the survival of the species.[6]

Competition 

Competition generally denotes the rivalry of organisms or species for limited resources. In this process, species mutually limit each other's birth rates and/or growth rates, and/or increase each other's death rates. Ecology considers competition to be a highly significant motor in forest dynamics; the competition for light was possibly one of the strongest selective factors in the evolution of land plants and the development of upright trunks.[7]

It has to be noted that analyses of the plant sociology and strategy of yew are inevitably based on its *present* status. The forests of Central Europe, for example, cannot be called primeval or virgin forest after millennia of intense usage by humans. Hence, an assessment of the ecological behaviour of a tree species has to remain incomplete.

In Europe, *Taxus* is a climax woodland species, 'climax' being an ecological term denoting the final stage of succession of a plant community under the local environmental conditions over time. The species composition of a climax community remains the same because all the species present successfully reproduce themselves and most outside species fail to invade. Inevitably, though, because of environmental changes (e.g. climate, soil) and invading species, even the climax stage will eventually change.

5.3 *An ancient yew in the Yenice Nature Reserve, Turkey, embraces an old beech tree.*
5.4 *(above) Point of contact between the two trees.*

Within the different plant communities, *Taxus* is a genus whose main strategy is safety. The slow growth, the effective storage of resources, the creation of interior roots in hollowing trunks, the immense potential for regeneration, the toxicity and high resistance of its organs, the low photosynthesis performance, and the extraordinary fastness of stomatal response – all these are safety mechanisms. Hence the ecological strategy of *Taxus* has been dubbed 'saving for safety'.[8]

The adaptability of yew in diverse and changing environmental conditions is very high, in particular its shade tolerance: even in an understorey with less than 5 per cent of the light of the open, yew is still able to produce flowers and seeds.[9] Most forest trees respond to lack of light – as long as they do not die of it – with an increased growth in height. Some species even invest in height increase when it implies jeopardising the diameter–height ratio and the development of the root system, and hence will make the tree unstable, for example, in high winds.[10] Not so the yew: even under poor light conditions it invests predominantly in the root system, a measure that provides not only a firm anchorage in the ground but also gives access to new resources.

The high adaptability of yew is a result of its large genetic variety, which makes possible its survival in the most diverse ecological conditions.[11] This, in connection with the aforementioned safety investments, results in *Taxus'* high life expectancy as a proof of the success of this strategy. The extremely advanced age that *Taxus* can reach,[12] 'means for the population that it can rejuvenate in especially favourable periods in-between large interims. And at the level of the individual, it is an expression of a sufficient adaptability and adaptedness to changing conditions during this long time' (Pietzarka).[13]

5.5 Yew and beech at Wakehurst Place, Sussex.

Only to a very small degree does yew take part in the 'competition' dynamics of the forest. Its regeneration rate, as well as certain growth rates (girth, biomass, but not height increase), is influenced by other species only to a limited extent. And to an even lesser degree does *Taxus* inhibit the regeneration, growth and death rates of other tree species.[14]

Taxus is a genus whose ecological strategy differs entirely from all other tree species in the temperate zone. Its strategy is based on an optimum adaptedness to its individual habitat while at the same time maintaining the highest possible adaptability (see box p. 23).[15] Thus the most appropriate name for its ecological strategy would be adaptation strategy.[16]

All in all, the view of *Taxus* as a 'weak competitor' does not hold ground. *Taxus* proves to be a forest tree that is perfectly adapted (to climax woodland as well), and one that has been capable of surviving in most forest ecosystems – morphologically almost unchanged for 140 million years.

The various safety mechanisms of yew are also the energetic prerequisites to ensure long-term survival under less favourable conditions, a crucial characteristic for an understorey species.[17] Yew, holly and box are indeed the only European trees that are shade-tolerant enough to survive under a dense beech canopy. But even *Taxus* has its limits – extreme lack of light inhibits the development of a full crown and also the production of flowers and seeds. On the other hand, it is the shelter that a mixed wood and its upper canopy provide that protects yew from high winds, lightning and, more importantly, from intense frost and other temperature extremes. Yew thrives in *medium* light mixed forest and can even eventually develop a crown almost as wide as that of a solitary tree.

Although a mature yew, like the beech, can create a dense canopy and hence a deep shade in which nothing much can grow, the ground of a mixed yew forest is not necessarily bare. Frequent species, as well as a variety of ferns and also mosses, include dog's mercury *(Mercurialis perennis),* wild strawberry *(Fragaria vesca),* ground ivy *(Glechoma hederacea),* ivy *(Hedera helix),* blackberry *(Rubus fruticosus),* common nettle *(Urtica dioica),*[18] and violets *(Viola).* Mixed yew-beechwoods (Taxo-Fagetum) may also offer a habitat to wood melick *(Melica uniflora),* woodruff *(Galium odoratum),* moor-grasses *(Sesleria),* cyclamen and various orchids. Mixed yew-oakwoods (Querco-Taxetum) yield cowslip *(Primula veris),* peach bell *(Campanula persicifolia)* and bracken *(Pteridium aquilinum),* among others.[19] In the Alps, *Taxus* can be found accompanied by, for example, mountain valerian *(Valeriana montana),* Alpine honeysuckle *(Lonicera alpigena)* and mountain rose *(Rosa pendulina).*

PURE STANDS

Because of its immense shade tolerance and the dense shade it can itself create, *Taxus* can eventually outgrow or outlive surrounding species (e.g., five to ten whole generations of beech) and become a dominant member of the plant community, or even develop into a pure yew woodland. These rare places are rather species-poor but nevertheless remarkably impressive. They remain *in situ* by means of the yew's extreme longevity rather than by

5.6 *Yew and juniper at Borrowdale, Cumbria, England.*

5.7 *A yew emerging from a hawthorn, Kingley Vale, Sussex.*

a successful regeneration through seeding. Usually, spaces vacated by fallen aged yews are quickly invaded by other species (beech, ash) rather than by its own offspring. Hence, regeneration occurs *at the edges* of yew woods, which has given rise to the idea that pure yew groves might be single generation stands that 'move' across the landscape.[20] But neighbouring vegetation, pasture land and human habitats rarely allow this to happen, forcing the yew grove to stay inside its enclosure. However, 'the circumstances under which yew develops from being an occasional component of woodland to becoming the dominant species remain poorly understood' (British Ecological Society, 2003).[21]

OPEN AIR

Hawthorn, sometimes blackthorn[22] and other wild shrubs of the rose family, but particularly juniper *(Juniperus),* prove to be great 'allies' of the yew when it invades grassland. Juniper is the most accommodating nurse plant as it grows in locations suitable for the yew (shallow soils on steep or exposed sites). The fruits of juniper or hawthorn attract birds that will digest the yew arils beneath them. The yew seedlings find shade and effective protection from herbivores among these thickets, and eventually a yew will outgrow its nurse shrub and one day might even stand among its dead woody remains.[23]

Taxus is generally absent from wet soils such as wet clay or wet acidic peat. In northern Europe (England, Germany, Poland), however, it can occur upon calcareous peat with oak, ash, pine, birch or alder. Although it can tolerate temporary flooding, it is nevertheless susceptible to poor drainage.

5.8 *Yew covered with hanging lichen, Duezce, Turkey.*

5.9 Lichen (left), liverwort (centre) and algae (right) growing on yew bark.

EPIPHYTES

Epiphytes are plants that grow upon other plants merely for physical support, *not* interfering with their metabolism in a parasitic or symbiotic relationship. They obtain water and minerals from rain, and also from organic debris that collects on the supporting plants. Lichens, mosses, liverworts and other bryophytes, and algae are all epiphytes of temperate regions.

Its smooth and flaking bark makes the yew generally inhospitable for epiphytes, except in areas with particularly high humidity. Reenadinna Wood near Killarney, Ireland, for example, is famous for its abundance of lichens and mosses.[24] In a research project (1994) on Inchlonaig, an island in Loch Lomond, Scotland, 60 out of 791 yews had epiphytes[25] and 28 had other *trees* growing on them as epiphytes. These trees were birch *(Betula pubescens* ssp. *carpatica)*, rowan *(Sorbus aucuparia)* and holly *(Ilex aquifolium)*, which had started to grow in humus pockets on the yews and could later grow roots down through the hollow yew into the soil. Eventually, those trees rooted in the ground were enveloped by the growing yew trunk, and went on to become 'partner trees'.[26] Other reports mention oak,[27] ash and *Rhododendron ponticum* growing as partner trees on yew in south-west Wales.[28] In Newlands Corner, Surrey, a mature whitebeam *(Sorbus aria)* emerges from an ancient yew.

MAN-MADE HABITATS

Doubtless the activities of man have often had detrimental effects on the expanses and richness of woodlands. But human impact has also created new habitats for various trees, including yew. The yew woods in the South Downs, for example, established themselves in the eighteenth and nineteenth centuries as a result of the Napoleonic Wars and the subsequent economic crisis with its abandonment of sheep pastures and outbreak of myxomatosis (a fatal viral disease of rabbits).[29] In woodlands used for coppicing – an ancient form of woodland management in which suitable trees such as hazel, lime, elm or ash are cut back every four to ten years – yew used to find a habitat too. But since yew grows so slowly it was often left untouched for centuries. During the last millennium, however, the most successful and long-term man-made habitat for the yew has been the churchyard; Britain in particular has a story to tell in this respect. In Asia, Buddhist and Shinto shrines serve a similar purpose.

In the British Isles, yews are also frequent in hedges, gardens and parks. On the European continent, too, the planted *Taxus* population might distract from the fact that in the wild the yew is still a threatened species.

6.1 This tree germinated in a deep rock fissure at over 1,300m altitude; Mt Limbara, Sardinia.

CHAPTER 6

THE ROOTS

THE GROUND

Taxus grows on almost any soil, typically on deep, well-drained, humus- and base-rich soils of variable pH. Moisture is very important, and the vicinity of springs is most favourable.[1] But *Taxus* also thrives on shallow, dry rendzinas on limestone, often rich in downwashed flints and poor in earthworms; and also on thin, warm, chalk soils; and on calcareous peat in fens. *Taxus* can also be found on sandy soils or loamy sand if there is enough moisture, and on siliceous soils derived from igneous and sedimentary rocks. The two ground conditions that *Taxus* avoids are those with poor drainage and (with exceptions) high acidity.

A few examples from Western Europe may suffice to illustrate the full spectrum of soils chosen by *Taxus*. In south-west Ireland, yew grows on a carboniferous limestone pavement,[2] and individuals are found on Devonian sandstone (Killarney woods).[3] In south-east England, it grows on the sandstones of the Lower Greensand and the Central Weald, and on the chalk of the North and the South Downs. Here, the rock is soft and free-draining but holds much water in a deep aquiferous layer.[4] In Germany, yews are usually found on chalk humus soils (rendzinas), for instance Jurassic chalk (at Kelheim), shell limestone (near Göttingen and in the Thuringian Basin), as well as the chalk tuff at Paterzell. In the Black Forest, the Bavarian Forest, and in the mountains of Schlesia, yews stand on gneiss and granite, and in the Harz Mountains (Bodetal) on quartzites, clayslate or gneiss. Talus slopes of marl, loess or chalk offer another habitat.[5] Further south, in the Mediterranean, the ancient yews of Sardinia thrive on schist, granite, chalk and basalt.[6]

Taxus grows on the widest variety of soils when located within the region of its optimum climatic conditions, but near the borders of its climatic distribution range keeps more to chalk soils.[7] As is the case with other plants, low soil temperature in spring can restrict the assimilation of minerals such as phosphorus, nitrogen and potassium. It has high requirements in regard to mineral nutrition, in particular such elements as potassium, phosphorus and calcium. One of the reasons for the disappearance of the yew from European forests could be soil degradation and the shortage of mineral components.[8]

Trees do not just adapt to their immediate locality, they also change some of its characteristics. A comparison of soil physio-chemical conditions under oak and yew trees growing on the same soils has shown that humic acids were more oxidised under yew; the mineral content was less than under oak; and that total amounts of carbon, nitrogen and calcium were significantly higher under yew than under oak, which is thought to be because of the absence of large earthworms under *Taxus*.[9]

6.2 *Fighting erosion on a steep slope in Alapli, Turkey.*

6.3 *This old or even ancient yew clings to a talus slope at Low Scawdel, Borrowdale, Cumbria.*

CLIFFS

In the cultural landscape, pastures, fields and human dwellings effectively occupy the deeper, richer and moister soils. Hence difficult terrains such as poor soils, steep slopes, gorges, rocky terrains and cliff faces have always been natural strongholds for wild yews.

Steep and vertical cliff faces are often inaccessible for human activity and even for browsing animals like goats. Today, vertical cliffs harbour 'some of the most ancient and least-disturbed wooded habitat types on Earth … even when they are situated close to agricultural and industrial activity, which has destroyed or altered most other natural habitats' (Doug Larson 1999).[10]

Many of these yews have little nourishment and water and therefore grow extremely slowly, but they have time on their hands, and, although of a stunted and much smaller shape, they can be as old as or even older than some of the large specimens in favourable soils (see figure 6.3).

6.4 Yew roots penetrate a vertical limestone rock face at Kentchurch, Herefordshire.

THE ROOT SYSTEM

The yew has an extensive yet dense root system that facilitates an effective penetration of the soil. It is very efficient in supplying the tree with water and minerals and also lends excellent mechanical support to the tree, even in difficult terrain such as rock pavements, steep slopes and vertical cliffs. The composition of the fine roots is, like so much in *Taxus*, variable.[11]

Already as a seedling, *Taxus* begins to invest heavily into its root system. Even at low light levels and hence limited growth, the strengthening of the root system has priority over height and girth increase.[12] This occurs as one of the safety measures employed by *Taxus*, because if lack of light gets too severe and photosynthesis performance fails to maintain root growth, yew actually dies of dehydration. The tree meets this danger with narrow xylem tubes as well as improving the root system even under poor light conditions.[13]

6.5 The extraordinary root system of the ancient yew at Bridge Sollars, Herefordshire.

Bio-electric studies of trees confirm that *Taxus* employs the root system with the highest vitality among trees (see Chapter 17). This results, among other things, in the ability to produce numerous shoots directly from the roots, even after the complete loss of the trunk (see figure 20.1). An intense root system that is perfectly adapted to its habitat is the crucial fundament for the unique regenerative abilities of yew.[14] *Taxus* roots are also capable of penetrating compressed soils, and only the most extreme cases of compression have been shown to lower its rooting intensity. Under the surface, too, competition does not seem to be an important factor for yew: no influence of the roots of other trees on the density of yew roots could be shown, which contradicts the (established) assumption of a possible inhibition of yew roots by root competition from other tree species.[15]

Root anatomy

Generally, the roots of higher plants absorb water, which passes through a complex filter system that allows only the required minerals to enter into the central cylinder of the root where the vascular bundles, consisting of xylem and phloem, are located. The xylem is the transport system that takes the water right up to the foliage where it is needed for photosynthesis. The nutritious sap produced inside the leaves is distributed by the phloem downwards to feed the living tissues in the branches, the trunk and the roots.

During the active growth of the *Taxus* root, the zone of elongation and the adjoining root-hair zone are situated close to the root tip (apex). When the absorbing function of the outermost layer of the root (rhizoderm) is completed it dies off, and the

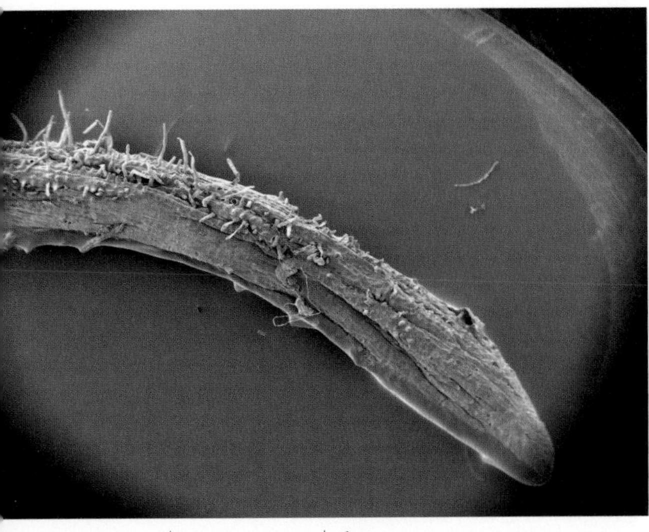

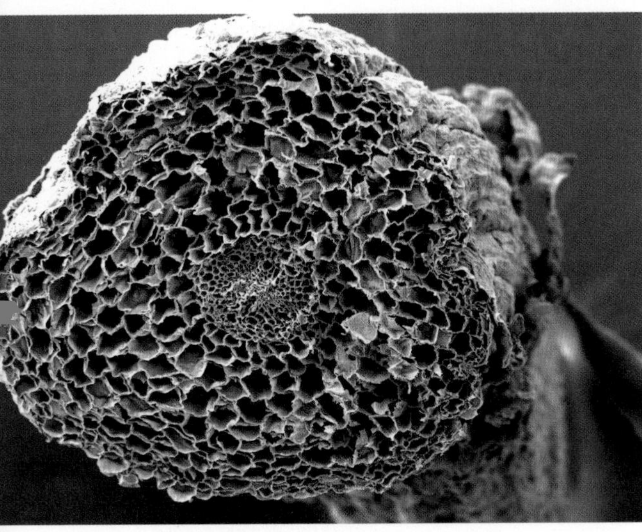

6.6 *Root tip (apex) with root hairs. Here, the greater part of water and nutrients is absorbed.*
6.7 *Cross-cut of root tip. Clearly visible are the vascular bundles in the centre.*

external layers of the bark (cortex) become suberised (i.e. cork-like), creating a protective outer layer called exoderm. When root growth is completed towards the end of the vegetative season, the inner cortex (endoderm) also becomes suberised, and another protective layer (metacutis) is formed around the root interior to protect it during winter dormancy. Its thickest layer covers the sensitive root tip. In spring, when the root commences to grow, this layer of suberised cells falls off.

The structure of the young yew root is diarch, meaning it contains two bundles of xylem and phloem. The phloem bundles are situated vertically beneath those of the xylem. Thin lateral roots as small as 1mm already begin to form cambium cells between the phloem and xylem strands. In the cross-cut of older roots, the concentric layers of cambium and secondary phloem and xylem resemble the structure in trunk and branches above ground. Yew roots have no resin ducts but single cells that contain resinous substances.[16]

Mycorrhizas

Generally, higher plants live in symbiosis with certain types of fungi whose network of tubular filaments (hyphae) is a significant extension to their root system. In this mutually beneficial relationship, the fungus is able to transform organic chemical substances in the soil and prepare inorganic plant nutrients (minerals) for the plant to absorb. In return, the plant supplies the fungi (which cannot perform photosynthesis) with carbohydrates and amino acids. The symbiosis of tree and fungus is called mycorrhiza; many tree species depend on it.

Mycorrhizas are most frequently classified into three structurally different groups: ectomycorrhizas, endomycorrhizas and ectendomycorrhizas. Ectomycorrhizas have formed a hyphal mantle and the hyphae grow into the intercellular spaces of the primary root bark of the tree. Endomycorrhizas do not have a hyphal mantle and the hyphae grow intracellularly. Ectendomycorrhizas have formed a hyphal mantle and the hyphae grow inter- and intracellularly. Only endomycorrhizas are known in yew.[17] *Taxus* connects with a number of different fungi: for example, research on four trees in Iran found seven species from the genera *Glomus*, *Acaulospora* and *Gigaspora*.[18]

CHAPTER 7

THE FOLIAGE

LIGHT

The foremost function of the leaves of higher plants is photosynthesis, that is, the use of solar energy to convert water and carbon dioxide into energy-rich organic compounds (carbohydrates). Unlike water, sunlight is generally abundant in all climates. But the light quantities available to a plant vary with the length of daytime during the seasons and with its individual position in relation to shading objects such as rocks or other plants and trees.

The great shade tolerance of yew does not imply that it prefers shadow. Like any other plant the yew needs a sufficient amount of light. It is, however, rather flexible in both directions. Its ability to bear intense sun exposure is demonstrated by specimens occurring on exposed rocks and cliff faces.

As a shade-tolerant plant, however, yew is photosynthetically active at comparatively low light intensities. The yew's light compensation point (the light level below which a plant cannot perform effective photosynthesis[1]) varies from 175 to 3,200 lux, correlated to season and temperature. At low temperatures, more light is needed to trigger photosynthetic efforts, for example 2,500 lux at –3°C and 3,300 lux at –4°C.[2] *Taxus* is the most shade-tolerant tree in Europe. If the shade tolerance of birch is given at the virtual value of 1, then spruce has 2.0 (i.e. it can still assimilate at only half the light value at which birch has to stop), oak has 1.6, beech 2.1, fir 2.2 and yew 5.8. This shows that *Taxus* can tolerate far more shade than even the 'classical' shade tree of Western Europe, the fir.

The widespread opinion that the shade-tolerant yew would grow faster under full light exposure[3] does not hold, and has recently been refuted by a detailed study.[4] In dense forests, light exposure can sink to 1 or 2 per cent of the values in the open. In places where medium-sized yews in the second tier bear foliage right down to the ground, the light exposure of yew seedlings underneath is reduced to an absolute minimum. It has been shown that even under such conditions yew seedlings are capable of growing in height at far less than 1 per cent of light exposure. Surprisingly, no statistical correlation between light exposure and height increase could be found,[5] despite the values being close to the vital limits for shooting plants.[6] *Young yews under higher light exposure do not significantly increase their growth in height.*[7] Thus, *Taxus* proves to be of low sensitivity, that is, highly adaptable, in this respect too, just as with the climatic factors. Once again, this reveals its ecological strategy 'according to which the safeguarding of a low but constant growth independent from exterior influences contributes to the ability of survival'.[8]

7.1 *The needles of a new year (apical growth) are distinctly lighter than the older ones.*

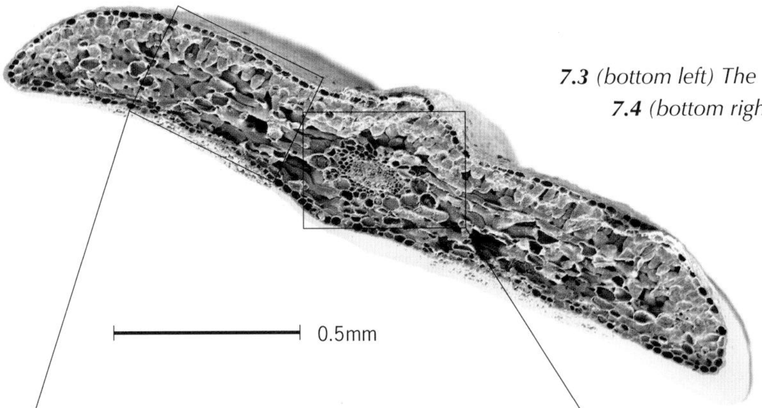

7.2 Cross-section of a Taxus *leaf.*
7.3 (bottom left) The two parenchyma types of the Taxus *leaf.*
7.4 (bottom right) The vascular bundle of the Taxus *leaf.*

1 epidermis with cuticle
2 palisade parenchyma
3 spongy parenchyma
4 vascular bundle
5 epidermis of the underside of the leaf
6 stomata

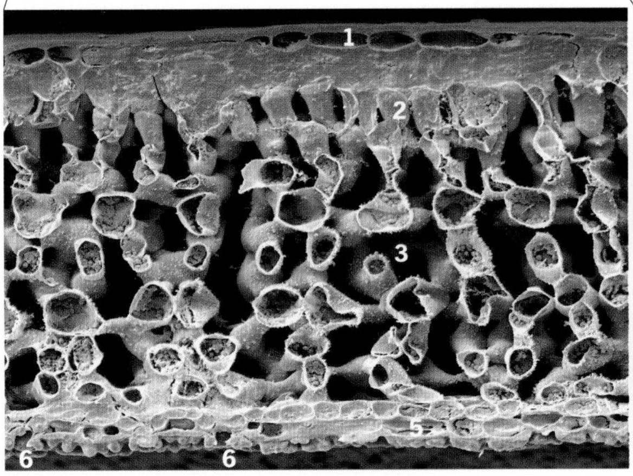

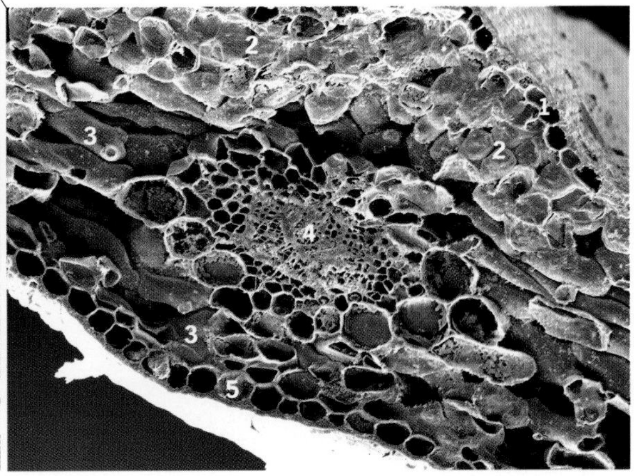

Even at well below 5 per cent relative light intensity, sexually mature trees are capable of producing sufficient numbers of seeds. However, yews, particularly female ones, produce more flowers under higher light exposure.[9]

Nevertheless, the half shade of deciduous woods such as ash or alder is quite favourable for yew because in their company it usually also finds a good water supply. Yews are very rarely found beneath conifers (such as fir or spruce) because the evergreen shadow of these trees would deny yew the important winter sun. Since the leaves grown in deep shade and those grown in sunlight differ structurally (see below), it takes a yew up to eight years to adapt from a shaded position to a fully exposed one, or vice versa. In this respect, the *sudden* disappearance of neighbouring trees (through storm, age or human interference) can impair the yew's vital balance, or even kill it.

THE LEAF

The needle-shaped leaves of yew are spirally attached to the twigs, but on lateral shoots they are twisted more or less into two ranks. They vary in length, usually 16–25mm, but in some individuals can be as short as 10mm or as long as 45mm. Their usual width is 2–3mm. The soft needles have parallel sides, short stalks, and end in a short, no-stinging tip.[10] *Taxus* leaves live for four to eight years before they are replaced on younger shoots.[11] Their photosynthetic ability, however, declines with age, to only 50 per cent in 7-year-old needles compared to young ones.[12]

The yew leaf is dorsiventral, meaning it is flattened and has a definite dorsal (facing away from the axis) and a ventral surface (facing towards the axis). The colour is dark glossy green above and clearly paler on the underside. The leaf surface consists of an outer skin, the epidermis, which is covered with a protective wax layer, the cuticle. On the lower surface of the leaf, the cuticle forms irregular, papillary thickenings, particularly near the stomata. The leaf has neither a hypodermis (lower skin) nor resin ducts. Underneath the epidermis lies a stratum of one to three cell layers called the palisade parenchyma, which contains the majority of the photosynthetically active chlorophyll grains. Moving inwards, the next layer of tissue is the spongy parenchyma (*parenchyma* is the name for plant tissue made up of undifferentiated and usually unspecialised cells). The relation between the two parenchyma types depends on the light conditions during leaf growth. Developing in light, a leaf forms more layers of palisade parenchyma and a more compact spongy parenchyma than a leaf growing in shadow.[13] Yew leaves do not have the mechanical support tissue *(sclerenchyma)* typical of conifers.[14]

Adapting to a variety of light conditions, yew brings forth light leaves and shadow leaves, the former with a clearly higher photosynthesis performance than the latter.[15] Light-adapted needles are thicker yet narrower than shade-adapted needles, which limits the evaporating surface. The cuticula, which gives additional water loss protection, is stronger too.[16] Shadow-adapted leaves only occur well below 1 per cent light intensity.[17]

Inside the yew leaf lies a single vascular bundle comprising xylem and phloem. The vascular bundle is surrounded by large, thin-walled parenchyma cells. These two tissues are linked by a transfusion tissue composed of living, thin-walled parenchyma cells and dead lignified tracheids. The inner surfaces of these tracheid cell walls have helical thickenings and bordered pits on their internal surface (see also Chapter 18).[18]

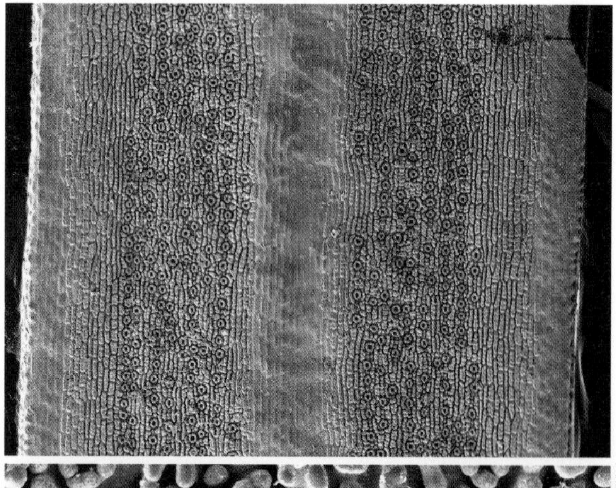

0.5mm

7.5 The bands of stomata on the lower surface of the yew leaf (c. 2mm wide). Their number varies in the different Taxus *species.*

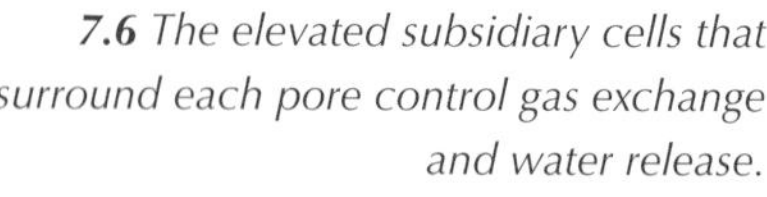

7.6 The elevated subsidiary cells that surround each pore control gas exchange and water release.

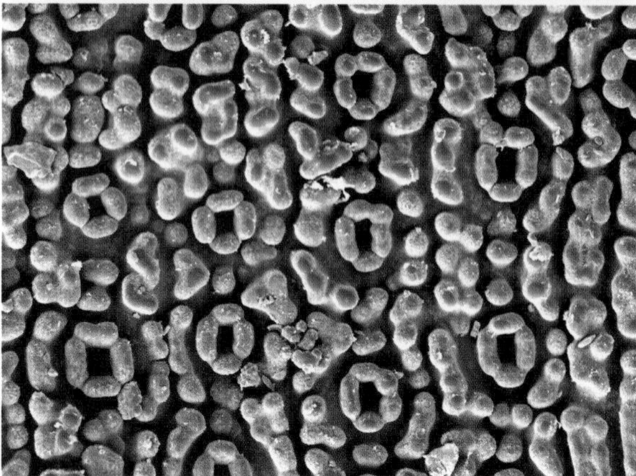

0.05mm

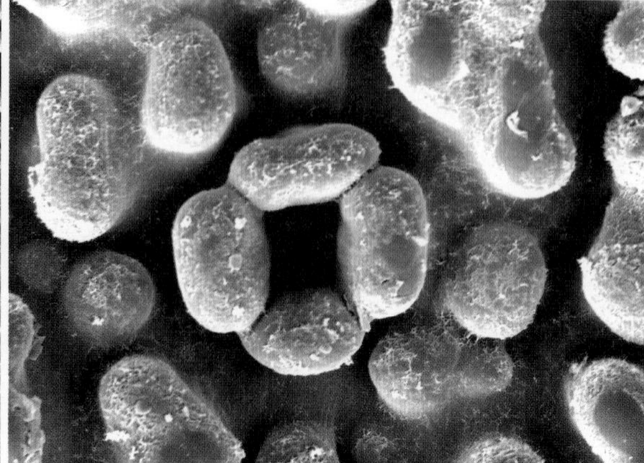

0.02mm

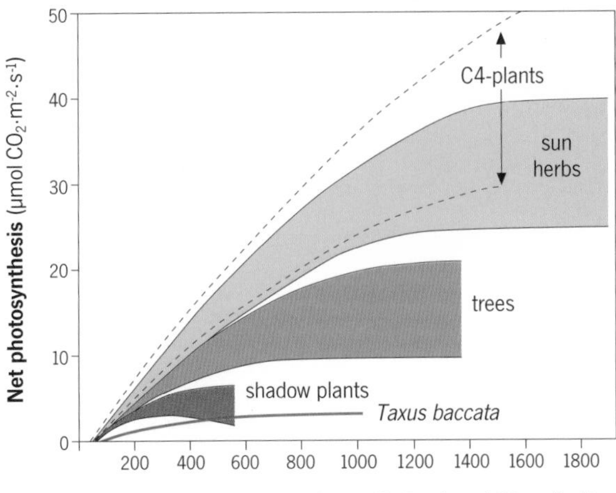

Diagram 1 Net photosynthesis of leaves under vertical light in dependency of photon flux density*

* at optimum temperature and normal CO_2-range. (From Pietzarka 2005, after Larcher 2001, modified.)

The minute pores in the epidermis of a leaf of a higher plant are called stomata. Through these, gases (carbon dioxide CO_2 and oxygen O_2) are exchanged and excess water is released. The yew's stomata are found on the underside of leaves only, within two pale stomatiferous bands, but they are not arranged in lines within the bands, as in some other species. The stomata are sunken into the epidermis, which minimises air movement and evaporation. But cuticular waxes, which assist the regulation of water transpiration in the other conifers, are absent in the yew.[19] Instead, its stomata are surrounded by elevated subsidiary cells. Stomata density in European yew varies, with counts of 59, 92 and 115 per square millimetre[20] (for comparison: pine usually has about 100). The density range of twenty-five different varieties of yew (82–119 stomata/mm²) is not wider than that of European yew alone.[21] A comparison of Pacific and European yew, too, revealed no difference in stomatal density, but there was a significantly higher number per leaf (4,684 vs.1,604) as the examined leaves of European yew were generally larger.[22] The leaf stalk has no stomata.

During the lowest temperatures in January, the osmotic pressure in yew reaches its value peak at *c.* 35 atmospheres. It diminishes from February onwards and remains constant from the end of April, at a level of *c.* 20 atmospheres. These values are comparable to other conifers, such as Scots pine.

Water shortage during needle growth decreases the leaf size and the density of stomata.[23] Several conifers, including yew, have a lipid layer (Greek *lipos* = fat) that surrounds the cytoplasm of the photosynthetically active cells and thus increases drought resistance.[24]

PHOTOSYNTHESIS

The ability to perform photosynthesis at low light and temperature levels enables yew to take advantage of the favourable aspects in the winter half of the year (more moisture in the soil and more light because the deciduous trees of the upper tier are bare of leaves) and create stores of excess energy.[25] The net performance of photosynthesis of *Taxus baccata*, however, is extraordinarily low even under optimum conditions. Yews are not capable of using a higher light exposure effectively. In periods of great summer heat, *Taxus* loses a lot of energy because of increased respiration and its inability to balance these with a more effective photosynthesis.[26] This missing plasticity is a characteristic of those shade-tolerant woody plants that are adapted to late succession stages.[27] Yew's photosynthesis performance is among the lowest of all plant species and has to be compared with that of distinct shade plants.

At 14–25°C, the yew's optimum temperature for photosynthesis is higher than of other conifers. The summer temperature minimum for photosynthesis is 3–5°C, and the winter minimum is –8°C. The summer temperature maximum is 38–41°C.[28]

The carbohydrate products of photosynthesis (sugars such as fructose, glucose and sucrose) can be found in *Taxus* all year round. The sucrose content is higher than in other conifers investigated, namely pine, larch and hemlock. A major part of sugar reserves is stored in the form of hemicellulose in the previous year's needles and released in spring for the development of new needles and shoots. Over half the yew's nitrogen and nitrogen reserves exists in the form of argenine, just as in apple and pear trees.[29]

CHAPTER 8

THE FLOWERS

Taxus has very effective means of *vegetative* reproduction (see Chapter 19), which can extend the lifetime of an individual and entirely renew its 'body' or create clones (e.g. via layering). But these methods do not make for genetic diversity. The great importance of sexual reproduction is that it creates *new* individuals with a unique genetic make-up and thus enhances the species' diversity of genotypes and its chances of long-term survival.

Generally yew is a dioecious tree, that is, male and female reproductive organs are borne on different trees. Sex ratios in *Taxus* stands are often half and half, but variations can be found too; for example, 44 per cent female in Kingley Vale, Sussex, 67 per cent female in certain stands in the Caucasus, and even 70 per cent female in the Sierra Nevada Mountains of Spain.[1] Bisexual trees exist but are rare, the highest known frequency reported as being 1–2 per cent of trees.[2] Usually, bisexual trees bear only single branches of the other gender. Change in sex of an individual tree is also possible,[3] particularly in extremely isolated specimens. However, in the Bakony Forest in Hungary, where yews are abundant but fruiting yews are rare, the staff of West-hungarian University in the autumn of 2002 marked a few female yews with exceptional aril crops. Returning in the following spring, the scientists, to their surprise, found these trees bore almost 100 per cent male flowers.[4] Nevertheless, the real frequency of bisexual yews is probably significantly higher than currently believed. Because of imprecise observation methods, at the first glimpse of a few scarlet spheres a tree is often prematurely documented as 'female'. In other seasons, too, gender identification usually consists of identifying the first few flowers.[5]

Sexual maturity usually starts at 30–35 years of age in single trees and open stands, but can begin as late as 70–120 years of age in dense woods. Maturity often begins slightly earlier in male than in female trees.[6] Light exposure has a strong impact on productivity: forest shadow might reduce the production of reproductive organs to about a third,[7] and many yews do not flower at all under dense beech canopy.

Small, green flower buds form in the leaf axils during the second half of summer. The flowers open

8.1–8.3 *(opposite) Macro shots of single female flowers with clearly visible micropylar drops.*
8.4–8.6 *(above) Macro shots of single male flowers.*

in the following spring – in mild climates in February/March, and in locations where winter snows or coldness endure, as late as April/May. The early flowering facilitates optimum pollen distribution because during this time the deciduous upper tier of the forest has no leaves to obstruct pollen flight. The opening of the flower buds is induced by a number of different exterior factors (air temperature, light intensity and others), but also by interior control mechanisms such as nutrient intake and distribution. For example, a high carbon/nitrogen ratio stimulates the flowering of trees: intense flowering and mast years can be expected when the weather just before the production of the flower buds is predominantly sunny (high carbon assimilation through photosynthesis) and dry (low nitrogen intake with the groundwater). A year of intense flower and seed production leads to a reduction of flowering in the following year because of a lack of resources (reduced carbon content), and hence most tree species pass through significant annual fluctuations in this respect. Yew, however, under favourable conditions usually shows an even annual flowering and seeding activity.[8]

The numerous male flowers are spread quite evenly in the leaf axils near the ends of branchlets, particularly in the front half of last year's growth. The amount of available light determines how far down the shoot flowers are produced. The tiny flowers are 2–3mm in diameter and consist of six to fourteen short-stalked stamens (microsporophylls), each with four to nine pollen sacs (microsporangia). Once they have opened, stimulated by warmth, the slightest breeze will stir the pollen grains for release in abundance. The pollen grains are wind-borne but this does not stop honey bees occasionally visiting the flowers. Wind pollination is a method that naturally implies a high loss of pollen, compared with the precision of pollination by insects, which fly from flower to flower. *Taxus* counterbalances this by producing and releasing extraordinarily large amounts of pollen; the pollen count per flower is higher than in any other conifer and suffices to secure a 100 per cent pollination of the female flowers.[9]

Small isolated populations of a species often run a high risk of losing part of their genetic information, in particular the rare varieties. In the long run, the consequence of a low population size is a decrease in the range of genotypes (i.e. of the special constellations of inherited characteristics). But it is this very variety of genotypes that is an important factor in dealing with environmental challenges. Epidemics, for example, often strike certain genotypes while

8.7 Swollen male flowers just before pollen release.

8.8 Male flowers after pollen release.

others are more tolerant. Genetic research has shown, however, that yew populations, compared with other conifers, are remarkably diverse. Yews succeed in passing on to their offspring not only their own genetic information but also a high amount of 'new' information brought in by pollen from outside the stand. The genetic variation within yew stands is particularly high and clearly above other gymnosperms. This applies not only to large populations but even to small isolated groups.[10] This is possible for two reasons. Yew pollen grains are not only numerous but very light. In still air they sink at a speed of only *c.* 2cm per second, which is even slower than those of the pioneer tree, birch, and twenty times slower than fir. Hence *Taxus* pollen grains can travel far indeed, showing a perfect adaptedness to survival in small and isolated groups. Still, all females in one cluster could run the risk of being fertilised by one 'alpha male', the one that sheds its pollen first or just stands closest to them. *Taxus* overcomes this danger by having individual trees (or genotypes) flowering at slightly different times. Hence, even in a close yew community, the females are pollinated at different times by pollen clouds from different male trees.[11] With its characteristics in terms of population genetics, *Taxus* is perfectly adapted to living in small isolated stands.[12]

The female flowers appear solitary or paired in leaf axils on the underside of shoots. They are even smaller than the male ones, about 1.5–2.0mm long. They consist of several overlapping scales of which the uppermost is fertile, bearing a single ovule,[13] rarely two.[14] At the tip of the flower, the outer layer (integument) forms a micropylar canal, which is sealed at the outside end with a drop of sticky, sugary liquid. The function of this micropylar drop (or pollination drop) is to catch pollen grains: an excess of 1,000 of them can hit its surface – 260 times the amount necessary to secure fertilisation.[15] Subsequently, the pollination drop and the pollen therein are reabsorbed by the nucellus and sink down into the pollen chamber where the pollen begins to germinate. The actual fertilisation will occur six to eight weeks after pollination (see next chapter).[16]

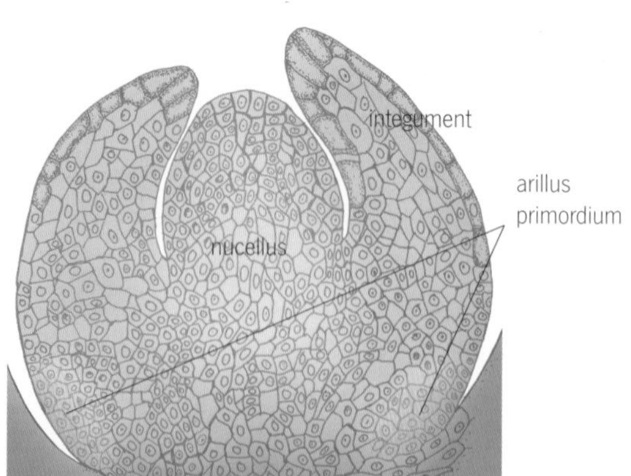

Diagram 2 Diagram of a longitudinal section of the developing ovule

After separating from the leaf bud formation, the female flower develops into a generative shoot apex. Its very tip (which otherwise would have become the tip of a leaf bud) changes into a nucellus. The primordium of the aril develops from the base of the integument during autumn and winter.

CHAPTER 9

POLLINATION AND FERTILISATION

Genetic basics

Chromosomes are thread-like parts of the cell that carry hereditary information in the form of genes. Every genus and species has a characteristic number of chromosome pairs (e.g. human forty-six, horse sixty-four). Each chromosome consists of two **chromatids**, which are separated in mitosis, a process bringing forth two cells with one chromatid of each chromosome. In the following phase of chromosome replication, the 'missing' chromatids are synthesised, which leads back to a complete diploid set of chromosomes.

Mitosis is the process of cell duplication, during which a mother cell gives rise to two genetically identical daughter cells. Between cell divisions the genetic material (chromatin) is diffused in the form of long threads throughout the nucleus in a tangled network of long filaments (**chromatids**). In mitosis, the filaments coil up and are surrounded by protein sheaths, thus forming the **chromosomes**. The chromosomes line up along the midline of the cell and each chromatid pair (chromosome) separates. The single chromatids are pulled to opposite ends of the cell. The cell divides, and after the replication of the chromatids both of the daughter cells have a complete (diploid) set of chromosomes. Within the new cells, the chromosomes uncoil again, forming the diffuse network of filaments.

Meiosis, also called reduction division, denotes the double division (similar to that of mitosis) of a diploid (2n) germ cell into four gametes, each possessing half the number of chromosomes (haploid, 1n) of the original cell.

The **gametes** are reproductive cells with a single (haploid = 1n) set of chromosomes. They are created by meiosis. The **microspore** is an organ that produces a male **gametophyte**, the mother cell for male gametes (pollen). The **megaspore** produces a female gametophyte, the mother cell for female gametes (eggs). During fertilisation, two gametes of opposite sex (egg and sperm) unite to produce a **zygote** (fertilised egg), a single cell that has a double (diploid) set of chromosomes again.

PREPARATIONS IN THE FEMALE FLOWER

The female flower has a single ovule sheathed in a single integument. The aril will not develop from the integument but from a region at its base; otherwise *Taxus* would be an angiosperm like the deciduous trees and not a gymnosperm.

In the early stage of development, when the ovule has just appeared, one or a few megaspore mother cells form in the subepidermal layer. After the meiotic divisions, one of them produces four megaspores of which only one, usually the innermost, reaches maturity and becomes the mother cell of the gametophyte. The internal layer of the nucellus that surrounds the developing megaspore acts as a tapetum, i.e. a tissue responsible for the nourishment of

9.1 Taxus *releases vast amounts of pollen.*

the megaspore. But a haploid tapetum, as is so characteristic for conifers, does not form in yew.

The megaspore grows rapidly and develops into a (haploid) prothallium with a large number of free nuclei, 512 in the yew (about 2,048 in pine). Inside the prothallium, the archegonium (egg cell) begins to develop. The initiating cell first divides into a neck cell and a central cell. The neck cell multiplies, while the central cell directly becomes the archegonium. But a canal cell, as typical for conifers, does not develop in yew. The archegonium is surrounded by cells of the primary endosperm, which will feed the egg cell (embryo) after its fertilisation.

PREPARATIONS IN THE MALE FLOWER

Already in the autumn, the mother cells of the pollen are contained in the microsporangium. They go through meiosis either in the autumn or at the end of winter.[1] The optimum temperature for meiosis in yew pollen mother cells is between 1°C and 10°C; temperatures above or below this, if lasting for several days, cause partial or complete sterility.[2]

The pollen grain of yew is irregular, tetrahedral or oval in shape, and has a rough surface. It has no air sacs and no pores. Sizes range from 22 to 30 μm. Its wall is composed of a two-layered, non-porous outer layer (exine), and an inner layer (intine) consisting of a gelatinous, cellulose-like substance. The intine causes the pollen grain during germination to swell and burst open the external layer.[3]

COMING TOGETHER

The gametes of *Taxus* have twelve chromosomes (n = 12).[4] Ten of these have the primary constriction close to the central part (metacentric), and two have it closer to one of the arms of the chromosome (submetacentric), and in one of these two, the smallest, the primary constriction occurs almost at the end.[5]

At the time the pollen reaches the micropylar drop at the tip of the female flower (see Chapter 8), the female prothallium is still undeveloped and the megaspore mother cell is in the process of meiosis. The pollen travels down the micropylar canal of the female flower and comes to rest in the hollow of the nucellus where it begins its germination, which will take ten to twelve days.

The pollen grain swells and eventually causes its outer wall to split (in a predestined place), creating an opening for the pollen tube to be released and to grow towards the (female) megaspore. As it grows, it secretes substances that stimulate the development of the megaspore. Midway to the (female) gametophyte, the pollen tube temporarily stops growing and the activity focuses on the inside: nucleus division. Two unequal cells are created, the larger is the vegetative pollen tube cell and the smaller one the generative cell. The generative cell divides into the stalk cell and the body cell. Shortly before fertilisation, the body cell divides into two gametes of which only one will take part in fertilisation. Ten days after its release, the pollen tube reaches the

9.2 *The* Taxus *pollen grain is extremely light.*

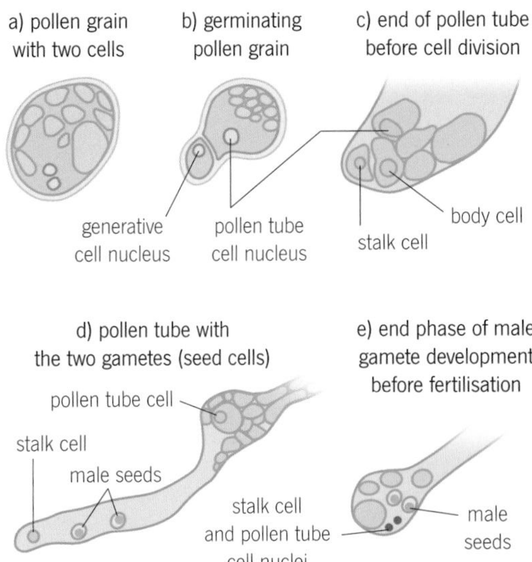

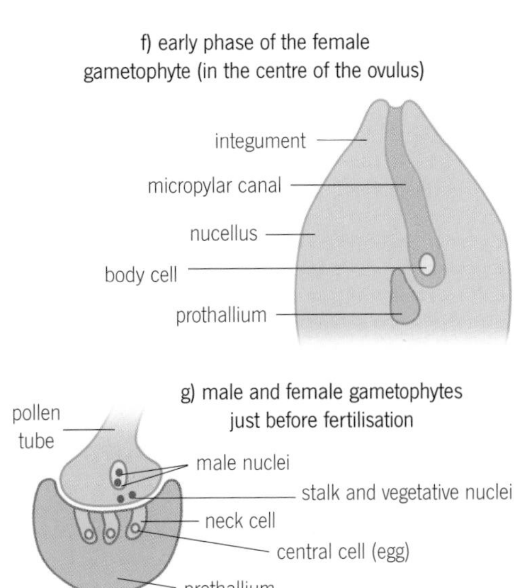

Diagram 3 The fertilisation process

female gametophyte. When it moves into the archegonium, the neck cells are dissolved. The contents of the pollen tube spill into the archegonium. The nucleus of the egg cell and the one male cell (i.e. one of the gametes) that will participate in the fertilisation wrap themselves in a mutual coat of cytoplasm and then merge.

The time from pollination to fertilisation is usually six to eight weeks (rarely up to three months).

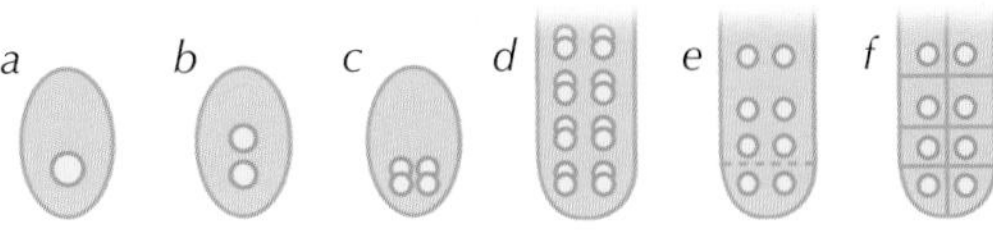

Diagram 4 The development of the proembryo begins with free divisions of the nucleus of the fertilised egg (a). Cell walls do not develop before the proembryo has eight or sixteen, sometimes thirty-two, nuclei (e).

THE EMBRYO

The development of the embryo takes about three months from the moment of fertilisation. With a length of 1.2–1.5mm it is rather small, compared with the entire seed that measures about 5mm in diameter. The embryo occupies the upper part of the seed (the apical, which is opposite the stalk). The embryo is not fully developed in the mature seed; it comprises two cotyledon primordia but no root or plumule ('feather', an apical shoot between the two cotyledons).

The whitish tissue surrounding and nourishing the embryo is the endosperm, which is rich in reserve substances, mainly proteins and fats. The embryo digests these by means of the cotyledons, which will perform this function until the first leaves develop. The cotyledons emerging from the seed are covered with a fine cuticle, and stomata begin to develop in the epidermis. The parenchyma cells are not yet differentiated as in the mature yew leaf. A cotyledon has one single vascular bundle, but the root of the seedling is already diarch (i.e. it contains two bundles of xylem and phloem) as in the mature yew.[6]

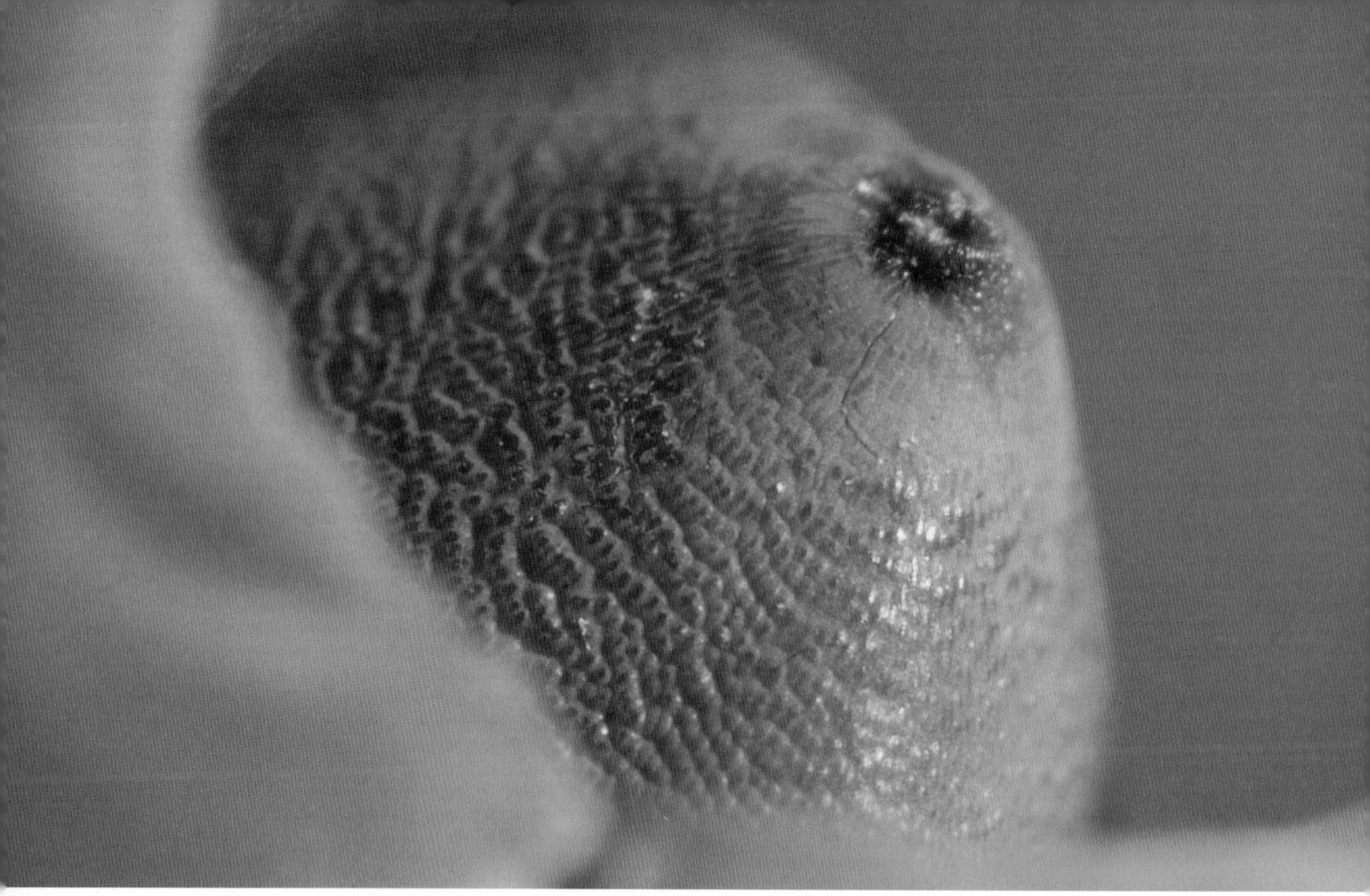

10.1 The seed of Taxus *is surrounded by the juicy and nutritious pulp.*

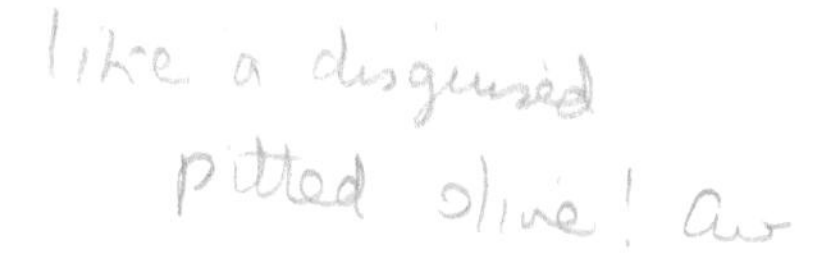

CHAPTER 10

THE SEED

THE ARIL

The 'fruit' of the yew comprises the seed and the surrounding red, fleshy aril, which has a typical size of about 7–9mm. Despite the tree's botanical name (*baccata* = berry-bearing), yew produces neither a fruit (gymnosperms do not have fruits) nor a berry (because it does not develop from the integument, see diagram p. 38). The aril's juicy, sweet-tasting, somewhat slimy pulp is the only non-poisonous part of the tree (*if* the aril is ripe – the green aril is toxic too!). In Europe and the north-eastern USA, the arils ripen between August and October, extending into November in Russia and southern Ireland. In a favourable, mild climate, seed is produced abundantly in most years, but every two to three years in less optimal climates.[1]

The main function of the bright red, nutritious aril is to attract birds, which are the main agent of seed distribution (see Chapter 14). The sheer scale of the yew–bird relationship can be illustrated with an investigation of the Killarney woodland in 1975 and 1976, when the number of arils was estimated at 2.6 and 6.2 million/1,000sqm, respectively. This corresponds to about 96 and 308kg of arils per 1,000sqm, with a total energy content of 0.6 and 1.9 billion calories, respectively. It has been estimated that the birds ate about 35 per cent and 43 per cent of the weight of the arils in each of the two years.[2] It is therefore no mystery why woodlands rich in yew trees attract more birds than areas without yews.

Measurements of the moisture content of the arils vary from *c.* 31 per cent (Poland)[3] to over 60 per

10.2 The green arils of Japanese yew.

cent (Spain).[4] The dry pulp consists of 94 per cent carbohydrates, 2.6 per cent fibre, 2.3 per cent protein, 1.4 per cent ash (i.e. minerals), and 0.2 per cent lipids. The mineral contents[5] are, in g/100g: calcium 0.2, magnesium 0.1, phosphorus 0.4, potassium 6.0, sodium 0.2, and in mg/100g: iron 25, manganese 1, zinc 5, copper 1. The energy content has been variously measured: dry aril pulp at 3.88 kcal g^{-1}, ripe arils at 5.07 kcal g^{-1}, and the embryo and cotyledons at 8.41 kcal g^{-1} (kilo calories per tenth of a gram).[6] A healthy feast for the winged ones!

THE SEED ITSELF

Unlike the aril surrounding it, the seed[7] in its woody shell is highly toxic.

10.3 Longitudinal cut through the seed.

The average weight of the fresh yew seed has been repeatedly measured across Europe and the United States. In America, the weight ranges from 59 to 76mg (European yew). In England, the average is 56.5mg, but it was found that seeds from the south and west of Britain, with a more oceanic climate, were significantly heavier than seeds from more central and eastern locations. Thus, seeds from Surrey and Sussex weigh over 69mg ±7mg while those from Overton Hall, Derbyshire, weigh 45mg ±7mg. Yew Barrow in Cumbria, however, was shown to produce seeds with a mean weight of 61mg ±7mg.[8] No data is available for Scotland.

10.4 Arils in the evening sun at Borrowdale.

Across continental Europe a similar trend occurs, with 77mg in the Netherlands and 70mg on the Iberian Peninsula, down to 43–59mg for the seeds from stands in Poland. It is reasonable to assume that these weight differences are caused by the effects of different climatic conditions, which become more maritime towards the west of Europe.[9] The surprising observation, however, that seed in Britain is, on average, lighter than in western continental Europe could be explained with genetic properties of the trees, annual variation in seed production intensity and possibly different water content at the moment of weighing.[10]

CHAPTER 11

NATURAL REGENERATION

GERMINATION

The mature seed consists of the dormant embryo, embedded in remnants of the female gametophyte and megasporangium, and the seed coat that surrounds it. The embryo inside the seed is still immature when the aril is ripe. *Taxus* seed very rarely germinates in the first year; generally germination occurs in the second or even the third year. The long dormancy of the seed represents – despite the prolonged risk of fungal infection or rodent predation – an expression of the yew's ecological strategy: it increases the chances of germinating in favourable (weather) conditions, and also creates an assured seed bank in the ground that is accessible in the case of a poor flowering year. The viability of *Taxus* seeds can be extraordinarily high, approaching 100 per cent, but germination rates of 50–70 per cent are more usual (and still high). The seed can remain fertile for up to four years.[1]

11.1 *Seedling on a tree stump.*

The seeds of most conifers of cool temperate climates actually require a winter period of cold, moist stratification before they will germinate. The absorption of water, the passage of time, chilling, warming, oxygen availability and light exposure all take part in initiating the process. For the yew, there is another component: passing through the digestive tract of birds. Primarily, this frees the seed of the aril, which must be removed for successful germination to take place (otherwise, viability can go down to c. 2 per cent).[2] It has also been assumed that it may chemically stimulate the seed coat.[3]

Germination is not stimulated by an increase of light in spring, but by warmth. The embryo begins to grow and absorb water, and a seedling root breaks through the seed coat and grows down into the soil. The stem below the cotyledons elongates and lifts them above the ground. The stomata bands are located on the upper side of the cotyledons (in adult leaves only on the lower side). As the cotyledons begin photosynthesis, they produce the energy needed for the early growth of the young shoot. Three to four opposite pairs of leaves emerge above the cotyledons, followed by a few alternate leaves around the terminal bud. By the end of the first year, the seedling is usually 2–8cm high, and has a strong taproot with laterals. Growth for the next few years is slow, often less than 2.5cm in height annually.[4]

THE SEEDLING

As in most plant species, the vast majority of seeds and seedlings never succeed – the annual mortality rate of yew seedlings is about 10 per cent on average. Over the years, then, more than half the seedlings will not make it beyond this stage (during the sapling phase the losses decrease significantly).[5] Yew seeds are lost on ground that is, for example,

11.2 Germinating Taxus *seed.*

11.3 Yew and beech seedlings in winter.

thickly covered with grasses, dog's mercury *(Mercurialis perennis)*, bracken *(Pteridium aquilinum)*, brambles *(Rubus fruticosus)* or rhododendron. Seeds germinate well on a thick moss cover, but the seedlings soon topple over because the young root is too far from the ground to anchor the plant. And some areas are contaminated with a root pathogen such as *Phytophthora cinnamomi*, which causes benevolent symbiotic fungi to retreat.[6] The main reason for high mortality rates, however, is deer browsing, followed by damage by mice, hares and rabbits. Apart from this, some plants die as an expression of an early selection of weaker individuals (and their genotypes), which actually represents a positive selection process for the survival of a population.[7]

The height of a seedling increases slowly but constantly (see Table 6).[8] As the young plant increases in height, the rate of annual height gain increases too, and does not culminate before a total height of 4m or more. The height increase of *Taxus baccata* is clearly below that of all other European trees – even including the other species of the lower tier, namely box and holly – and can only be compared with that of juniper *(Juniperus communis)*.[9] It is worth noting that even during its seedling stage yew displays its tendency *not* to give priority to height increase over horizontal expansion or over root growth.[10] The budding of the apical shoot occurs only about one week after that of the side shoots, and their growth is even faster than that of the apical shoot for the first four to six weeks.[11]

Table 6 Early life phases of yew *(after Pietzarka 2005)*

height	*annual height increase*	*phase*	*approx. age (estimated)*
up to 10cm	-	**seedling phase** seedling, usually no side shoots yet	1-2 years*
10-50cm	1.3-1.8cm	**sapling phase** beginning of side shoot development; still in competition with ground vegetation	2-20 years
50-200cm	4-8cm	**juvenile phase** strong height increase, but still in reach of deer browsing	10-60 (70) years**
over 200cm	9-17cm	**mature phase** sexual maturity;*** crown out of reach for deer	50-200 years****

* Up to 8-year-old seedlings without side shoots have been observed, however.
** Single yews are known that reached 2m height in only 10 years.
*** Some yews flower already in the juvenile phase.
**** With the height increase of *Taxus* reaching its peak at about 160 years, the mature phase (uninhibited growth and flowering) can last explicitly longer than stated here.

Apart from the normally branched long shoots, *Taxus* also produces short shoots and linear shoots. **Short shoots** bear leaf scales only and are exclusively produced to bear female flowers. The female inflorescence of yew is borne on three types of twigs: normally branched shoots, 1-year-old

11.4 *Seedling with prominent sideshoots.*

11.5 *Competing apical shoots (compare page 4).*

short shoots, and older short shoots. Older short shoots (up to 7 years old has been known[12]) can be found on almost every flowering female yew. The term 'short shoot' relates to their extremely slow growth in length, which often is well below 1mm/year.[13]

Linear shoots, on the other hand, are side shoots that develop normally in length but fail to fork out into side twigs. The occurrence of linear shoots is often interpreted as a symptom of a decreasing vitality in growth potential; ash and horse chestnut trees, for example, produce linear shoots as a normal reaction to too much shade,[14] and in old trees it is an effect of a general drop in vitality.[15] But these reasons can be excluded for *Taxus*, even though yew seedlings with linear shoots display a clearly lower height increase than those without.[16] A vitality drop corresponding to ageing does not apply to seedlings. And there are old yews such as that at the monastery of Fonte Avellana in Italy, or at the church of Wormley, Essex, that have an abundant and vital crown, and even stand in full sun exposure as well, and yet display a clear tendency to linear shoots. Some such yews give the impression of a 'hanging' type, reminiscent of weeping birch or weeping willow. Thus, the reasons for linear shoots in *Taxus* are not understood – another area waiting for future research.

A delay of, or damage to, the budding of the apical shoot may easily result in a **multi-stemmed** crown. Already during the first three life stages (compare p. 75), 10–35 per cent of young yews have *competing apical shoots*, on average three to six.[17] The annual height increase of yews with competing apical shoots is even bigger than that of yews with a single dominant one! *Taxus* responds fast to the damage or withering of a dominant apical shoot, which occurs in about a tenth of the seedlings, even in fenced forest plots free of deer browsing. An important cause of this is the bud gall mite *Cecidophyes psilaspis,* which overwinters in the bud and can contribute significantly to the damage of apical shoots if it returns annually.[18] Yew is capable of balancing the loss of its apex by raising one or more of the upper side shoots to take its place. This produces the entire array of simple or multiple multi-stemmed shapes, right down to a shrub-like appearance.

The percentage of multi-stemmed yews continually increases with age – a study from 1996[19] showed 63 per cent multi-stemmed yews, with an average of 2.2 stems per tree – but varies enormously, mainly because of different browsing pressure and genetic differences between populations (sometimes also local coppicing practice). Hence, it is not correct to interpret mature and old multi-trunked yews generally as being advanced root suckers or even the result of 'trunk fusion' (see box p. 81). A multitude of leaders is a natural part of the appearance of *Taxus* from an early age.

CHAPTER 12

A POTENT POISON

All parts of the yew apart from the ripe (red) aril are poisonous. This is the result of a toxic alkaloid called taxine B[1] that is rapidly absorbed by the digestive tract of humans and other mammals. It strongly irritates the digestive tract and affects the nervous system and the liver, but its chief action is against the cardiac muscles, resulting in heart failure and death. Yew trees contain very different amounts of this poison, with the exception of the Himalaya yew which does not contain any taxine B and hence is not poisonous.

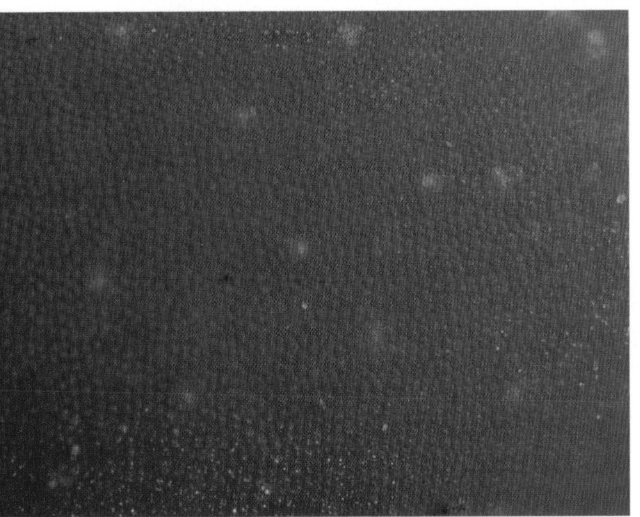

12.1 *Despite its alerting red colour, the red aril is the only non-poisonous part of the yew.*

Other poisons in *Taxus* are the taxanes, which are not soluble in water and cannot therefore be absorbed in the digestive tract, and hence do not lead to poisoning. Taxanes are highly sought-after for cancer chemotherapy (see Chapter 25). They occur in the bark and mainly in the leaves, with highest values in August. The 10-DAB content is highest in young buds and decreases with age (of the leaf).[2] Taxane levels among individual trees of any species *(T. baccata, cuspidata, brevifolia* and *floridana*) differ greatly.[3] An investigation of ninety-two cultivars (mostly of European yew) in the United States, for example, showed the taxane contents of the leaves varying from 0.0135 to 0.1471 per cent (in weight)[4] – a factor exceeding 10.8! A pharmaceutical study of ancient yews in Sardinia yielded even more puzzling results: the needles of male trees with a girth larger than 4m were collected from eleven different sites and investigated for their contents of 10-deacetyl baccatin III (DAB-III), paclitaxel and taxine. The levels of DAB-III varied in the extreme, and with no apparent correlation to soil type or altitude, which means the DAB-III content is genetically regulated. Not one of the samples contained paclitaxel; only two samples contained taxine, and in much lower levels than expected, about 300mg/kg of dried needles as opposed to 10–30g/kg. Furthermore, several trees were completely devoid of all three substances.[5]

HOW POISONOUS ARE YEW TREES?

Between 50 and 100 grams of yew leaves are considered fatal to adults, and an even smaller amount for children. For dogs and chickens, 30 grams is considered fatal, for horses and pigs, 100–200 grams, for sheep, 100–250 grams and 500 grams for cattle.[6] Goats, cats and guinea-pigs are less susceptible; and wild animals like deer, hare and rabbits are widely immune to yew poison, in fact, they take the foliage as a delicacy (being much softer than spruce needles, for example) and possibly use it for self-medication against parasites of the digestive tract. Yew seems

A worldwide investigation in 1998 shows 11,197 records of yew poisoning (from all *Taxus* species) in humans (96.4 per cent in children less than 12 years old) and found no deaths.[7] A 1992 article in the *Forensic Science International*, however, states that ten authenticated cases of fatal yew poisoning in humans had been recorded in the previous thirty-one years, but all had been deliberate.[8]

to be the favourite food especially for deer – a huge problem in *Taxus* regeneration attempts.

The symptoms of *Taxus* poisoning are manifold and equally horrifying. A fatal case may progress as follows: 30 to 90 minutes after intake vomiting occurs, with intense internal pains, diarrhoea with colicky pain, dizziness and numbness. Breathing speeds up at first but then becomes continuously slower and shallower. Circulatory disorders, and possibly suffocation cramps, are followed by fading consciousness, collapse, coma and finally death by paralysis of the heart and the breathing apparatus. Death occurs within twenty-four, sometimes even in one and a half hours.[9] However, around the turn of the last century, a number of authors expressly mention a 'particularly happy facial expression' on *Taxus* corpses.[10]

First aid is to call an ambulance, induce vomiting, and to give 10 grams of carbon powder and a dose of sodium sulphate. Hospital treatment includes the pumping of the stomach, anti-epileptic drugs to reduce cramps, blood plasma expanders, control of blood coagulation and of the liver and kidney values. A survivor is likely to have liver damage.[11]

Animals poisoned by yew first become very agitated, and subsequently their breathing becomes increasingly slow and shallow; they dribble, suffer cramps and paralysis of the breathing apparatus, and collapse. Horses undergo increased urination, colic and paralysis, and can be dead within half an hour, sometimes only minutes.[13] Throughout history, yew poisoning has led to a number of deaths of domesticated animals. Horses and even cattle have died, particularly when let into winter fields with no other green food present, or when given access to cut branches. *Yew foliage can be even more poisonous when wilted or dried* (which probably relates to the water content).[14] However, it is the sudden excess that is fatal: cattle, sheep and goats (and, to an extent, even horses) can build up an immunity, the key is habitual access to small quantities at a time.[15] As a matter of fact, yew foliage was a traditional part of animal fodder in many regions all over Europe and the Caucasus (until the introduction of artificial fodder supplements in the twentieth century), not only in times of scarcity but also for disease prevention.[16] Often, the yew leaves were mixed with other fodder.[17] This practice still persists in some regions, for example in Albania, where the foliage is even dried before feeding.[18]

Table 7 Oral lethal doses of yew foliage
in grams per kg of animal body weight[12]

horse	0.2–0.3
donkey and mule	1.6
pig	3
dog	8
sheep and cattle	10
goat	12
rabbit	20

12.2 *A simple fence keeps both horses and yews safe.*

In the early 1970s, five New Forest ponies in a 17-acre paddock near Kingley Vale ate yew leaves from time to time, and, keeping to small doses, suffered no ill effects. Equally wise were three donkeys in 1976 in the same area who did not even touch the yews despite a bleak, wet winter, but opted instead to eat the (supposedly poisonous) privet.[19] *Nobody should take these sorts of risks with their animals, but should keep them safely separate from yew trees (sound fencing).*

The toxicity of *Taxus* has been known since prehistory and was first committed to writing by Theophrastus (371–287 BCE).[20] Many other classical Greek and also Roman writers subsequently commented on this subject, not necessarily without exaggeration or superstition. Corsican honey could be tainted, if not poisoned, by *Taxus* flowers, wrote Virgil (70–19 BCE), for example.[21] Wine from yew wood containers would poison travellers, wrote

Pliny the Elder (23–79 CE).[22] The Greek physician and pharmacologist Dioscurides (40–*c.* 90 CE), whose work *De materia medica* remained the leading pharmacological text for no less than sixteen centuries, stated that 'the yew growing in Narvonia [Spain] has such a power that those who sit or sleep under its shade suffer harm or in many cases may even die!'[23] This might have been a useful hint about a local genotype especially rich in taxanes (the kind the research pioneers in chemical cancer remedies in the late twentieth century would have been desperate to find), but soon enough after Dioscurides's remark, sleeping in the shade of any yew was whispered to bring certain death;[24] and variations of this idea, and other superstitions, permeated European lore and physics for the nineteen centuries to follow.

As the great medieval physician Paracelsus (1493–1541) says, '*Dosis sola fecit venenum*' ('the dose makes the poison'), and hence, the yew's reputation to yield strong medicine lived on along with its folklore. But unfortunately, the know-how of handling it did not, and in the eighteenth and nineteenth centuries, the victims of lethal doses are legion. Yew derivatives were given for various ailments. A recommended small dose to increase menstrual flow, for example, was never quite a problem, but desperate attempts to abort killed countless mothers along with their unborn babies in England, France, Germany and Austria.[25]

A HALLUCINOGEN?

Alkaloids ('alkali-like') are organic nitrogen-containing bases, naturally occurring in certain families of flowering plants and not at all in conifers – except for the yew, that is. Many alkaloids, when taken internally, have diverse and important physiological and psychological effects on humans and animals. Well-known alkaloids include nicotine (from the tobacco plant, *Nicotiana tabacum*), morphine (a powerful narcotic from the opium poppy, *Papaver somniferum*), ephedrine (a blood-vessel constrictor from *Ephedra* species), cocaine (a very potent local anaesthetic from *Erythroxylon coca*), and hallucinogenic drugs such as mescaline (in the peyote cactus, *Anhalonium* species).

Apart from the taxine complex, *Taxus* also contains small quantities of ephedrine,[26] but despite the illustrious alkaloid 'neighbourhood' mentioned above, psychoactive effects of the yew are not known in literature; or *were* not, until Dr A. Kukowka, a retired medical professor, published his 'experience' in 1970.[27] At the age of 71 he had been doing some gardening underneath four yew trees for about two hours when he was overcome by dizziness, nausea, headache and a sudden restlessness. He became disorientated and lost his sense of time, and began to hallucinate. A spell of visions of vampires, vipers and 'diabolic scenes' was accompanied by a fearful cold sweat and somewhat paralysed limbs, but was soon followed by visions of a paradisiacal realm, heavenly music of the spheres under a huge dome, and a euphoric and 'indescribably happy' mood. When he finally emerged from his otherworldly experience, he called his doctor who could find no physical signs of poisoning, nor salt deficiency or overheating. These effects of the mere presence of yews proved to be reproducible, but Kukowka admitted fear and ceased his experimentation.

In his article, Kukowka himself mentions other people who have had similar experiences, but pharmacology has not yet confirmed any psychoactive effects of *Taxus*. However, a number of yew researchers and visitors[28] have reported 'funny things' happening to them underneath certain yew trees. Such experiences are usually too subtle to be measured or 'proven' clinically, but nevertheless suggest that a diffusion of taxoids from the leaves in certain weather conditions (hot and dry, no wind) might be possible. A tree, therefore, with high taxoid values could very well have a chemical influence on humans and animals underneath its crown, which would then reveal a core of truth in some of the ancient Greek texts on the airborne emanations of yew.

13.1 Red deer in winter.

13.2 Browsing damage by deer.

CHAPTER 13

MAMMALS

HERBIVORES

Despite its poisonous properties, *Taxus* is highly susceptible to browsing and bark-stripping by red deer, rabbits and hares, as well as domestic animals such as sheep, goats and sometimes even cattle. The yew is tolerant of repeated pruning though, and established young trees are able to continue to (re)grow even under (moderate) browsing pressure.[1] However, high populations of herbivores (mainly red deer) worldwide cause very serious problems in forest regeneration, not only for yews.

13.3 After over twenty years of being browsed, this yew still does not exceed 30cm in height.

Roe deer *(Capreolus capreolus)* especially are seen as a main factor inhibiting successful yew regeneration in many regions. Coming upon a yew woodland, deer browse the foliage of those trees at the wood's edge (up to the browsing line, which for deer is about 1.35m), because here the foliage reaches further down than inside a grove. Additionally, the male deer cause bark damage when they rub off the velvet skin from their annual antler regrowth each summer. But the main damage caused by roe deer is the destruction of the vast majority of seedlings and saplings: because deer have no incisors in the upper jaw, food is usually torn rather than cut (as with the hare and rabbit), hence yew saplings disappear entirely, including their roots. Deer browsing varies, of course, with season and with overall food resources. In Norway, for example, roe deer grazing on yew is highest when bilberry *(Vaccinium myrtillus)* is in low density, or when the ground is snow-covered.[2] Fallow deer *(Dama dama)* cause damage too, but are not as common as roe deer.

***13.4** Browsing line (sheep, goat) in old yew in Arzana, Sardinia.*

Hares and rabbits also browse yews regularly without harm,[3] that is, no harm to themselves. High grazing pressure can drastically affect the net growth rates of the yew, slowing down its already slow growth even more: one small yew on the South Downs, heavily visited by rabbits over the years, was found just 18cm high but with approximately fifty-five growth rings.[4] While rabbits prefer the shelter of the grassland (where they browse on pioneering yew seedlings, among many other things), hares like a clear view around them and so the rather bare ground of a pure yew stand is very convenient for them. They build their overground forms (i.e. a hare's daytime nest) in the woods where they lie all day. At dusk they come out of the woods to feed on the grassland, or on ivy, low hawthorn and blackthorn buds or bark, and twigs of oak, bramble and yew.[5]

Grey squirrels *(Sciurus carolinensis)* were introduced to the United Kingdom from North America in the late nineteenth/early twentieth century and have become a major concern for many forest keepers. They browse on leaves and twigs of many trees and even strip the bark of trees such as hornbeam, beech, birch, willow, Scots pine, sycamore, and also yew.[6] Sometimes half a tree is girdled, and the callus that covers the wound is again attacked in the

***13.5** Hare.*
***13.6** Young rabbit.*
***13.7** Wild boar.*

following year. Wounds can render a tree prone to fungal infections or parasites. Richard Williamson, in his time as warden of Kingley Vale, 'saw yew trees of about 100 years of age stripped of bark all down one side from a height of 16ft right down to the ground'.[7] Even greater damage, according to Williamson, was caused by the grey squirrels cutting hundreds of yew twigs to make their dreys in summer, wastefully leaving many of them cut but unused on the ground. Grey squirrels also eat yew arils, regurgitating the seed later.[8] Additionally, they rob birds' nests, particularly the first nests of mistle thrush, when the young have hatched. Mistle thrush is a great benefactor of *Taxus* (see the next chapter).[9]

Badgers *(Meles meles)* eat large numbers of yew arils from the ground, and therefore large quantities of seeds and undigested arils can be found in badger faeces.[10] The wild boar *(Sus scofa)* eats arils off the ground as well, while small mammals such as the edible dormouse *(Glis glis)* and forest dormouse *(Dryomis nitedula)* pick them from the trees.[11] In the fir forest Ghomaran Rif in the Moroccan mountains, the arils are part of the diet of wild monkeys (Barbary macaques, *Macaca sylvanus)*.[12]

Yew woods support higher densities of rodents than comparable deciduous woodlands, particularly the wood mouse *(Apodemus sylvaticus)* and the bank vole *(Clethrionomys glareolus)*, and to a lesser extent the yellow-necked mouse *(A. flavicollis)*.[13] In autumn and winter, yew seeds become the main diet of various rodents, which remove and eat vast numbers of yew seeds; for example, up to 60 per cent of seed fall for the woodlands of Co. Durham, in north-east England,[14] and *c.* 87 per cent in the Andalusian highlands in south-east Spain.[15] Unlike most birds, which eat the aril and pass the seed unharmed through their digestive tract, rodents open the seed shell and eat the toxic seed. They might also eat the energy-rich aril, but at times leave the pulp behind.[16] In all fairness it needs to be said that rodents often stash more seeds than they eventually eat, sometimes resulting in close groups of ten or more seedlings on the forest floor later on.[17] Voles, however (probably the water vole, *Arvicola terrestris*, in particular), prey on yew roots, and seedlings up to about 2m in height have reportedly been killed by this activity in Bavaria.[18]

13.8 *Mouse.*
13.9 *Rodents open the seed shells and eat the embryo.*
13.10 *Young yew root damaged by vole (*Arvicola *sp.) activity.*

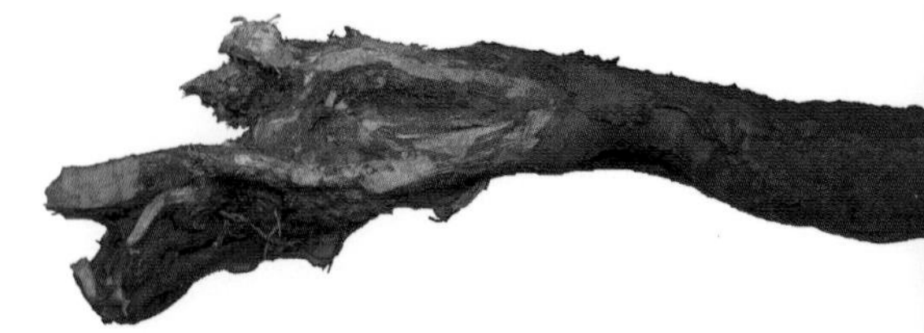

CARNIVORES

Foxes *(Vulpes vulpes)* have been observed delicately licking yew arils from the ground, and such meals of up to two pounds of arils later lead to the dropping of an excremental red 'haggis' that can contain over 200 yew seeds![19] But no doubt for the fox the greater attraction of the yew wood would be the numbers of rodents, rabbits and hares feeding there. Similarly, martens *(Martes martes)* have been reported eating yew arils (in Germany and in the Caucasus) and distributing the seeds with their faeces;[20] but like the stoats (ermines) and weasels *(Mustela* sp.*)* in Europe and America, their greater interest would have been in the rodents, and in birds and their eggs. Williamson describes stoats and weasels rapidly climbing up large yew trees to prey upon almost all woodland birds. Of the fox, he relates a story of the hunter having become the hunted: the fox-hunters around Kingley Vale told Williamson that the hunted foxes always run for the ancient yews and then disappear without a trace, most likely climbing up inside hollow yews and hiding there where they can never be found.[21]

On the ground, the rodents might attract snakes, for example, the common adder (European viper, *Vipera berus*). Obviously, all these predators help to regulate the populations of herbivores and hence decrease the damage they cause to forest plants. The greatest contributors to the natural regeneration of woodland are the wolf and the lynx, or were, before they became extinct in wide parts of Eurasia and America. The wolf *(Canis lupus)*, the largest wild dog-like carnivore, feeds primarily on deer (in the north also on moose or elk), and in their hunting activities the packs perform an important natural function in controlling the numbers of large herbivores. Hence a Carpathian proverb goes, 'Where the wolf walks, the forest grows.'[22] The various species of lynx are long-legged, large-pawed cats with tufted ears, hairy soles, and a broad, short head. They live alone or in small groups, hunt at night and feed on birds and small mammals, but occasionally also on deer. Animals such as wolf and lynx are not the 'enemies' of their prey. On the contrary, by continuously selecting the individuals that are less fit for survival they maintain an excellent gene pool for the hunted species.

Another visitor to the yew wood used to be the bear. In Eurasia, the various species of brown bear *(Ursus arctos*, in America also the grizzly) feed on small mammals, fish, vegetable materials, fruits (including yew arils),[23] and honey. Before the introduction of man-made beehives, honey bees *(Apis mellifera)* made their homes exclusively in rock crevices or trees, in hollow old yews for example (see figures 33.16–19).

13.11–13.13 *The greatest mammal allies of trees: lynx, wolf, and bear.*

***14.1** Mistle thrush on yew.*

CHAPTER 14

BIRDS

SEED DISPERSERS

Mixed woodlands containing yew populations attract a higher number of birds than other woods.[1] At least eighteen species of birds, one-third of them in the thrush family, eat yew arils, and some of these birds are the main agents of *Taxus* seed dispersal. They consume arils and seeds and later disgorge the seeds or pass them intact through their digestive tract.[2] This removes the fleshy aril that otherwise would inhibit germination. It is possible that the digestive juices of the birds act on the outer seed coat as a chemical germination signal, though there is some debate about this.[3] Additionally, those yew seeds dropped directly underneath birds' nests or roosting places receive nutrients from them for years to come. It is birds' wings that enable yew seeds to travel great distances, and also to reach isolated and sometimes inaccessible locations such as steep cliff-faces and walls.[4]

The eating of yew arils is limited to those birds that can swallow this size of seed, primarily starlings *(Sturnus vulgaris)* and, as mentioned above, various thrushes, particularly the song thrush *(Turdus philomelos)*, blackbird *(T. merula)* and mistle thrush *(T. viscivorus)*. Other aril-eating thrushes include fieldfare *(T. pilaris)*, redwing *(T. iliacus)* and ring ouzel *(T. torquatus)*. Other seed dispersers are robin *(Erithacus rubecula)*, blackcap *(Sylvia atricapilla)*, waxwing *(Bombycilla garrulus)*, jay *(Garrulus glandarius)* and pheasant *(Phasianus colchicus)*.[5]

Among all of these, the song thrush is the bird most fond of yew arils; they constitute about a quarter of its autumn diet. For mistle thrush and starling, the arils make up 16 and 18 per cent respectively of their food totals; for blackbirds only 7 per cent; and for redwing, robin and blackcap only 1 to 4 per cent.[6]

Most birds excrete the seed. The mistle thrush, however, eats large quantities of yew arils and later regurgitates a red mass of partly digested pulp with brown yew seeds. Up to twenty-three seeds have been found in one spit.[7] The fieldfare too can eat quite a few arils in one meal (up to thirty),[8] but usually feeds mainly on hawthorn fruits (haws) in semi-open habitat, only entering woodland when winter weather becomes more severe (by which time the more easily accessible yew arils are largely gone anyway). Generally, however, birds eat smaller numbers at a time. In Great Britain, for example, most thrushes take on average eight to ten arils per visit.[9] A study conducted in Germany,[10] however, shows thrushes having slightly bigger meals (ten to twelve arils per visit). It also shows a few more robins, fieldfares and even jays *(Garrulus glandarius)* feeding on yew than in Britain. Other German studies also mention wagtails *(Motacilla)* and spotted nutcracker *(Nucifraga caryocatactes)* eating arils.[11] After a yew meal, birds need to drink frequently and they can be observed washing their bills to remove remains of the glutinous pulp.[12] With no water in the immediate vicinity they may wipe their bills clean on branches.

For the tree, small meal sizes ensure a better seed dispersal as only a few seeds will be deposited at one site at any given time. Small meal sizes may also explain why so many yew seeds are left on the tree to fall to the ground. However, seeds transported by birds often accumulate underneath birds' resting places.[13] Shrubs like hawthorn and particularly juniper are able to provide a suitable microclimate for yew seedlings (see Chapter 5), as well as protection from deer, but often are heavily frequented by rodents which feed on the seeds. Groups of yew seedlings also occur underneath old spruce *(Picea)* trees in which thrushes or blackbirds sleep.[14]

In southern Britain, Williamson observed the seasonal bird activities at Kingley Vale. In the early autumn, various birds begin to arrive from their foreign summer habitats. Large groups of thrushes, partly from the more continental climate of Europe, flock in and begin to feed on yew arils. In October, vast flocks of fieldfare, arriving from Scandinavia or

14.2 *Blackbird on yew.*
14.3 *Washing off the sticky pulp; a blackcap.*

14.4 *Blackbird's nest on yew trunk.*
14.5 *(right) Robin's nest on yew trunk.*

14.6 Yew arils last well into the new year.

Russia, and redwing, on their way from the Baltic to Ireland or Spain and Portugal, join the other thrushes. After midwinter, when aril stocks are largely depleted, blackbird becomes the commonest bird. The only yews that still have a lot of arils in January are those defended by resident mistle thrushes, which claim their territories in late summer and keep them well into spring, driving off intruding birds. Their guarding work pays off in late winter when there is little other food available. Song thrushes have territories too but do not seem to defend 'their' yews as much as mistle thrushes.[15] Interestingly, the starling, which in Britain and Saxony is a keen feeder on yew arils, does not seem to eat arils at all in Eastern Europe (Hungary, Czech Republic).[16] Pheasants eat arils but also yew leaves, which seems to be detrimental to their health: around yew stands, these birds tend to be smaller than average (poisoning by yew leaves is also known in domestic poultry).[17]

The great significance of birds for the seed dispersal of yew reaches back far into earth's history. The commonly accepted predecessor of all birds is *Archaeopteryx*, a winged and feathered dinosaur, fossils of which have been found in strata 159 to 144 million years old. *Taxus jurassica* is 140 million years old, and it seems possible that its arils could have been eaten by some of the earliest bird species. If so, this would be another example of the co-evolution of species, similar to the mutual development of the first flowering plants and the insects that pollinated them. However, we do not know enough about the evolution of the aril and its possible precursors, largely because its watery, thin-skinned characteristics render conservation next to impossible.

SEED PREDATORS

While the seed dispersers eat the juicy pulp of the aril and leave the seed itself behind, other birds want just that. The main yew seed predator is the greenfinch *(Carduelis chloris)*, but seeds are also eaten by the great tit *(Parus major)*, bullfinch *(Pyrrhula pyrrhula)*, hawfinch *(Coccothraustes coccothraustes)*, nuthatch *(Sitta europaea)*, green woodpecker *(Picus viridis)* and the great spotted woodpecker *(Dendrocopos major)*, and occasionally by the blue tit *(Parus caeruleus)* and marsh tit *(P. palustris)*. The chaffinch *(Fringilla coelebs)* also feeds on yew seeds but only takes the seed fragments left over by the greenfinch.[18] The overall impact of the seed predators on yew regeneration, however, is quite small

14.7–14.9 Various aril meal deposits.

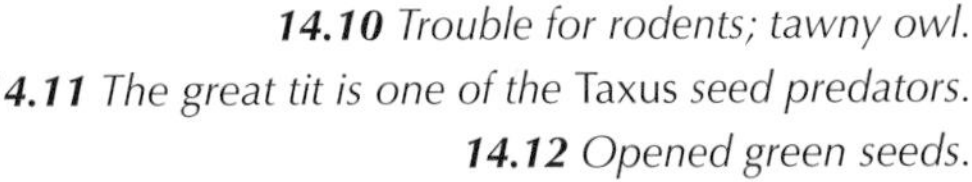

14.10 *Trouble for rodents; tawny owl.*
14.11 *The great tit is one of the* Taxus *seed predators.*
14.12 *Opened green seeds.*

compared with that of the seed dispersers (see Table 8).

The greenfinch 'chews' the aril until the pulp falls away, then rotates the seed in the bill, stripping off the brown seed coat (which contains cyanogenic glycosides) before eating the rest of the seed. Poisons in the seed coat might also explain why the woodpigeon *(Columba palumbus)*, generally a seed predator, does not eat yew arils, and crows do so only very rarely. The woodpecker and nuthatch wedge the seed into a crack in a tree, rock or wall, to stabilise it and hammer it open with their beaks. The nuthatch rubs the aril off on the tree bark before wedging the seed into a crack.[20] Unopened seeds can germinate in such crevices and, however slowly, develop into healthy plants.

Tits also use the wedging method, especially in cracks in the yew bark, but alternatively are able to firmly fix the seed with their feet. They also feed on other tree fruits such as acorns, beechnuts, haws and ash keys, but largely hunt insects in the canopy of trees. *Taxus* canopy has little in the way of insect life to offer, but its peeling bark holds moderate populations of small spiders for the tits.[21]

Table 8 Avian consumption of yew arils[19]

	Jul	Aug	Sep	Oct	Nov	Dec	Jan	Feb	**total**
seed dispersers									
starling	-	32	230	445	127	3	-	-	837
song thrush	6	55	120	109	124	53	43	-	510
blackbird	1	95	150	135	85	11	3	-	480
mistle thrush	2	26	68	44	54	28	10	1	233
redwing	-	-	-	34	19	1	1	-	55
ring ouzel	-	-	3	-	-	-	-	-	3
robin	-	2	9	5	4	1	5	-	26
blackcap	1	3	-	-	-	-	-	-	4
all dispersers	10	213	580	772	413	96	62	1	2,148
seed predators									
greenfinch	-	15	42	31	87	32	4	-	211
great tit	-	-	1	-	-	6	-	-	7
all predators									218

ANIMAL PREDATORS

The great numbers of birds and small mammals feeding well on yew arils occasionally attract birds of prey, for example, hawks *(Accipiter)* and buzzards *(Buteo)*. The culinary interests of crows *(Corvus)* and magpies *(Pica)*, although they eat insects and the eggs of other birds, are not primarily focused on the yew wood either, but they might take up residence and overwinter in it.[22] Night predators, in Europe particularly the common barn owl *(Tyto alba)* and the tawny owl *(Strix aluco)*, are interested in the mice, too. Although the tawny owl is rarely observed in coniferous woodlands in Europe (because of the lack of suitable holes in the straight and mostly under-aged trees) it seems to have a preference for nesting in yew woodlands.[23] Indeed, old hollow yew trees offer perfect homes for owls of various kinds.

14.13 *Woodpecker damage on a yew trunk at Kelheim, Germany.*

WOODPECKER DAMAGE

In Switzerland, Austria, Germany, the Czech Republic, Poland and the USA (south-western Oregon), many woodland yews bear strange marks – small, evenly spaced holes in horizontal rows or rings around the trunk, often arranged in vertical columns or in spiral patterns to produce characteristic arrays in the bark of trees. The holes are usually 3–8mm in diameter and 3–4.5cm apart from each other, while the distance of the horizontal rings measures about 9–11cm.[24] This phenomenon damages the living tissues (cambium, phloem and xylem) underneath the bark, and thus impairs the production of new wood and bark, the water transport, and especially the sap transport of the tree. In a few single cases, young yews have even died of this.[25]

Since these small wounds are already slightly weathered in summer, they are believed to be inflicted in the early spring.[26] But as to who and how, there have been no human witnesses as yet. The prime suspects in North America, however, are the acorn woodpecker *(Melanerpes formicivorus)* and four species of sapsuckers *(Sphyrapicus)*; and in Eurasia, the great spotted woodpecker *(Dendrocopus major)*, northern three-toe *(Picoides tridactylus)* and green woodpecker *(Picus viridis)*.[27] These birds are known to drill patterns in the bark of selected trees (oak, lime, beech, pine, larch, spruce, yew, wild service tree), where they function as 'community feeding stations'.[28] Other birds, mammals and insects visit woodpecker holes to consume sap and inner bark, and glean associated insects. But why the woodpeckers should prefer the rare yew tree remains to be seen. Also, at times the birds seem overly keen: some single inflicted yews have not just hundreds of holes, which would be sufficient, but thousands.[29] Assuming that the woodpeckers themselves consume some yew liquid or tissue,[30] and taking into account *Taxus*'s toxicity, the birds' unusual behaviour could be an instinctive act of self-medication to get rid of parasites in their digestive tract.[31] However, some trees show many holes not penetrating deeply enough to the living tissue,[32] and so the mystery continues.

CHAPTER 15

INVERTEBRATES

Compared with other trees of the temperate zone, the number of insects living on or from the yew is small. The wood is not readily attacked by woodworms, and the leaves only by a very few insects. This is mainly a result of the effects of the taxoids (see Chapter 12) contained in *Taxus*, particularly 10-deacetylbaccatin III and V.[1]

The most noteworthy insect species engaging with *Taxus* is the yew gall midge, *Taxomyia taxi* (see box on next page), cause of the 'artichoke' galls so often seen on yews in Europe (not in the US and Japan). The gall is a dense needle cluster formed by the tree in response to stimulation by the newly hatched larva.[2] Inside the gall, the plant cells adjacent to the gall midge larva are characterised by young, undifferentiated cells. Their epidermis is permeable; in order to feed, the larva probably uses its mandibles to cut a minute slit[3] into the plant tissue and, owing to the osmotic pressure in the tree, a tiny constant flow of plant sap flows and feeds the larva.

15.1 *Wasp nest in Japanese yew, Hakusan Jinja, Nagano prefecture, Japan.*

The life cycle of the yew gall midge, *Taxomyia taxi*

By Dr Margaret Redfern, University of Sheffield

Taxomyia taxi is an unusual gall midge in that it can develop from egg to adult in either one or two years, with these two life cycles producing quite different galls. Most individuals develop in two years, and cause the large artichoke gall (figure 15.2a) in their second year. The one-year life cycle, which contributes only 5–10 per cent of a population each year, induces enlarged buds only about 5mm tall.

In both galls, the gall midge develops through the same stages: an egg, three larval instars, a pupa and an adult, but the first instar lasts about fourteen months in the two-year galls and only two months in the one-year galls. Eggs (fig. b) are laid in late May and early June on the leaves of new shoots and hatch in about ten days (or twice that if June is cold and wet). The tiny hatchlings crawl towards the tip of the shoot and burrow into the buds – they prefer the terminal bud although they grow and develop normally in the lateral buds too. One- and two-year galls cannot yet be distinguished. These are tiny dormant buds that, if not infested, would produce the next year's shoots. The larva sits on the meristem (cell-producing tissue) which becomes flattened and slightly enlarged, but otherwise has little effect on the bud at this stage.

15.2a *The artichoke gall of* Taxomyia taxi *on yew.*

In August, when the galls are two months old, development of the one- and two-year galls diverges. In buds that will become two-year galls, leaf production speeds up so that by October they contain twice as many new leaves as in ungalled buds and the artichoke gall is recognisable. In buds developing into one-year galls, the meristem erupts into a fleshy pad consisting of nutritive tissue, and this event signals the end of leaf production. The pad develops in the two-year galls, too (fig. c), but not until one year later and after many more leaves have been produced. Eruption of the pad stimulates the larva to moult into its second instar and to grow very rapidly – larval weight increases seven- or eightfold during September. In October, the larva moults into its final instar and grows until prevented by cold weather. It grows again in the spring, when its weight doubles (fig. d), and pupates (fig. e) in late April. Adults (figs f, g) emerge a month or so later, with females outnumbering males by about three to one. They live only about a day, so must mate and lay eggs straightaway.

The larva can suck the sap with the aid of pumps located in its mouth and throat, and it can live on sap alone. Salivary gland secretions may reduce starch inside the plant cells, and because the plant supplies sugar from other tissue areas, a continuous supply flows to the larva.[4]

However, the yew gall midge is not a dangerous 'parasite'[5] of the yew, it is a specialised herbivore whose activities have little effect on the growth of the host trees, or vice versa.[6] Obviously, the growth of single shoots is inhibited by the formation of galls, but any healthy tree can tolerate this small percentage of loss. And the gall midge populations are kept within boundaries that are 'safe' for the yew populations by the effect of two (true) parasites that prey on the gall midge.

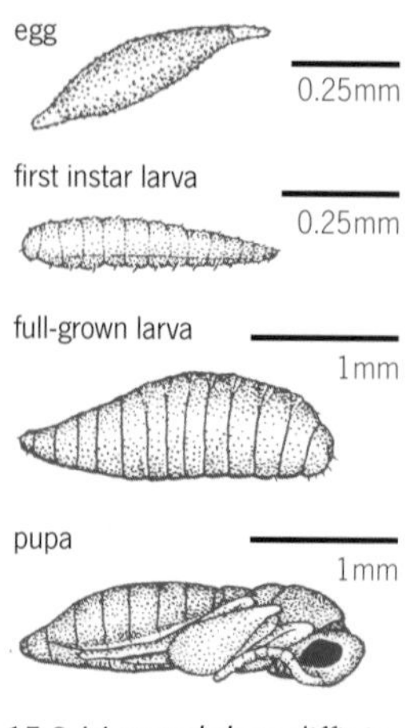

15.3 *Mesopolobus diffinis.*

The first of these two (true) parasites is the parasitic wasp *Mesopolobus diffinis,* which attacks the third instar and early pupal stages of

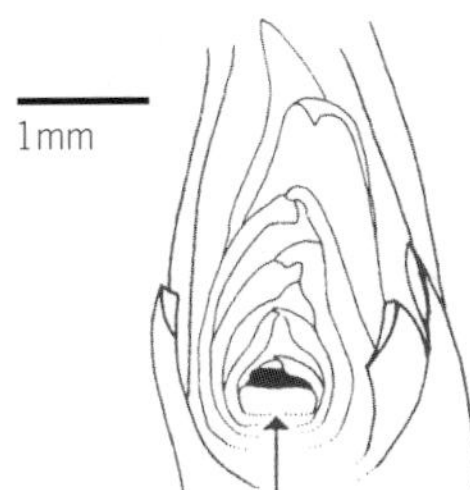

***15.2c** Section through the inner part of an artichoke gall showing a larva (in black) on the erupted pad.*

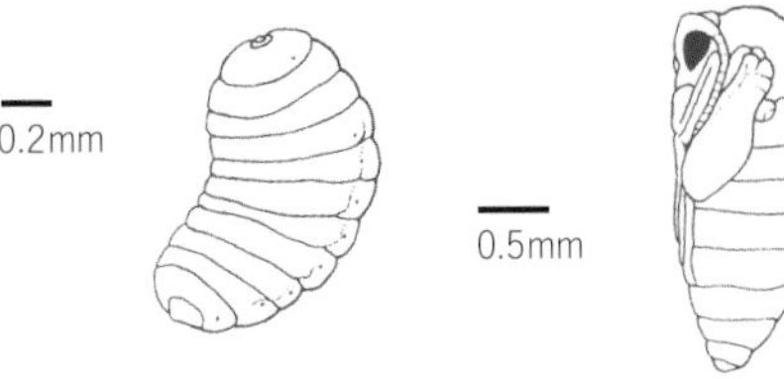

***15.2d** The full-grown larva of* Taxomyia taxi.

***15.2e** The pupa of* Taxomyia taxi.

Despite the different galls, development is identical in the one- and two-year life cycles after the first instar; development just occurs one year later in the artichoke galls. The artichokes have a greater effect on the tree, though: they stimulate a greater production of leaves and prevent the shoot from elongating so that the leaves remain in a compact cluster.

Other gall midges also have varying life cycle times of one, two, or more years – a strategy that has probably evolved to avoid or reduce parasitism. Usually, the dominant life cycle is annual: a species capable of developing in one year will normally do so. But not in *Taxomyia taxi.* Although females would produce twice as many offspring if they developed in one year, many more individuals would die from parasitism. Parasitism (caused by *Mesopolobus diffinis*) is much higher in the one-year galls, so two-year parents leave more descendants than the one-year adults, a strategy that clearly benefits the species. But even though few one-year individuals survive to breed, those that do provide an important link between the two-year life cycles of odd and even years, allowing genetic exchange between them – otherwise they might evolve into two distinct species.

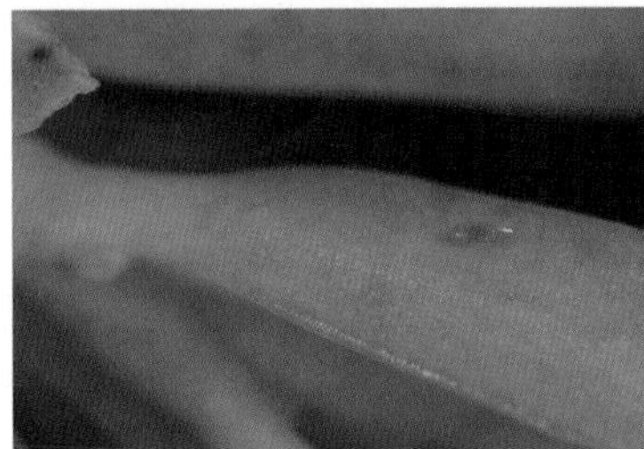

***15.2b** An egg of* Taxomyia taxi *on a yew leaf.*

***15.2f** Adult female.*

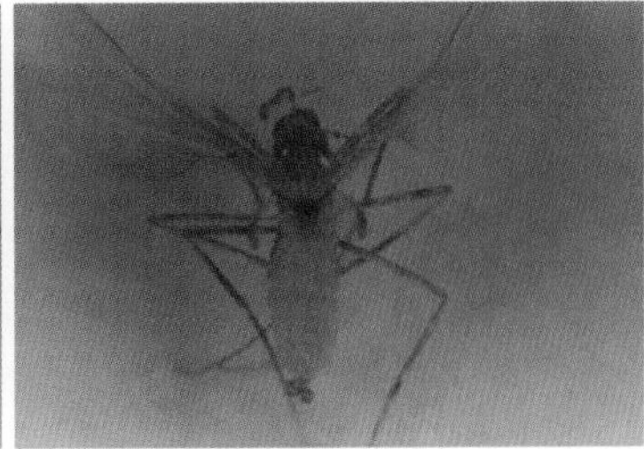

***15.2g** Adult male.*

Taxomyia. *Mesopolobus* adult insects appear on the galls from September to November and lay their eggs on the host larvae in both the two-year and one-year galls (see box). The *M. diffinis* larva soon kills the *Taxomyia* larva, grows quickly, overwinters with the shrivelled skin of the gall midge larva, pupates, and emerges as a new adult in March. Each winter and spring, *M. diffinis* has three generations on one generation of its host (the gall midge). This enables it to build up its numbers quickly within a single host generation when the host is common, and hence it might be able to act as a kind of population control of the gall midge.[7]

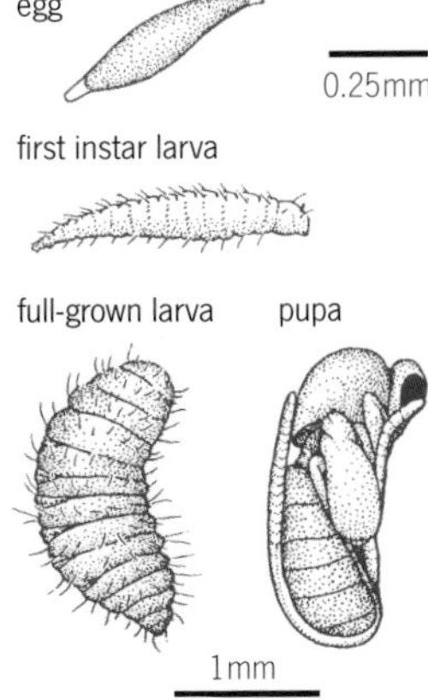

***15.4** Torymus nigritarsus.*

The other *Taxomyia* parasite is also a parasitic wasp. *Torymus nigritarsus* lays its eggs on fully fed larvae and pupae of the yew gall midge in late April and May, and only attacks hosts in the two-year galls. The parasitoid larvae are already fully grown by July; they overwinter in the gall,

15.5–15.8 *Occasional visitors …*

and adults emerge in late March to April. In the life cycle of the yew gall midge, *T. nigritarsus* is the last major peak in the mortality rate.[8]

The caterpillars of two butterfly species (order Lepidoptera), *Ditula angustiorana* (red-barred tortrix) and *Blastobasis lignea*, are commonly seen feeding on yew leaves (including gall leaves and gall midge larvae). They are not restricted to yew, however. Their larvae feed from late July to autumn, and pupate in May and June. The caterpillars themselves are prey to birds, especially tits.[9]

Other insects on *Taxus* include the larvae of the house longhorned beetle (*Hylotrupes bajulus*), and those of the death-watch beetle (*Xestobium rufovillosum*) that attack the sapwood of yew.[10] A dangerous beetle is the black vine weevil *(Otiorhynchus sulcatus)*, which 'ringbarks' yew shoots all around the stem, causing them to turn red, wither and die, though they remain on the branch for some time. It also attacks the roots of seedlings as well as the apical buds,[11] the latter leading to multi-stemmed (and later multi-trunked) yews. Rather benevolent to the yew are ants, as they only eat the sugary aril pulp of some arils on the ground,[12] thereby exposing the seeds and thus improving seed germination to a humble extent. Last but not least, wild honey bees *(Apis mellifica)*[13] enjoy perfect nest-building conditions inside hollow yews (compare figures 33.16–19).

Commonly found in the vicinity of *Taxus* is the woodlouse (*Porcellio scaber*, also the pill bug, *Armadillidium vulgare*) which is not an insect but belongs to the ancient biological class Crustaceae. Most of the animals in this class are aquatic (e.g. shrimp, lobster) and the woodlouse, albeit terrestrial, still breathes through gills (pseudotrachea). The woodlouse does not have a waterproof waxy cuticle on its exoskeleton, as insects do, and is therefore prone to dehydration (desiccation). For these reasons, woodlice are usually nocturnal, and prefer a dark, damp and cool habitat.

15.9 *Woodlouse.*

16.1 *Wood decay inside a hollow yew trunk.*

CHAPTER 16

PARASITES

FUNGI

Wood on the forest floor takes about half as long to be decomposed by fungi as dead wood in the crown region,[1] mainly because of moisture conditions. Fallen trunks and dead branches on the ground also harbour a greater number of wood-decomposing fungi. But the wood of living trees is not safe from fungi attack either.

Compared to other forest trees, *Taxus baccata* is affected by only a few serious fungal diseases. In stands in Poland, for example, *Taxus* regeneration is inhibited by a hostile microbiological environment including the pathogenic fungi *Nectria radicicola* that kills yew seedlings.[2] In the Fürstenwald near Chur (Canton Grisons),[3] Switzerland, every fourth tree of the yew population manifests widespread stem canker. Fungal infection is thought to be the cause, and some of the tree wounds show fruiting bodies of the golden spreading polypore *(Phellinus chysoloma)*. This fungi also attacks spruce, fir and larch, but never pine. While trees normally cover their non-pathogenic wounds with wound material (callus), the failure to do so usually indicates the presence of an infection. *P. chysoloma* enters the heartwood via broken branches or other deep wounds, and attacks the cambium. It later spreads into the sapwood as well, and if the decay is faster than the new growth it will reach the bark surface

16.2–16.4 *Various fruit bodies of the sulphur bracket.*

where cankers will develop as a consequence. The living part of the cambium is increasingly pushed away from the bark and ceases to be able to seal the wound. The fungal growth is very slow, however: dendrological research at this location reveals that the first cambium damage occurred as early as 1948. After decades, single trees die from this infection, while others manage to eventually cover the canker with layers of healthy bark.[4] Similar stem cankers in yew are known from the Uetliberg near Zürich, Switzerland, and from Derbyshire and Sheffield in Britain.[5]

Taxus is not attacked by the common dry rot fungus, *Serpula lacrymans*.[6] It is also very resistant to honey fungus (*Armillaria* spec.),[7] a mushroom genus whose individuals can live for centuries to produce the largest fungal underground networks (mycelium) known, and live on nutrients from decaying wood or living roots of hardwood trees and conifers alike. There are no authenticated cases of yew death as a result of honey fungus.[8] The only known fatal disease of yew in Britain is an infection with the root pathogen *Phytophthora*.[9] Another study of wood-consuming fungi[10] conducted over four European countries confirmed the extreme slowness of fungal decay in *Taxus*. Eighty species of fungi[11] were identified in this study, only one of which is specific to yew, *Aleurodiscus aurantius*.[12] In total, 258 species of fungi and slime moulds have been identified so far on *Taxus baccata* or on the soil beneath (for comparison: 2,200 species have been found on beech or oak).[13]

Fungal growth in wood is strongly bound to the permeability of the wood, which allows the hyphens to penetrate but also affects the availability of water and gases. This makes yew a difficult subject for fungi. The longitudinal (along the xylem fibres) permeability of yew heartwood is extremely low – much lower than in all other European trees – and its radial permeability is even lower. To give an example, the longitudinal permeability of the wood of the Western red cedar *(Thuja plicata*, a tree native to America) is about 1,000 times higher than that of yew. But the wood of this tree also decays rather slowly, because of its high contents of terpenoids. Hence it has been proposed that the high fungal resistance of yew wood probably owes more to its physical characteristics than its wood chemistry.[14]

In old yews, however, fungi play a crucial part in the hollowing of the trunk and branches. Most commonly observed in this activity is the sulphur bracket, *Laetiporus sulphureus*, also known as 'chicken of the woods'. Although at first glance an attack on the structure of the tree, the decay of the old heartwood is a vital step in the life stages of yew (see p. 75).

16.5 *(left)* Ganoderma carsonum *is rare on yew (this one was found in Kent).*
16.6 Ganoderma resinaceum.

16.9 *Wood breaking down into dark brown to black cubes is characteristic of the work of the sulphur bracket.*

16.7–16.8 *Two examples of stem canker; healing has begun in the tree on the left.*

MITES

The yew big bud mite *(Cecidophyopsis psilaspis)*[15] is known as a pest of yew in northern and Central Europe. Its infestation causes abnormal elongation and the swelling of the bud scales, which leads to extensive and chronic bud mortality throughout the canopy and an irregular branching pattern.[16] Another mite, *Eriophyes psilaspis*, causes tumour-like growths and deformations as well as discolouring of leaves and buds. In the long run, leaf buds and male flowers dry up and fall off.[17] Altogether, 199 species of mites have been sampled from yews in England and Wales. The majority of these mites are most likely free-living predators and not strictly confined to yew. Interestingly, there are distinct biological differences in endemic mite species associated with solitary yews (thirty-nine species have been found solely within the environment of solitary yews) and those in stands of forest yews (twenty-five species of mites).[18]

CHAPTER 17

VITALITY

In trees and humans alike, health is more than the absence of disease. Before a tree gets damaged by parasites, root pathogens or other fungal infections, there is a weakening or imbalance in the tree itself. At present, one of the methods offered by conventional technology in studying the vitality of biological organisms is to measure their electrical currents. The electric activities of trees were first measured in 1925,[1] and since the 1960s it has been known that electric tree potentials reflect the rhythms of day and night, of the seasons, and of the moon.[2] They interact with the air electricity that directly surrounds the tree, and even with the earth's magnetic field.[3]

Since his pioneering studies of geo-phyto-electrical currents (GPEC) began in 1969, the Czech scientist Vladimír Rajda[4] has become an international expert in this field. For this book he kindly agreed to conduct a twelve-month study of *Taxus* in the Czech Republic, from June 2004 to June 2005.

The environment in the Czech Republic is in the slow process of recovering from the effects of heavy industry during the period of Communist rule. The majority of trees are ill, with vitality rates never higher than 68 per cent.[5] Yew trees are known to be quite tolerant of chemical pollution, but given the

Electrodiagnostics of trees

Most of the physical and chemical properties of all atoms and molecules derive from their electrical charges. Chemical reactions between molecules, for example, or the absorption of mineral nutrients by a living cell are governed by the laws of electricity. Hence the measurable electrical activity of an organism can indicate its state of vitality. Long before visible symptoms of illness occur on the outside there will be an irregularity in the electrical currents.

The method of electrodiagnostics of trees uses the existence and the laws of electrical currents and the tension voltage between the soil and the tree. The GPEC are present in all tree parts, above and below ground. Vladimír Rajda's method of electrodiagnostics uses a mobile measuring device for direct current (DC), and two special metallic probes. One probe is inserted 20–60cm deep into the ground at a distance of 0.2 to 40m from the tree; the other probe is much shorter (*c.* 10cm) since it only has to penetrate the cambium and phloem layers beneath the bark (the xylem has only *c.* 65 per cent of the strength of the currents of the cambium and phloem). Measurements take place at the base of the trunk because here the currents are strongest. They decrease with height, and at 6m above ground are only half as strong as just above ground level.

The strength of the GPEC mirrors the vitality of a tree, and a decrease in the electrical activity is inevitably followed by a decrease of water uptake and therefore nutrient supply. This causes the electrical currents to decrease further, while the electrical resistance of the tree grows exponentially (from 30–50Ω in a healthy tree to 30,000–60,000Ω in a severely ill tree). The balance of the nutrient distribution collapses, and after a phase of malnutrition the tree is too weak to ward off parasites and pests. The close relation between the bioelectrical and biochemical metabolism of plants has enabled Rajda to develop an early warning system for forestry commissions, detecting unhealthy trees before physical symptoms appear.[6]

Furthermore, Rajda's studies have shown that an increase of light raises the GPEC, as does an increase in air temperature (1°C/1.8°F equals a change of 2.8 microampere in electrical intensity). Results of earlier American studies[7] could confirm, allowing for the above variables plus differing soil and water conditions, that each tree species has its own distinct characteristics[8] in the GPEC, characteristics that are identical for every specimen regardless of elevation or geographical location. Within the limits specific to the species, the GPEC of each tree follow two patterns: during the juvenile stage a continuous rise of the GPEC activity with increasing trunk diameter, and throughout its entire life an annual rhythm, with a peak in summer and a low point at midwinter.[9]

17.1 *The ancient yew (male, girth 788cm at ground level in 1999) at Ankerwyke, Buckinghamshire.*

present environmental condition of the Czech Republic it was still surprising for Rajda's team to find all examined yew trees with high vitality rates, some of almost 100 per cent.[10]

The *Taxus* study[11] that followed this discovery shows the yew once again in an intermediary role between the native leaf trees and conifers of Europe. Deciduous trees generally have clearly higher GPEC than evergreens, and the yew is in-between. The highest levels have been measured in horse chestnuts (*Aesculus hippocastanum*, 410μA/cm^{-1} and 0.936V) and oaks (*Quercus robur*, 370μA/cm^{-1} and 0.830V) with 50cm trunk diameter. One of the least GPEC-active trees is Scots pine *(Pinus sylvestris)*,

Diagram 5
Comparison of the vitality levels of different tree species

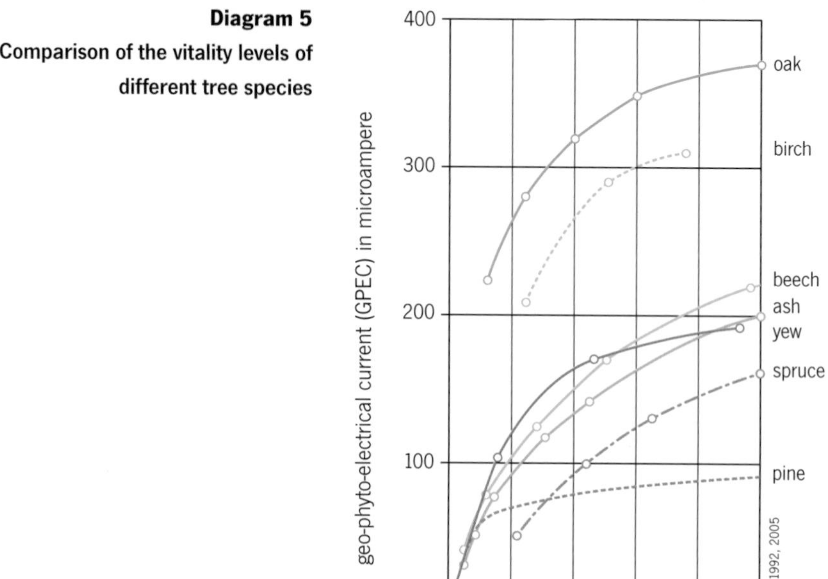

Diagram 6
The annual vitality rhythms in different tree species

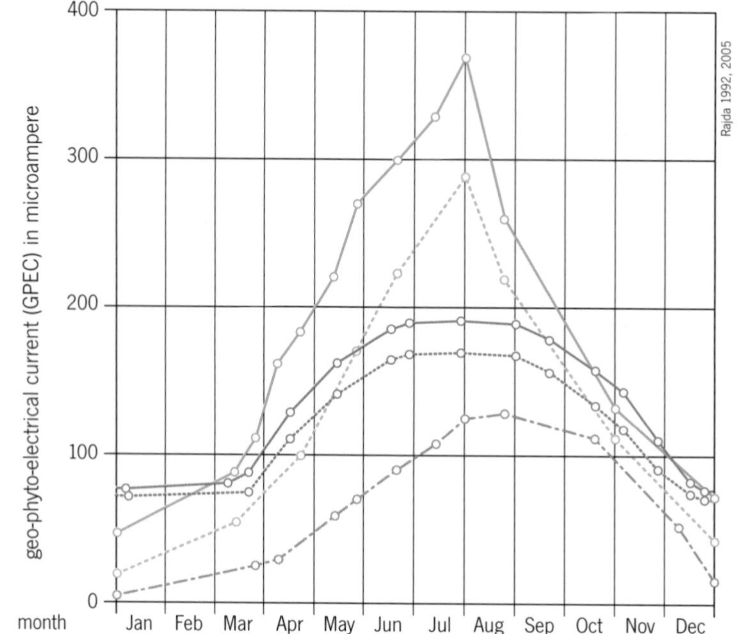

17.2 Thriving yews at Conswell Scar, Borrowdale.

which lives on less than a third of the currents produced in oak (see Diagram 5). Spruce *(Picea abies)* values are only slightly higher than pine. The cedar of Lebanon *(Cedrus libani),* however, during its juvenile stages has similar GPEC values to the yew; but it is not a native tree of Europe.

The annual rhythm of our flora shows all investigated plants sharing an annual absolute low point at the turn of the year, and from there an uninterrupted continuous increase of metabolic activity (in healthy plants) until the peak is reached at the end of July. Conifers such as spruce might even (slightly) increase well into August, but before the end of the month join the common decrease that continues until the end of the year. The curves of deciduous trees rise and fall more sharply than those of conifers because deciduous trees reach much higher levels than the evergreens in summer.

The yew, however, is different. Its annual GPEC curve resembles a mighty dome that spreads from mid-March to mid-December. While the short-lived summer maximum values of deciduous trees only last for a day or two, yew holds its top level (195μA/cm-1 and 0.833V) for a full two months. This unique dome shape for *Taxus* indicates that, although the GPEC in oak around 1 August are about twice as high as those of yew, the annual averages of both trees are fairly similar. In other words, the annual average energy level of yew is one of the highest among European trees, despite the tree being an evergreen.

Another striking observation is that yew reaches its annual minimum between 20 and 25 December, which is out of sync with all other trees investigated. Exactly during these shortest days of the year, the phloem and cambium layers of yew are at their driest. And yet, even during this low point of general vegetative hibernation, the GPEC in yew never fall below 70μA/cm-1 and 0.833V while those of other conifers sink to a fraction of this value. Zero, by the way, is only reached at the death of a plant.

An unexpected observation was made at two old multi-stemmed yews. Generally, the strength of the electrical currents in the base of a multi-stemmed tree equals the sum of the different currents measured in all its single stems above. But in yew, apparently, the single trunks maintain their individual GPEC characteristics right down to ground level. This shows an independence of the metabolism of the single stems that is not known in other trees. And since the full strength of the GPEC must unite somewhere, it must be the root system where this occurs.[12] Hence we arrive at yet more evidence that the root is of outstanding strength and importance in yew.

As for health, about half of the yews measured in the Czech Republic showed 100 per cent vitality; the other half, with 83–87 per cent, must be considered physiologically weakened but still range above the country's average of 68 per cent for the other tree species. The study confirmed the high vitality of *Taxus baccata* and its significant resistance against environmental pollutants. It also suggests that the high life expectancy of yew is at least partly a result of its high vitality as shown in the GPEC rates.

17.3 The set-up for measuring geo-phyto-electrical currents (GPEC).

CHAPTER 18

THE WOOD

The structural wooden parts of a tree are covered by a protective layer of bark. The bark of the yew is usually reddish-brown, thin, odourless, and has a slightly bitter and astringent taste. With expanding girth, old layers of bark are irregularly shed in thin, flat scales on the outside; while on the inside, bark regrows from the bark cambium.

18.1 Cross-section of young trunk.

Under the bark of a tree there is only a thin layer of living tissue, the cambium. Towards the outside, the cambium produces phloem cells, the other long-distance transport system of the tree, responsible for the distribution of the nutritious sap of assimilates created by the leaves. Unlike the xylem cells, those of the phloem are alive. The phloem cells or 'sieve cells' form continuous channels (sieve tubes), and their living substance is confined to a thin layer to make space for the passing sap. In case of injury, the sieve plates interrupt the sap stream in order to avoid unnecessary loss of the precious liquid. The phloem cells, too, eventually die and create the bark, older layers of which are shed and fall to the ground.

18.2 Annual rings in the fluted section of a 160-year-old trunk.

Towards the inside of the tree, the cambium produces the xylem, the tube system responsible for transporting water from the roots into the crown of the tree. When the xylem cells have reached their destined shape and size, they die. The inner core and everything inside the cell wall vanishes, leaving only a hollow but strong tube of cellulose, additionally strengthened by the incorporation of substances such as lignin. After functioning, for a year or longer, as part of the water transport system (the sapwood), old xylem cells eventually become blocked and retire. Chemical changes convert them to heartwood, a mechanical support structure for the trunk and crown. The single rings are distinguishable because the xylem cells produced in spring (earlywood) are larger and have thinner walls than those produced towards the end of the growth season (latewood). In *Taxus*, the reddish heartwood is distinct from the pale sapwood, which is usually about ten to twenty annual rings thick.[1]

Generally, the annual rings of yew are comparatively narrow, mirroring its slow growth. However, in this respect too, a large spectrum of variety exists in yew. Wood grown in low altitudes sometimes has broader rings than wood from mountainous areas, and the especially fine grain and immense load-bearing capacities of the Alpine yews were highly sought after during the age of the British longbow. *Taxus* annual rings in cross-cut are often wavy, which, after annual amplification of the impulse, results in fluted trunks and branches as in hornbeam

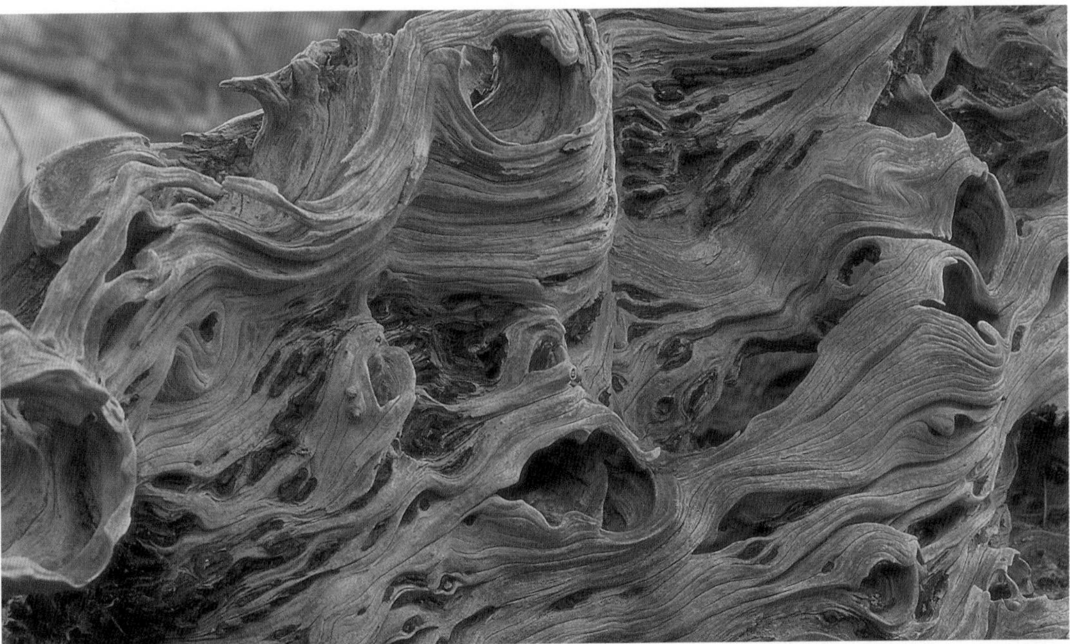

18.3 *The flowing shapes of secondary growth.*

(Carpinus betulus). But those in yew are often far more accentuated, so much so, in fact, that their appearance frequently leads to rash assumptions of 'trunk fusion' (see box on p. 81).

Within the annual rings of yew, the transition from early (spring) wood to late (summer and autumn) wood is gradual as compared to many trees that show two very distinct annual growth phases under the microscope. Rainfall from February to July has a positive effect on the growth of yew wood, mild temperatures in late winter (January/February)[2] and late autumn (October) can extend the growth season, while high summer temperature (particularly in July) can inhibit growth.[3]

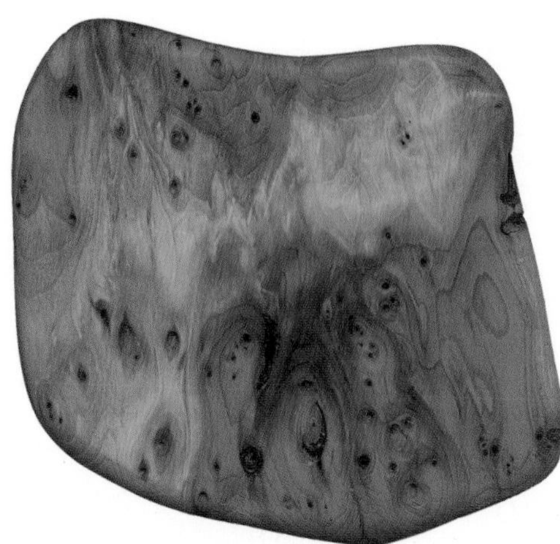

18.4 *Always full of surprises …*

While tree rings grow concentrically outwards from the pith, they do not necessarily grow equally in all directions. A slanting tree, for example, exerts a pressure on one side of its trunk, and a pull on the other. The root system, too, has to adapt to that, and so does the trunk architecture. Compression wood is built for stabilisation and shows wider rings, a higher lignin content, and a darker colour than normal wood.[4] *Taxus* is extraordinarily slow in shedding its dead branches, and this causes additional disturbances to the annual ring pattern (and prevents long straight sections of timber that commercial forestry would wish to see).

The extremely slow growth of *Taxus* makes its wood hard, heavy and durable. Its high density is evident in its weight: 640–800kg/m^3 (compared with redwood 420, pine 510, beech and oak 720).[5]

The rate of girth increment of *Taxus* culminates at an age of about 110–120 years, and height only at about 150–160 years,[6] both much later than those of all other trees.

18.5–18.6 *Adventitious growth covers the trunks of these old yews in Wales: Garthbeibio (Foel), Powys (left), and Llangeitho, Ceredigion (right).*

REGENERATIVE GROWTH

The hollowing of the trunk of ageing yew trees is a general characteristic of *Taxus*.[7] As the core of the trunk decays, a process of vegetative regeneration (fully discussed in the next chapter) occurs. During the various stages of a yew tree's growth, the load-bearing stresses on the supporting trunk repeatedly need readdressing, and as a result, new growth is laid down in various areas around the increasingly thin-walled trunk. This regenerative growth is usually even less straight-grained than the original wood; it is of a rather curvy, winding and flowing nature.

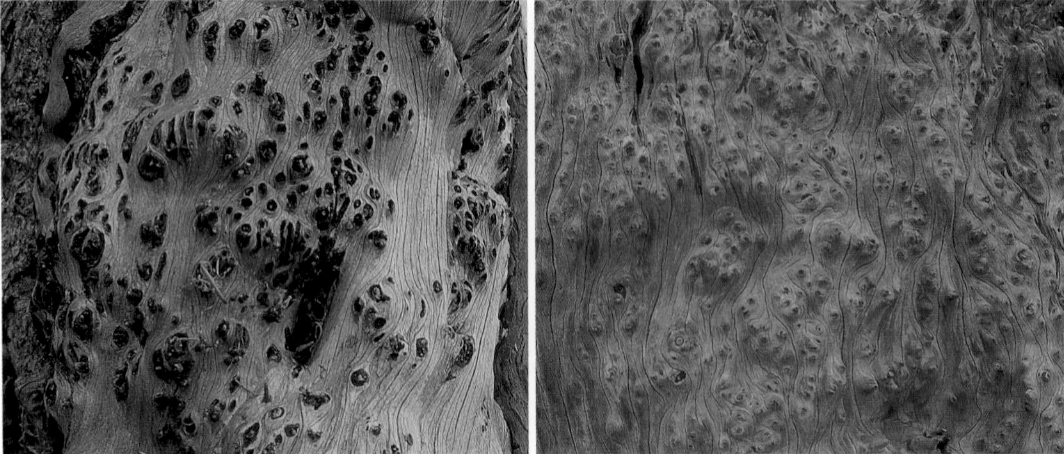

18.7–18.8 *Wood with the imprint of adventitious (epicormic) growth.*

Furthermore, yew cambium is able to produce green shoots from anywhere beneath the bark, and often the lower trunk of light-exposed solitaries is abundantly covered with such adventitious growth.[8] Most of these shoots are quite short-lived, as they either wither away or get eaten by browsing animals. The swelling of many old yew-boles into burrs, with convoluted, intensely gnarled and irregular structure ('grain') beneath, is probably the result of generations of adventitious growth, stimulated to regrow by prolonged exposure to browsing (compare figures 13.2 and 13.3).[9]

Another form of unusual growth characteristic for *Taxus* is the interior roots, which are discussed in the next chapter.

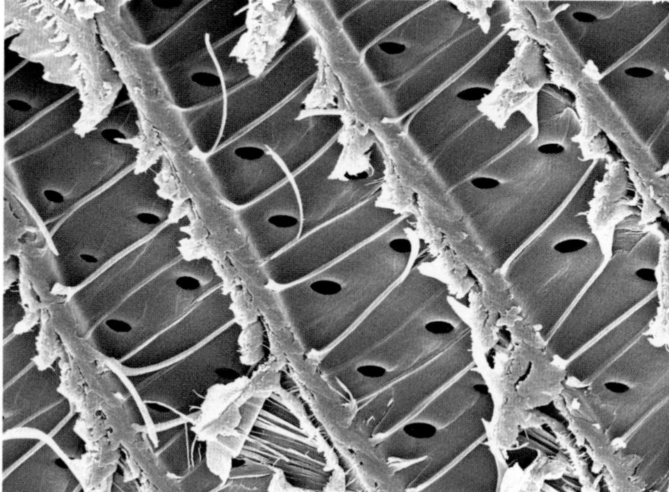

18.9 *The electron microscope reveals the spiral thickenings in the xylem cells.*

WOOD ANATOMY

Yew wood has neither axial parenchyma nor resin ducts, although resin ducts may occur after injury. The phloem of yew is composed of sieve and parenchyma cells, crystal-containing (crystalliferous) fibres and sclereids. These cell types occur in regular bands in the tangential order: sieve cells–parenchyma cells–sieve cells–crystal-containing fibres. During one vegetative season, a slow-growing branch might produce two such layers, the trunk four, and a young, fast-growing shoot four to five. The internal cell walls of the crystal-containing fibres are covered with tiny crystals of calcium oxalate (calcium sand). The crystal-containing fibres originate from parenchyma cells which in older (inactive) phloem change into fibrous sclereids with thickened walls.[10]

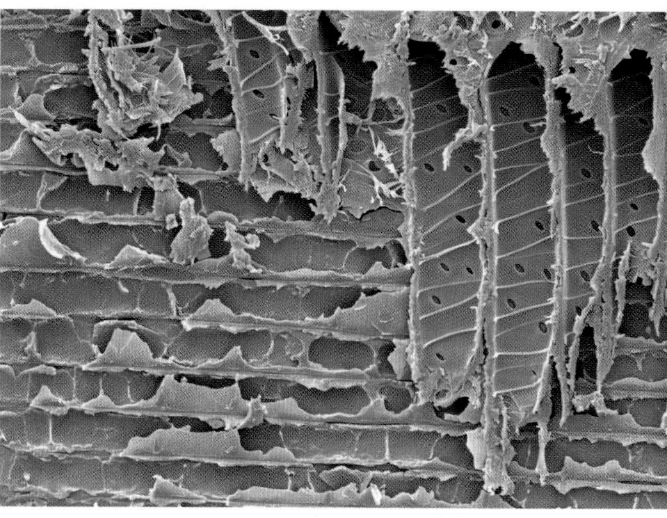

0.5mm

18.10 *This tangential cut shows a layer of rays behind the tracheids.*

In the xylem, the tracheids are extraordinarily narrow; their mean diameter of 18.4 μm is the smallest among all European tree species.[11] They are arranged in regular, radial rows, separated by a single row (in some sections two rows) of rays. The sapwood also has rays, consisting of one row of parenchyma cells that are rich in starch and resinous substances. The rays of the sapwood and heartwood facilitate radial (horizontal) water transport. They are usually one to fifteen cells high (rarely up to twenty-five),[12] and consist entirely of parenchyma cells, with no radial tracheids. As in the other Taxales, an important anatomical feature of yew wood are the spiral thickenings of the tracheids. In *Taxus*, however, they are particularly strongly developed, and this characteristic feature is believed to contribute significantly to the remarkable elasticity of yew wood. Where the radial walls of rays and tracheids meet, the wood has so-called crossfield pits, small indentures in the cell walls whose membranes facilitate the (horizontal) exchange of aqueous solutions between the various cell groups of the heartwood.[13]

19.1 *Adventitious shoots can be produced anywhere on the trunk and branches of* Taxus.

CHAPTER 19

REGENERATION ABILITY

Trees in the temperate zone generally have *three phases of life*: (1) a 'formative period' during which the increase of trunk girth reflects the increasing crown size; (2) the 'mature state', when the optimum crown size is reached (in most temperate species after 40–100 years) and the annual increment of new wood stabilises – which implies that the annual rings become *increasingly thinner* as the trunk circumference increases;[1] and (3) 'senescence', when the tree has outgrown its limits of feeding itself and parts of the crown die back and hence reduce the photosynthesising foliage and thus the annual wood production and annual ring width. Most species can barely survive when the mean annual ring width of the trunk is reduced to 0.5mm[2] and will die in due course. *Taxus*, however, is different. It can survive long stretches of time with much less growth than that, with the effect that such minute girth increments through microscopically thin rings (so-called 'missing rings') will escape most tape-measuring attempts. In some old yews repeated girth measuring after a number of years even shows an apparent decrease; this is usually due to loss of branches or part of the crown and trunk having broken away. The only scenario where an ancient yew possibly truly halts wood growth of the main trunk can be observed, for example, in the tree at Totteridge, Hertfordshire, which seems to have kept the same outer girth for 314 years (1677–1991).[3] The reason is that all cambium growth now takes place in the mature internal stems (see below), which have taken over the support of the crown.

Apart from considerably slowing down its growth rates in older age, *Taxus* is also able to 'return to formative (i.e. vigorous) rates of growth at almost any stage in its very long life. It may be stimulated by a boost of plant food from branch layering,[4] or by vigorous regeneration after catastrophic damage' (J. White 1998).[5] But most significantly, by means of vegetative regeneration an individual yew tree can completely renew itself as part of its natural life process.

THE LIFE STAGES OF THE YEW

How exactly does yew carry on from the point where other tree species die? The increases and decreases in the girth measures and growth rates of yews have fuelled controversial discussions ever since the Swiss botanist Augustine de Candolle 'discovered' the meaning of tree rings in the 1830s (i.e. the fact that the rings visible in cross-sections of wood represent *annual* increments and hence the age of that part of a woody plant). A fresh start in this field was provided by Toby Hindson in his Alan Mitchell Lecture 2000. With his proposal of seven life stages of *Taxus baccata* he was the first to fully acknowledge the complicated growth rhythms of this species.[6]

Stage 1 – The seedling. During the first few years, the rate of girth increment is slow.

Stage 2 – The juvenile. The tree has a fast growth rate; annual rings are usually a few millimetres thick.

Stage 3 – The solid tree. Full size is reached and the tree is in its prime, the core is intact. The amount of wood produced each year stabilises, which means, however, that with the increase in girth the actual ring width decreases accordingly.

Stage 4 – Hollowing. Trunk girth continues to increase, but very slowly. Core rot begins inside the trunk, but trunk and crown will remain functional for a long time to come. As more and more of the heartwood disappears, however, weight pressure on the outer trunk areas, namely the sapwood, increases. Because the sapwood is not designed to cope with compression, the tree begins to put down new growth ('secondary growth', see Chapter 18) in the sections under stress, and with increasing vigour. Annual girth increment therefore rises.

Stage 5 – The hollow tree. Although (almost) fully hollowed out, the trunk and supporting structures maintain a full crown. The girth increment rate remains high, particularly at the base of the trunk, to re-distribute weight pressures. In the last phase of this stage the tall empty cone of the trunk cannot support itself anymore and begins to collapse, loosing parts of the canopy as well.

Stage 6 – The shell. The short, hollow tube has little canopy and hence little weight to support. Fast wood production is not required anymore (and would not be possible anyway with such reduced foliage). Growth slows down considerably, sometimes even to a temporary halt.

Stage 7 – The ring. In the final stage of the cycle, (semi)circles of vertical fragments of the old shell may begin to take the appearance of independent trees – though most probably continuing to share the same root system and hence be one tree – with primary and secondary wood slowly giving their trunks a round, 'finished' look. If the old shell disappears completely, new shoots may appear from the stump or root system. In any case, the life of the tree continues, making it impossible for us to determine an end or beginning of its life, let alone its age.

REGROWTH

At almost any stage in its life, the yew is capable of vigorous regeneration. Great numbers of new shoots can rise from the cambium of branches and trunks, even from stools, and as 'suckers' from roots close to the surface (above or below). In the case of a partly broken crown, vertical shoots can appear on the neighbouring branches and soon fill the gap. *Taxus*'s ability to produce new foliage anywhere enables it to cope with even severe mechanical damage such as storm damage.[7] Fallen yew trees will continue to grow as long as some contact with the root plate

19.2 *Six healthy stems rise from this fallen yew at Hawkwood College, Gloucestershire.*

Diagram 7 The effect of the established girth on its growth rate

Please note that this graph only illustrates the principle of the different life stages. Any individual yew can deviate enormously from the ratios given here. Also, the dotted lines which indicate the transition points from one life stage to the next are approximates and naturally vary between individual trees

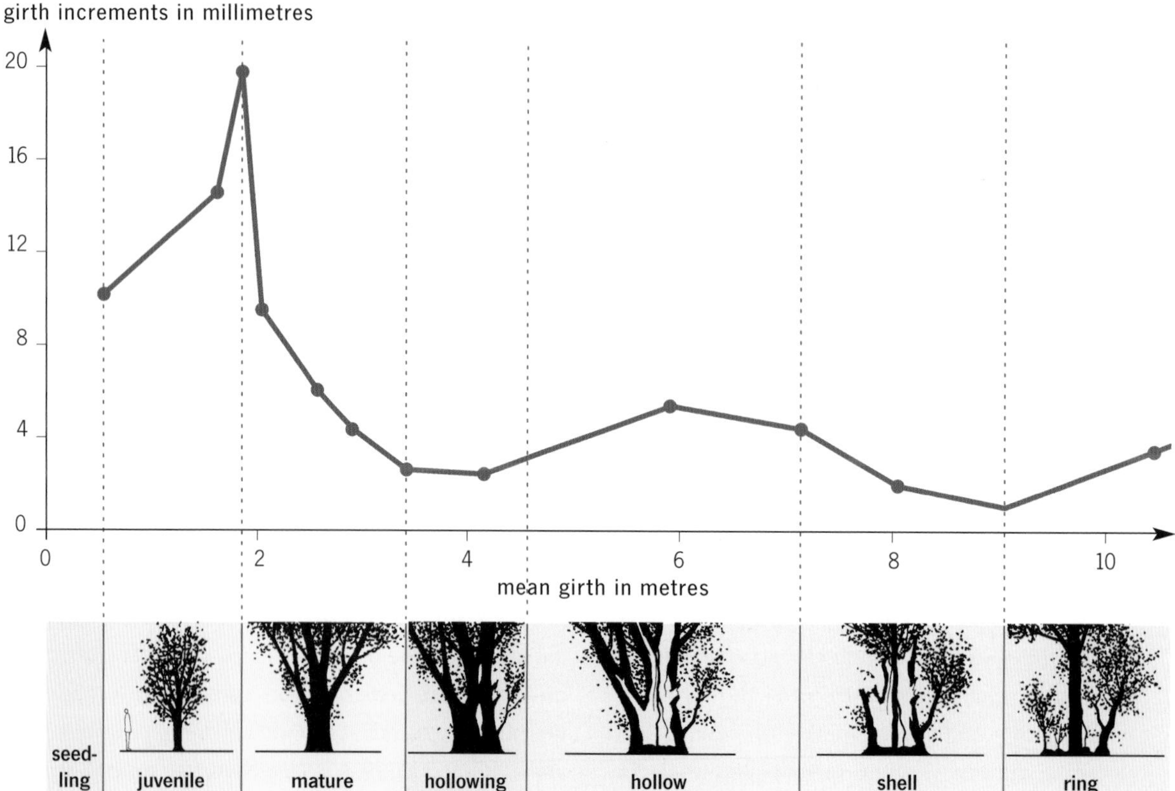

Diagram 8a Trunk appearances during the life stages *(after Hindson 2000)*

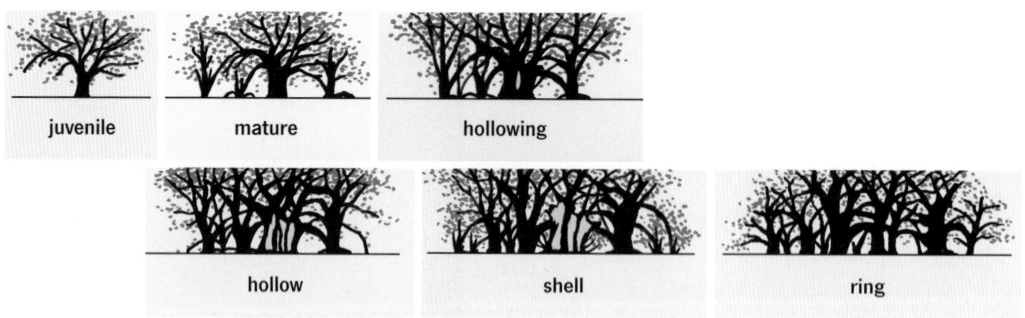

Diagram 8b Trunk appearances in combination with intense branch layering *(after Meredith n.d.)*

'There is no theoretical end for this tree, no need for it to die.'

Alan Mitchell (1922–95)[8]

is maintained, with lateral branches becoming new leaders (see also figure 37.16). And, surprisingly, age does not seem to weaken this capacity for regrowth.

However, there is a fine line between severe but recoverable damage, and a catastrophic event. In various cases, yew trees have been completely killed by the loss of important parts of the crown and upper trunk, for example through unnecessarily excessive pollarding.[9]

BRANCH LAYERING

One phenotype of yew (which seems most widespread in the British Isles) has a distinctly broad crown and lower branches that begin early to grow in the shape of an arch towards the ground and then take root. From these, new shoots appear, some of them developing into stately new trees. For a long time they stay connected with the parent tree through the arching branch, which might deliver nutrients in both directions.[10] Thus an old yew can turn into a whole grove and anchoring itself at different points like this grants ultimate security against high winds.

In Britain, the typical low height and broad shape of the yews predestined for layering are often seen in yews growing in the open (e.g. in parks or churchyards; in other countries layering branches are 'cleaned up' even more often than in Britain) as well as in the primary pioneers that colonise juniper scrub in open grassland. Secondary colonisers that grow towards gaps in the forest canopy may sometimes have a narrower crown and taller, straighter boles.[11] But in regards to the modest light demands of *Taxus* (see Chapter 7) one should resist drawing conclusions from this. There are examples showing exactly the opposite (i.e. straight tall trees where there is enough light, and broad trees under dense forest canopy). The secret of branch layering might rather lie in the immense genetic diversity of *Taxus*, as discussed earlier.

THE HOLLOWING

The trunks of very old yews invariably begin to hollow, with the aid of specialised fungi. This is a very slow process (centuries-long), and in its course will include the main branches as well. As every engineer knows, a hollow tube is more flexible than a solid one, and indeed, hollow trees have been shown to be less susceptible to high winds than solid ones.[12] The hollowing requires a perpetual redistribution of the crown weight, and the yew

19.3–19.5
Young yews 'planted' via branch layering by a parent tree (centre).

__19.6–19.7__ (top row) Internal roots develop from the cambium layer, usually a few metres above the ground.

__19.8–19.12__ (middle row) Early stages of internal roots, seen in Ortachis, Sardinia, and Alapli, Turkey.

__19.13__ (bottom left) Mature internal root at Ulcombe, Kent.

__19.14__ (bottom right) The internal stem of the yew at Bettws Newydd, Wales, is turning into a new trunk.

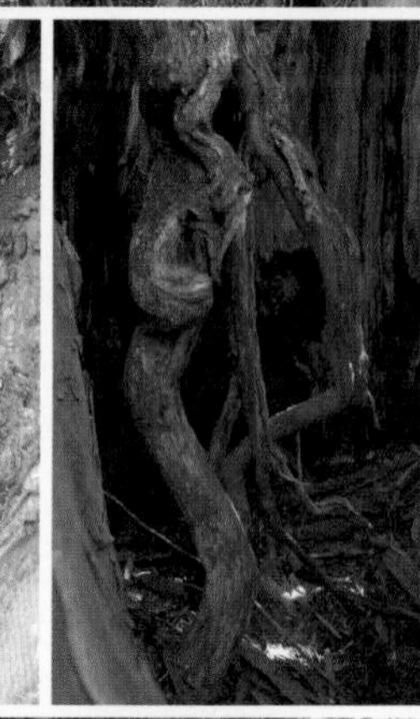

19.15–19.16 *The old shell of the yew at Ninfield, clearly visible in this photo from the early twentieth century (below), has gone, leaving the internal stems to carry the crown of the tree (right).*

meets these needs by laying down more 'primary wood' in the sections that demand it, and by additionally strengthening them with the production of 'secondary growth' (see Chapter 18).

INTERIOR ROOTS

In old yews, parts of the cambium at the top of the trunk are capable of developing roots that grow downwards through the hollowing centre of the tree. Thus the process of self-recycling begins – it has been suggested that the interior roots are capable of absorbing nutrients from the dead wood. Eventually, the interior roots reach the ground and penetrate it. Above the surface, they then begin to clad themselves in trunk bark and eventually turn into new trunks themselves, standing inside the space of the by now hollow old trunk or shell. Since they have always been connected to the living tissues at the top of the old trunk, they can progressively take over the role of the main stem, in terms of sap flow as well as mechanical support structure.[13]

The processes of branch layering and interior roots reveal how an ancient yew can entirely renew itself, with no part of the tree being as old as the living organism itself.

CHAPTER 20

DATING OLD YEWS

It is anything but easy to ascertain the age of a single old yew, or the growth rates of the species. Since old trunks invariably hollow out, there are no annual rings left to count, and radiocarbon-dating proves equally useless when an old tree has none of its early material left. Furthermore, growth rates, too, change during the life phases of a tree, rendering dendrochronological calculations very complicated. Nevertheless, precisely because of these changes in growth rates, modern dendrochronological investigations offer considerable refinement over age estimates based on tree girth alone.

One way to reach a somewhat reasonable estimate of the age of a hollow yew without knowing the planting date of the tree is to work out the most likely growth rates for this tree. This method involves taking measures of exposed annual rings of the tree in question and possibly of other yews within the locality as well. This data needs to be carefully interpreted with regards to light exposure of the tree, local soil conditions and climate. However, as a result of the weight of various flexible (and often unknown) parameters – for example, we know nothing about the tree's light exposure centuries ago – different researchers have suggested various methods of calculation, with widely differing results.

Measuring the girth of a tree trunk

Dendrochronology usually investigates annual rings either in a cross-cut (which implies that the tree was felled) or in a drill probe taken with a Pressler borer (which implies that the tree was injured). None of these methods is permissible for old or ancient yews (not to mention the fact that yew heartwood is very hard and often breaks the expensive hollow drill parts). What remains is the only non-invasive method to estimate the growth rates of a tree: to measure the girth (circumference of the trunk) and to compare it with existing data for the same species within the locality.

Foresters usually measure the diameter of a trunk, subtract a value for the thickness of the bark, and then multiply it by its height (also considering a factor for the upward tapering of the trunk because it is a cone and not a cylinder), resulting in a value of cubic metres of timber. Measuring the diameter works well in regards to tree trunks of a fairly even shape with 'parallel' sides, but when it comes to twisted or crooked individuals or a non-circular cross-shape, a girth measure is far more precise.

Both diameter and girth measures are generally taken at 'breast height', that is, 5ft/150cm above ground (in Germany at 130cm). When swelling branch bases or burrs would jeopardise the measurement, the trunk should be measured at the narrowest section (or directly above and below the burr to calculate the average of the two), with a reference to the measuring height. Such an additional note is also necessary when measuring trees on steep slopes.

Diagram 9 Measuring the girth

GROWTH RATES AND AGE ESTIMATES

The discussion about the growth rate and potential lifespan of yew trees began when Augustine de Candolle established the causal connection between the age of a tree and the number of annual rings of its trunk.[1] Many of the early dendrologists of the nineteenth century had a tendency to overestimate the age of yew trees,[2] but with the turn of the century this began to change, if not to reverse. People like John Lowe[3] in England, and later Prof. H. Eddelbüttel[4] in Germany, opposed the generalisation of the low growth rates of old trees and then went on to generalise the fast growth rates of young trees! Both Lowe and Eddelbüttel brought down their age estimates even further by adding the argument of trunk fusion with which yews supposedly 'pretend' to be of greater age than they are.[5]

'Trunk fusion'?

In his classic *The Yew-Trees of Great Britain and Ireland* (1897), John Lowe pleaded the case for applying the fast growth rates of young yews to old ones too. He also addressed exceptional cases and growth modifications such as some cross-cuts of youngish trees revealing a fusion of a number of young stems (or, as he assumed, possible fusions of root-suckers, or adventitious shoots as well). He incorporated this observation into his main argument by saying this phenomenon could further amplify an age over-estimation *where this phenomenon occurs.*

His argument was taken on by many experts, but also misunderstood by some. What was an exception for Lowe became an absolute rule for others. The German botanist O. Kirchner, for example, wrote in 1908 that yews had a single trunk only for the first 200–250 years maximum and that older yews 'always' had a 'feigned trunk', combined of many young ones.[6] Thus, the possible great age of yew was lost in translation for most of the twentieth century. Even the *Encyclopedia Britannica* generalised in 1984 that old yews consisted of a 'fusion of close-growing trunks, none of which is more than 250 years old'.[7] For more than 100 years now the misunderstanding of 'trunk fusion' producing 'fake trunks' *as a norm* has been copied by writers, and continues to be so.[8]

Indeed, it is possible but rare that branches or young stems of yews merge. Yet great caution should be applied when turning a phenomenon into a rule and thus declaring that all multi-trunked yews and all cases of inadequately observed interior roots as well as extreme flutedness (see figure 20.6) are mere further signs of a 'trunk fusion'.

Fusion, when it happens, is actually not an easy task for the tree; the cambium layers will finally have to merge but are firmly separated by layers of bark and bark cambium. A long period of rubbing against each other (more intense in fusing branches than in protected interior roots) also causes injury, which bears a high risk of infection with parasites. Wound material has to be produced, and the position of both wooden bodies in relation to each other will finally have to become fixed. Thus, fusion is a risky and energy-intense business not to be undertaken lightly. And indeed, different trees with interior roots – for example at Ninfield, England, and Jirobei, Japan (figures 19.16 and 20.7) – show that there seems to be no tendency to 'bundle up' early but rather to space out evenly within the hollow trunk.

Furthermore, a group of seedlings would comprise plants of both genders, generally about 50 per cent of each. Although the change of sex in single specimens of *Taxus baccata* has been observed, we cannot assume this to be a standard procedure just for the sake of supporting the trunk fusion idea. What complicated biochemical process would enable them to 'agree' which half of the group would finally change their gender? No such case is known to botany, and one wonders why some experts have preferred to accept such ideas instead of simply considering that *Taxus* might be able to live longer than 250 years, after all.

Lowe was plainly wrong to assume that every yew would inevitably lose its trunk at about 300 years of age. He also greatly overestimated adventitous growth. Adventitous shoots rarely make it beyond the size of mere twigs; they either dry up again within a year or two, or get entirely browsed by animals around the base of the trunk. If they regrow time and again they create intensely burred wood at the trunk base, which may also swell enormously. But hardly ever do they advance into full additional *stems.* Being a phenomenon most common in free-standing yews exposed to full sunlight, their function seems to be the shading and cooling of the living tissues underneath the thin bark. And for this, a multitude of recurring green shoots is more effective than a few up-shooting adventitious stems.

And last but not least: what difference does it make whether a tree is single- or multi-stemmed when it is one and the same root system that facilitates the annual growth?

20.1 *The felling of the ancient yew at Tomiyoshi, Japan, revealed a great number of interior roots.*

The dendrochronology of yew

By Andy Moir, Tree-Ring Services (www.tree-ring.co.uk)

Dendrochronology has been defined as 'the dating of annual growth layers in wood plants and the exploitation of the environmental information which they contain' (Fritts 1971). The fundamental basis of dendrochronology is the annual growth ring that forms inside the bark by division of the cambial cells. In suitable tree species capable of anaylsis, each ring is identifiable by the abrupt change in cell size between the winter-formed wood of one year, and the spring-formed wood of the next year. Variations in climate over a period of years lead to unique patterns of wide and narrow rings. By sequencing overlapping patterns from successively older samples (for instance, from living trees, building/boat timbers, peat bog wood, etc.), long reference tree-ring chronologies can be established. By matching particular ring sequences obtained from sampled timbers against dated reference chronologies as in diagram a, it is possible to precisely date in calendar years the sequences for some tree species. Currently the reference chronology for oak in the British Isles dates back to 5,452 BCE.[9]

Yew has some particular problems for dendrochronological application. Older yew trees commonly have irregular growth, which, combined with extremely narrow rings, can produce locally absent rings: rings 'missing' over part of the circumference of the tree. However, it appears that this problem can be generally avoided by sampling/measuring at lobe areas of growth. For larger yew trees, their hollow trunk typically makes complete precise dating impossible. But partial tree-ring sequences, which fall well short of a tree's full span, combined with girth measurement, offer the possibility of considerably refining an age estimate based on girth alone. Smaller trees neighbouring a large hollow yew can also sometimes provide additional information on the probable formative growth rate at a site, and be used to further refine a tree's age estimate. The indicated growth rate of a yew tree can vary considerably for different radii of the trunk. This is thought related to strip-bark growth, a growth characteristic that yew shares with the Methuselah bristlecone pines of California which can reach over 4,000 years of age (strip-bark's influence on radial growth is complex; one mechanism of its cause is thought to be the death of a root affecting

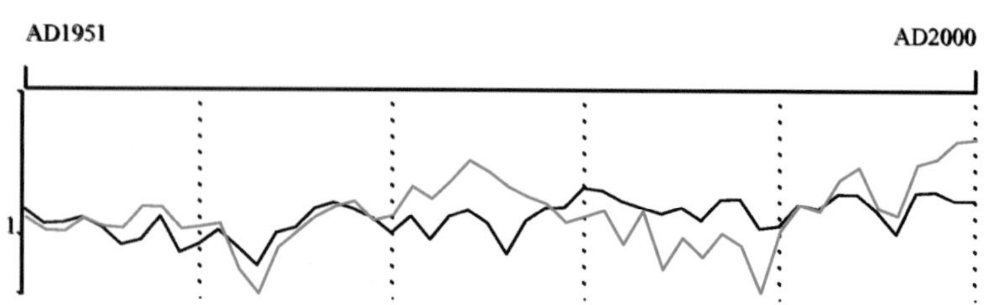

Diagram 10a Visual matching between tree-ring sequences

KEY: Parts of the UK yew reference chronology are in black and the Dunsfold yew sequence is in red. Note: The ring width (mm) is plotted on a (y axis) logarithmic scale, using common axis.

Thus, the lifespan accredited to *Taxus* was rather low during the larger part of the twentieth century, and popular botany books would rarely state a maximum lifespan of more than 800 years. In the 1980s, however, things began to change. Allen Meredith, a British naturalist who began his intense yew research in 1975, travelled the British Isles far and wide and accumulated an unprecedented database of ancient yews (mainly in churchyards). He measured and recorded the trees themselves and also hunted any available material regarding their history, including recorded planting dates and historical girth measurements. His gazetteer of 404 ancient yew sites was published in 1994,[10] and this took yew research into a new era. Meredith himself produced an age/girth relationship for *Taxus* using his empirical data from trees throughout the British Isles (compare diagram 11).

Contemporary attempts towards an estimation of growth rates and possible ages of yews are far more differentiated than those of the past. Senior forest research dendrologist John White, for example, describes the growth pattern for trees in general as a three-phase process (see previous chapter): a 'form-the cambium in the corresponding vertical section of the trunk). The variation between different radii can be problematic when estimating age. Currently, average growth rates are used, but as our understanding of yew develops methodologies may change.

In the British Isles, dendrochronological analysis has been applied to a number of yew trees, mainly in Surrey. The research has resulted in the establishment of a well-replicated yew reference chronology that extends from 1690 to 2004 CE. As well as refining the age estimate of the trees analysed (diagram b), change in phases of growth and periods of increased or reduced growth may also be established. This information in turn can help establish the effects of management, and the assessment of a tree's state of health.

To date, sequences of no more than *c.* 190 rings have been recovered and analysed from the trunks of large hollow yew trees (over 5m girth) in Britain. However, even these short tree-ring sequences can be useful in resolving instances where historical girth measurements suggest there were periods when trees did not increase in girth. The recent analysis of a fallen/felled branch of the Ankerwyke yew in Surrey produced a sequence of 317 rings. The sequence is significantly longer than those recovered from trunks, and highlights the future potential of branches, which occasionally become available for analysis through tree surgery or windfall. Dendrochronology provides an empirical method for refining the age estimates of large yew trees, and validating them as being as much a part of our heritage as our venerable buildings, which they often pre-date.

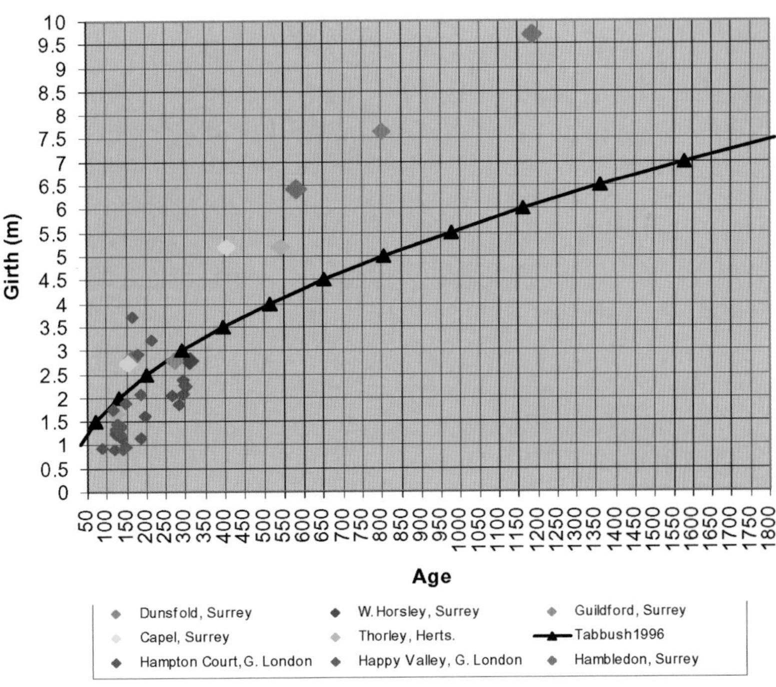

Diagram 10b The relationship between age and girth established through the tree-ring analysis of yew trees

ative period', the 'mature state' and 'senescence'.[11] Furthermore, White accounts for the effect of different environments: for the purpose of age estimation he suggests that the formative period of fast juvenile growth (at 4mm ring width) may last for about sixty years in yews grown in gardens and parkland, fifty-five years (at 3mm ring width) for yews in a churchyard, but only forty in poor ground, and thirty in shaded woodland. Practically speaking, this means that for a tree in a park or garden, the innermost 24cm of the radius can be expected to have grown in only sixty years at a juvenile growth rate (60 x 4mm = 240mm), and the remainder of the radius would be assumed to have grown at a slower growth rate (with ring width permanently decreasing with extending trunk circumference). Meanwhile, a forest yew would be calculated with thirty years of 3mm ring width (9cm for the core), outside of which the slower growth rate of the mature state would have to be applied.[12]

Investigating the ring widths of the ancient yews at Kingley Vale, Sussex, John White and Paul Tabbush soon found that this particular yew wood located on poor soil required a revision of White's growth assumptions as ring widths of 0.2mm and less (in old wood) clearly proved much slower growth rates than expected. In consideration of the diversity within this wood, Tabbush and White developed two curves for the ratio of girth and age for the ancient yews at Kingley Vale, stating that 'the true curve for any individual tree at Kingley Vale is likely to lie somewhere between these two'.[13]

As a result of rich soil and abundant light, yews (and other trees) in churchyards and parks generally grow somewhat faster than those in woodlands or poor locations. Clipping them, as often practised in such locations, has no significant effect on the growth increment.[14] Tabbush refined White's age estimation method in respect to churchyards, and suggests their age can be fairly estimated by the equation: tree age = girth^2/310.[15] This would make a tree with a girth of 7m about 1,580 years old (compare diagram 11).

The large variation spectrum among yew trees even at the same location was never as clear as when dendrochronologist Andy K. Moir examined the yew avenue at Hampton Court, London: twelve trees planted at the end of the nineteenth century vary in trunk diameter from 30 to 80cm (the generally fast growth rates were attributed by Moir to the warmth of the city).[16] Another example is Monnington Walk in Herefordshire, where yews planted in 1628 have varying girths from 1.47 to 4.42m.[17] The results of Moir's work at Borrowdale in 2004 surprised many experts when it became clear that the biggest yew, a male with a girth of 7.27m, achieved its volume in less time than its companion tree with a girth of 5.49m.[18]

Hindson's model of the seven life stages of *Taxus* (see previous chapter) combines his own empirical data with known planting dates for trees up to about 800 years of age and, with data from Meredith's gazetteer, creates an extended basis to relate age to girth measurements.

Naturally, most of the discussion of how to date old yews has occurred in Britain where most of the ancient trees can be found. But it is also worth taking a look beyond the insular viewpoint.

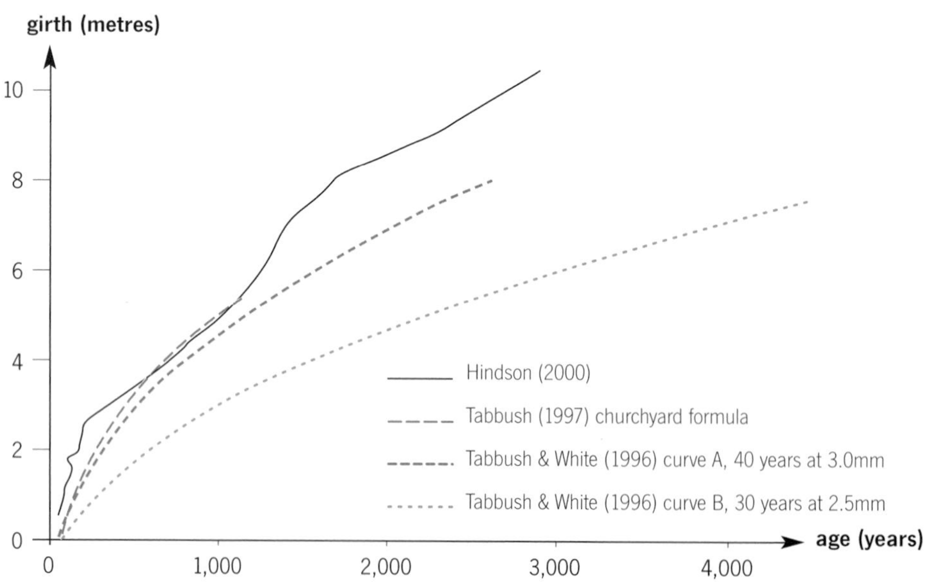

Diagram 11 Age to girth ratio *(after Tabbush and White 1996; Tabbush 1997; Hindson 2000)*
The curves in this graph can be used as a very rough guide to how old a yew tree of a given girth might be, but it must be noted that it is ***not possible*** to measure a tree and discover its age in any formula. The information contained in this graph is statistical and applies only to populations, not individuals. Any individual yew can deviate enormously from the average age.

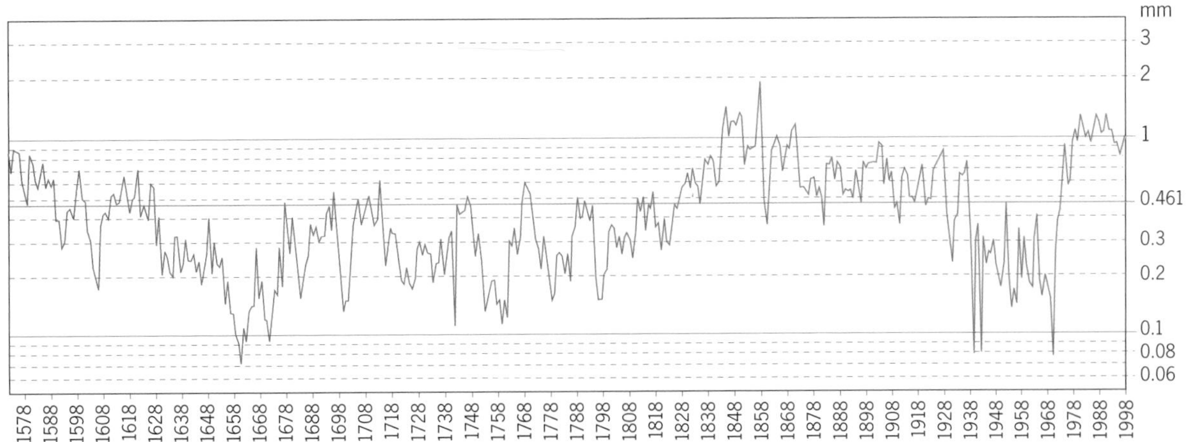

Diagram 12 Ring-width curve of the monumental yew at Bartin, Turkey *(after Kaya 1998)*

In a dendrochronological study from the Black Sea region, Zafer Kaya (1998) investigated one of the 'monumental yews' of Turkey, with a trunk diameter of 220cm, and at an elevation of *c.* 900m.[19] Kaya took a probe of about 20cm depth, spanning 426 years, from 1572 to 1998. His graph (diagram 12) shows an average annual growth rate of 0.461mm ring width, which even in single years hardly ever ventures into the 1mm region. The overall average ring width for more than four centuries indicates that an ancient yew south of the Black Sea can grow very slowly and densely, although the favourable climate of the area might have tempted us to have expected otherwise. Furthermore, it shows how periods of slower and stronger growth are spread over the entire period.

Doug Larson and his research team from the university at Guelph, Canada, have shown that vertical cliffs often support exceptionally old, deformed and slow-growing trees. All core samples from such natural bonsais show no difference in the growth rates of saplings and the mature stages of development. Cliffs can be said to harbour some of the oldest trees in the world. For example, a juniper *(Juniperis phoenicea)* from the Verdon Gorge, France, took 1,140 years to produce only 8cm of trunk radius,

20.2 *Ancient yew clinging to the vertical cliff at Whitbarrow Scar, Cumbria.*
20.3 *Original size of Doug Larson's dead wood sample from this tree, over 220 rings within c. 16mm radius.*
20.4 *Not even the macro shot reveals everything: some rings are only a single cell layer thick.*

which equals an average ring width of just over 0.06mm.[20] Additionally, with such low growth rates we can expect scattered single years when the tree does not put on anything at all. Yews of similar density and age have been found, for example, at Llangollen (Wales), Markland Grips (near Sheffield),[21] and Whitbarrow Scar (Cumbria). These trees must be among the slowest-growing woody plants on earth, each producing a total woody biomass of well below 0.1 grams on annual average.[22]

Professor Mikhail V. Pridnya, leading botanist and curator of the western Caucasian nature reserves (south-western Russia), states about *Taxus* that 'from living individuals nearby Khosta exceeding a diameter of 2 meters, one knows that they can exceed an age of 3000 years (referring to the number of rings of sample trunks with similar diameter)'.[23] Figure 20.5 shows a sample from a fallen monumental yew near Khosta. Pridnya also reports a circular cross-cut disk of a yew trunk with just over 1,000 rings in a just over 50cm radius, and still with a completely intact core. Unfortunately, this evidence fell victim to the flames that destroyed the contents of the Sochi State Museum in 1970.[24]

20.5 *This sample (original size at the top) was taken from a fallen old monumental yew in the Khosta Nature Reserve (western Caucasus) by H. Rössner in 2000. It has about 113 rings in 33.1mm, which resembles an average annual increment of 0.3mm in radius.* The trunk was c. 1m in diameter.*

* Counted by Dr Grosser, Institute for Wood Science of the Technical University of Munich. The sample was taken from *c.* 10m height.

Pitfalls

Many pitfalls (a frequent term in the age debate of yews) threaten to distort age estimates of yews. *Taxus* produces immensely different ring widths, wider in the fluted sections of a trunk than in the grooves, often wider on the south side of the tree than on the north side, and sometimes showing additional irregularities for reasons unknown. The annual rings are rarely circular and concentric, sections can be missing altogether (Lowe reported a trunk with 250 rings on one side and just 50 on the other),[25] or the trunk stops growth altogether for a period of time. The multitude of dynamics in the forest adds further irregularities. Grazing can keep a tree unexpectedly small. Old yews in the process of decomposition can produce new stems and appear as young trees. And, on top of all of this, there is the fluctuating climatic influence through the centuries.

CONCLUSIONS

This brief history of yew age estimation illustrates the vast challenge that *Taxus* poses to the human intellect. The work of more recent researchers such as Meredith, Tabbush, White, Hindson, and Moir, however, supplies a range of reasonable ratios of age and growth patterns, and we may expect the truth to be 'somewhere between', to quote White once again. Maybe one of the most valuable contributions to the discussion is Hindson's suggestion to consider seven instead of three successive life stages for *Taxus,* and also the effect of the shape of an old tree on its growth rate. Hindson's model explains much about the extremely varied results (and opinions!)

that have appeared over the last 170 years, and it gives us a new instrument with which to gain a deeper understanding of this fascinating species.

It seems inevitable that Tansley's warning from 1911 that the age of old yews in the British Isles may be 'often considerably over-estimated' should be counterbalanced with the notion that it can be considerably underestimated just as easily. The spectrum of variables and the timespans in which they act are simply too great.

A last point for consideration: whatever the result of an age estimation, every yew we examine that is not planted by man or undoubtedly sprouted from seed could be the result of layering of a far older tree, or an interior root that has replaced the original trunk, or a new shoot appearing from an ancient root system. Everything we see of a yew 'tree' might be young but it could represent the new physical form of an ancient living being. In such cases, all efforts of age estimation become superfluous. Could this still be the 'same' tree that stood there a millennium ago? Well, it is the same genetic code, for one, proving its profound ability to survive. It challenges and expands our view of life, as we learn that the living organism is more than what we can see and touch. In the words of Tabbush and White (1994): 'Age in this sense might be defined as years of continuous regeneration since germination.'

20.6 *(left) Some yew trunks are extremely fluted and can easily be mistaken for a fusion of coalescing suckers ('trunk fusion' theory). Ancient yew in West Tisted, Hampshire, girth 739cm at 30cm (1998).*

20.7 *(right) The trunk of the sacred yew* (T. cuspidata) *at Jirobei, Japan, is neither fluted nor made up of suckers but of a rich mass of what were once interior roots, the old shell having disappeared completely.*

CHAPTER 21

GREEN MONUMENTS

FOREST MANAGEMENT

Across Europe and Asia over the last three to four thousand years, yew has become locally extinct, or reduced to small isolated populations (mostly in mountainous areas). In the wild, it is now a rare and endangered tree, and prone to extinction even in the Mediterranean mountains of southern Spain, in Poland, Bulgaria, in the eastern Transcaucasus and Norway.[1] The decline has a number of reasons, among them heavy grazing by deer and other herbivores, adverse soil conditions (e.g. the build-up of pathogens), poor soil–water relations and changes in microclimate, as well as fungal diseases. One major contributor to yew decline has undoubtedly been excessive felling, or at least felling at a faster rate than regeneration. Indeed, the yew has been described as 'one of the most negatively affected European trees by human intervention'.[2]

For Europe, many archaeological records attest the manifold uses of yew wood from the Stone Age onwards, but no doubt the most dramatic impact on the yew population of a whole continent was the demand of the English armies for yew longbows between the thirteenth and sixteenth centuries; the European yew stands have never recovered. In post-medieval times, yews were increasingly destroyed to avoid poisoning of horses and cattle.[3] Recovery in the forests did happen, but only continued until the major shift in forestry management about two centuries later: until the eighteenth century, yew had had its place both in the understorey of the wood pastures (e.g. oak and beech feeding pigs in the autumn) and in the coppice woods; but since the woodland economy switched to timber production, there has been little room for the slow-growing yew.

In Germany, Dr Thomas Scheeder of the Eibenfreunde ('Friends of the Yew') society has proposed an integrated forest cultivation programme to restore yew populations in modern forests. In about 200 years, yew trees can attain a diameter of about 40cm on average, or less than 30cm with little light and poor soil, and close to 60cm in optimum conditions.[4] Harvesting would not allow them to reach their full age potential, but at least it would bring back the species to many areas where it has become extinct. However, in Germany and other countries where the yew is fully protected as a species, such endeavours are made impossible, paradoxically, by the species' very legal protection (once it grows it cannot be harvested, and so, as a potential hindrance to economical forest management, it is not planted in the first place). In Britain, where *Taxus* is not protected, the Forestry Commission (1994) recommended that rapid growth of yew can be promoted by felling the overstorey in beech-ash woods rich in yew (however, as of yet, the Forestry Commission in Britain does not include *Taxus* in its forestry plans and policies). Generally, *Taxus* is able to outcompete hardwood regeneration and dominate, but usually yew woodland after being cut is likely to regenerate to ash or birch, with yew as a scattered understorey, therefore forest management should be restricted to only very occasional harvesting of yews. Around 50 per cent of lowland yew woodlands in England are at present designated as Sites of Special Scientific Interest (SSSI).[5]

Today, the two main threats for yew trees worldwide are the destruction of habitat by **general deforestation** (particularly in eastern Siberia, China, Tibet, across South-east Asia, and North and Middle America) and the **hunt for taxanes**, the tumour-active molecules extracted from yew for an ever-growing global pharmaceutical market (see Chapter 25).

THE DASH FOR TAXANES

With the identification of an anti-cancer agent from the bark of Pacific yew in September 1966, this tree shot to fame and faced unprecedented demand. As 12kg of dried bark produce only 0.5 grams of the pure substance (a worryingly low yield of only 0.004 per cent[6]), the bark demands of the research institutes continuously increased during the following three decades: 106kg of yew material in April 1966; about 1,350kg of dried bark in spring 1967; 39,000kg in 1990;[7] 725,000kg in 1991, and the same amount again in 1992.[8] This quickly became a logistical problem, and eventually played a dramatic part in the political and ecological debate about America's old-growth forests and their destruction.

21.1 *Trunk of a Pacific yew.*

In the Pacific West of North America, the focus of the Forest Service is on those species of enormous commercial value, such as Douglas-fir, Sitka spruce or Western cedar. The smaller trees and shrubs of the understorey have no commercial value; they get damaged by the big machinery that hauls out the large timber, and after the clear-cut they are burned

The history of Taxol and the Pacific yew unfolded during the very same decades as the slow environmental awakening of Western societies. Citizen action groups began to voice protests, Friends of the Earth was organised in 1969, governments passed the first acts for protecting the environment,[9] and science began to change as well. In 1977, the Forest Service ordered a study of old-growth forests (about which next to nothing was known). The resulting Franklin report,[10] published in February 1981, was the first of its kind, representing not merely technical literature highlighting the commercially valuable species, but an ecosystem approach, describing the forest in its entirety. The authors challenged the attitudes prevailing at the time by pointing out that '*old-growth forests provide highly specialized habitats and are neither decadent, unproductive ecosystems nor biological deserts* ... An old-growth forest is much more than simply a collection of large trees. The dead, organic component is as important as the highly individualistic, large trees ...' (my italics).[11] Resisting a simple definition, the authors noted that it was the *complexity* of the ecosystem that was the key to understanding it. 'Old-growth' is a *process* whose characteristics begin to emerge after about 175 to 250 years under natural conditions. This report laid the foundations for what became 'New Forestry', the engagement in managing younger forests by recreating certain old-growth characteristics.

It was the third time that the yew tree had participated in the birth process of nature conservation:

In 1911, the English botanist Sir Arthur George Tansley (1871–1955) and his German colleague Professor Drude[12] stood on the hill above Kingley Vale. Drude remarked: 'You did not tell me that you were going to show me the finest yew forest of Europe.'[13] The two discussed how in Britain nothing was done to protect such sites. In the years after this meeting, however, Sir Arthur took action. He founded the British Ecological Society, was the editor of the *Journal of Ecology* for twenty years, and became a pioneer in the science of plant ecology – it was he who coined the terms *ecosystem* in 1935 and *ecotope* in 1939.

And as early as the 1880s, Hugo Conwentz (1855–1922), botanist and curator of the Museum of Western Prussia in Gdansk (modern Poland), had grown concerned about the yew populations of western Prussia and was the first to designate areas of nature conservation in continental Europe. He published fourteen essays and articles on *Taxus baccata* between 1891 and 1921, and in 1906 became the director of the first 'national office for the conservation of natural monuments'.[14]

***21.2** Yew stump after bark-stripping.*

in so-called slash-piles. Before taxol research made it a temporary priority species, the Pacific yew was nothing but such a 'trash tree', unvalued and unwanted by the timber industry. Even trunks more than 500 years old could be found on the slash-piles.[15]

Today, the burning of yews on slash-piles has greatly diminished with recent restrictions on the logging of old-growth forests in the Pacific north-west (at most, 10 per cent remains), but wasting yew wood has not ceased entirely as some industrial logging of old-growth forest still continues on public lands throughout the region where Pacific yews grow.[16] Pacific yew remains part of the ecosystem of the remaining Pacific rainforests, but the vast majority of these trees do not exceed *c.* 10–15cm in diameter, old trees being rare indeed. The bottom-line lesson of this American saga is that old-growth forests and their inhabitants harbour value to humanity that is often unanticipated, so that in addition to their inherent right to exist, we also have a stake in the preservation of the little that remains because we never know what will be important next.

THE PROBLEM GOES GLOBAL

The official approval of taxane production through semi-synthesis from leaves in October 1994 paved the way for the utilisation of yew trees anywhere in the world. By 1998, Taxol® had become the best-selling anti-cancer drug ever, with world sales of $1.2 billion;[17] in 2003, the annual turnover in the United States alone had grown to a reported $3 billion.[18] In response to the increasing ecological concerns being voiced in the USA, and resources running thin anyway, huge plantations in the Pacific north-west of the United States had been set up. But yews grow slowly ... and the harvesting of wild trees was shifted to Asia, primarily to India and China. Since 1992, enormous amounts of yew biomass have been exported for various Western pharmaceutical companies: Madras/Cochin and Delhi exported several thousand tons. In one month alone, the Arunachal Pradesh region in Assam exported 170,000kg of dried needles. In 1994, the Indian government applied to the Convention for International Trade in Endangered Species (CITES) to include the Himalayan yew in their list of endangered species, a request that was finally granted in 1999 but somehow too late: since 1992 India had lost 90 per cent of its yew trees. Exporting Himalayan yew is now forbidden[19] but illegal trade carries on: an Indian company, for example, sold material to a company in Marlborough, MA (USA), which processed paclitaxel for non-US markets in Canada, the European Union and Asia. Various organisations for conservation – among them the World Wide Fund for Nature (WWF), the International Union for the Conservation of Nature (IUCN) and the Species Survival Commission – presently consider Himalayan yew as *one of the ten most threatened species*.[20]

The situation in China is similar to that in India; the populations of the four yew species native to China – *Taxus chinensis, cuspidata, maírei,* and *wallichiana* – have been heavily depleted and what remains is now protected. The Himalayan yew *(Taxus wallichiana)* in China is on the list of endangered species (WCMC 1999), and so it is, together with *Taxus chinensis,* in Vietnam. The Himalayan yew in Tibet reached a critical low point in 1999 (WCMC).[21] There are no reports at all about the exploitation of *Taxus cuspidata* in eastern Siberia, but general large-scale deforestation, for example, in the region of Chabarovsk and along the Amur river, does not suggest good news. Furthermore, the new forestry law (since the late 1990s) of the Russian confederation enables the sale of exploitation rights for hitherto protected woodlands.[22]

Plantations

The way forward, it seems, is the plantation of yew trees on an industrial scale. Indeed, the first research programmes into yew plantations were in fact launched in 1987, and since then plantations for the extraction of anti-cancer compounds have grown in number and volume at extraordinary rates. In 1993, *Taxus* plantations in Oregon and Washington State were estimated to contain over 10 million young trees; three years later the estimates were already of 20–25 million plants. One of the biggest Chinese plantations is located in Sichuan province; the pharmaceutica yielded from the estimated 18 million yew plants benefit Chinese patients. The USA also have about 30 million plants (1995) in tree nurseries in Taiwan and in the Philippines. Indena, a phytotechnical company from Milan, Italy, that collected leaf material in northern India during the early 1990s, has set up a plantation in Mandi, Himachal Pradesh. Other countries, for example, Korea, have begun a national industry of yew pharmaceutica. During the 1990s, the number of plants in yew plantations worldwide grew from about 30 to 75 million, and by 2010 the 100 million threshold will be reached.[23]

However, with the spread of plantations arises an entirely new problem: genetic pollution. The seedlings for plantations are chosen from genetic strands for fast growth and high taxane content. Standard practice is to use hybrids – a favourite being *Taxus* x *media*, a cross of *T. baccata* and *T. cuspidata* – which are usually bred from species that are not native to a region. Multitudes of single plants are propagated from a small number of seedlings, via cuttings of these mother plants. Hence, the masses are – although not genetically engineered – effectively clones: the *genetic diversity* among these millions is incredibly small. Genetic diversity is one of the prime survival mechanisms of a species, and *Taxus* species naturally show an extraordinarily extensive range of genetic diversity, but millions of plantation *clones* will reach sexual maturity in the near future and are bound to interfere with the natural regeneration of the small numbers of remaining *native* yew trees, and within a few generations will flood the genetic material of these. Yew pollen travels far.

The obvious answer seems to be to focus on female plants. Another priority is to start plantations as far away as possible from native populations. Other ways out of the dilemma have been investigated since the 1990s, in particular the cultivation of *Taxus* cells in laboratories. One company in Schleswig-Holstein, northern Germany, has begun to cultivate yew cells in industrial-scale fermentation tanks under environmentally controlled conditions. Such methods might prove to be sustainable, cost-effective and environmentally acceptable – but to date, they do not achieve taxane levels as high as those of the trees, and it is a long way from the laboratory to clinical approval.

For the time being, plantations are growing in number, and native yew populations are still under threat. The WWF, IUCN, CITES and TRAFFIC are working hard to decrease illegal or other non-sustainable trade of *Taxus* resources in various parts of Asia.[24]

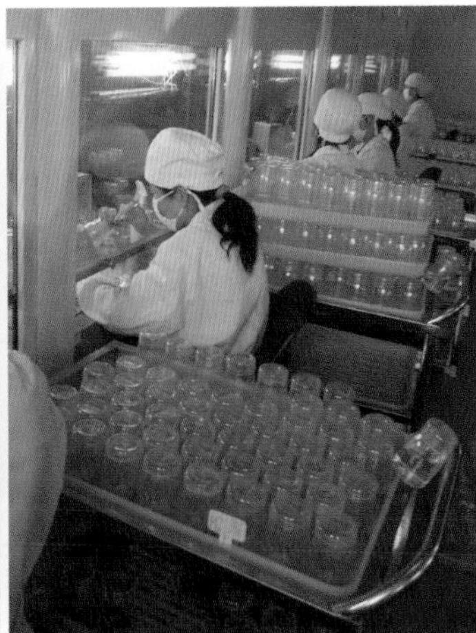

21.3 *A greenhouse containing tens of thousands of young trees, China.*
21.4 *Planting* Taxus x media *under sterile conditions, China.*

PROTECTING THE ANCIENTS

Where ancient yew trees have survived they occur singly or in small numbers scattered over a few European mountain sites, mainly in the Pyrenees, Spain, Italy, Sardinia, Corsica and the Alps; in larger numbers in northern Turkey and particularly the Caucasus; as small, stunted cliff trees (e.g. in Germany, France and the British Isles); and on consecrated ground, mostly in Japan (Shinto and Buddhist shrines) and England and Wales (Christian churchyards). Many of the ancient wild trees are by now on protected land, either national nature reserves or even (European) Biosphere Reservations. But although the Church has preserved many trees over the centuries, many non-woodland yews in churchyards as well as parks and private gardens endure unimaginable – and unnecessary – levels of neglect, mutilation, incompetent management and, above all, ignorance.[25]

As early as 1819, Alexander von Humboldt (1769–1859) coined the term 'tree monument' for the protection of trees. Are we, almost 200 years later, still too insecure about our economic strength to consider the exemption of the most venerable and ancient of trees from the iron rule of the stock-markets? Can the richest economies of the world not care for a few hundred ancient trees that are among the oldest living things on earth and irreplaceable for ecological, scientific, historical, religious and ethical reasons?

Interference

When it comes to yew trees in human 'care', the hollowing of the trunk of an old yew is the most fatally misunderstood aspect of all. It is still widely believed that the decay of primary yew wood as well as crown breakage are signs of old age and approaching death, and all too quickly tree surgeons are brought in who usually either pollard the tree and minimise its size drastically (which could kill the tree), or apply a host of props and crutches as 'safety measures' – if they do not cut it down straight away. What's more, metal chains and braces are impediments that interfere with natural girth increments. Even the inside of the tree is not safe. It was once the fashion to fill hollow trunks with bricks and cement. For example, in 1913 the hollowing yew at Warblington, Hampshire, was drastically treated in much the same way that a dentist deals with a hollow tooth. The cavities were rid of decaying matter, thoroughly disinfected with creosote, and then filled with cement – under advice from the experts of the Royal Botanical Gardens at Kew. The tree, however, stood up well to this treatment and still appears to be healthy.[26] Such measures show no consideration for

21.5 *Even yews that seemed dead for many decades have been observed to burst into new life again, Sheepwalks, Enville, Staffordshire.*

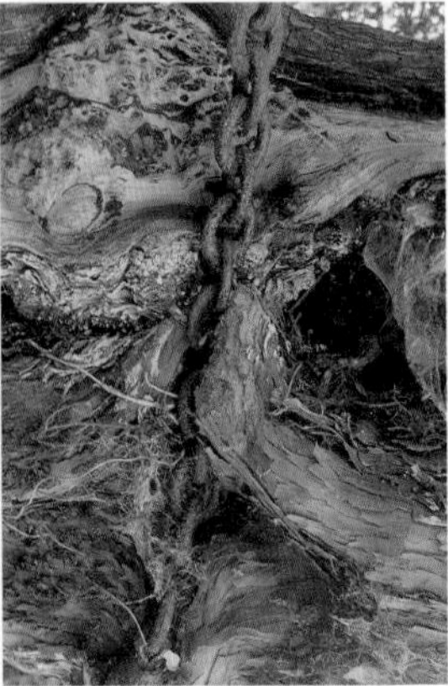

21.6–21.7 *Chain reactions, Stansted, Kent; Beltingham, Northumberland.*

21.8–21.9 *Props have to substitute for natural branch layering, Bentley, Hampshire; Wilmington, Sussex.*

the possibility of interior root growth – and no understanding of the tree's dignity either; they should be avoided altogether.

Storage

Compost heaps, garden sheds, dustbins, storage shelters for building materials, stacks of disused gravestones and even oil tanks can frequently be found beneath ancient yews in churchyards, parks and gardens. They constitute varying degrees of threat to the root systems and hence the trees themselves.

Tips for yew tree 'owners'

The following are guidelines given by Russell Ball, the one-time Senior Executive of the International Society for Arboriculture, European Office.[27] *Full guidelines are available at www.ancient-yew.org/management-consideration.shtml*

- In general, tree roots grow in the top 60cm of soil, and normally as wide as the crown of the tree extends: sometimes beyond. Tap roots extending vertically downwards are rare.
- Tree roots need oxygen, water and free space to grow; their function and development is inhibited by soil compaction as caused by the weight of (parking) cars, garden sheds, oil tanks, stacks of building materials, etc.
- Trees store 'food' reserves in the trunk, scaffold branches and woody roots. Any severe pruning/root loss can jeopardise the tree's health.
- Old/ancient yews in particular have an unfavourable mass–energy ratio, i.e. they have a large mass to support but only a relatively small canopy to 'feed' the tree via photosynthesis.
- If an old/ancient yew must be pruned it should be kept to the absolute minimum. Crown reductions of more than 20–30 per cent should only be done in staged pruning events over a period of years (see *mass–energy* above).
- Props for weighty branches may be better than pruning, but better still branch layering should be encouraged/facilitated. Braces and cables should be reviewed to ensure they are not damaging the tree and are still providing the required support.
- For advice on caring for old/ancient yews consult your local council Tree/Forestry Officer.
- *Before* pruning your old/ancient yew tree contact your local council to check if it is subject to a Tree Preservation Order or a Conservation Area designation. Permission/consent may be required.
- **The best policy is non-intervention. A yew tree will live longer with the minimal amount of human interference. Old/ancient yews have survived on this planet for hundreds if not thousands of years without the benefit of a chainsaw.**

21.10–21.11 *Filling hollows with stone and concrete, Bedhampton, Hampshire; Tisbury, Wiltshire.*

Insurance companies

The (registered and protected) oldest tree of the city of Hamburg, north Germany, is a yew standing on the dyke of the Elbe river. All the dead wood inside the hollow trunk has been meticulously scraped out, and the tree is supported by an elaborate system of metal poles and braces, outside *and inside* the tree. The braces are buffered with rubber against the wood, and adjustable to allow for future growth. There is no possibility of interior roots, wood-decaying fungi or insects. The sterility could only be further enhanced by a glass box around the trunk. And yet this masterpiece of German thoroughness was probably the only way to save the tree – from the insurance company. It is located only a few metres from an inhabited bungalow.

Nowadays, insurance companies pose a greater threat to some trees than developers, heavy storms or arson. Insurance agents scare tree owners with all the worst-case scenarios that their trees could bring upon humankind, supposedly leaving the tree owner (or the insurance company) liable to pay compensation in perpetuity. When the old yew on top of the Iron Age mound of Wychbury Hill near Stourbridge, West Midlands, was struck by lightning in 2001, it 'had to be' cut down because the lightning revealed a cavity in the trunk that rendered the tree 'unsafe'. A simple circular fence could have done the security job, with the tree in its centre, a radius of the tree's height, and with warning signs allowing entrance only at one's own risk (so if the tree were ever to fall it would do so within the fenced area and not be an insurance problem). But renting a chainsaw is cheaper than building a fence.

21.12 *Clockwork orange: Swiss precision in Hamburg, Germany.*

21.13 *Struck by insurance policies: old yew at Wychbury Hill, 2001.*

The large-scale destruction of yew populations in North America and Asia has achieved in little more than twenty years what in medieval Europe took

21.14–21.16 *How (in)accessible should a fence make the tree? Top left: Gresford, Wrexham; bottom left: Capel, Kent; right: La Haye-de-Routot, France.*

over three centuries: to push a species to the brink of extinction in many regions where it once was abundant. Furthermore, the small islands of surviving native trees with their naturally extensive genetic diversity are now about to face another threat, that of genetic contamination from many million yew 'clones' in plantations. In light of this global scenario, the ancient yews that have survived up to this point appear more precious than ever, and not only for their genetic codes that have proved a unique capability for survival over millennia. A small country full of ancient yew trees, such as Britain, is becoming a true Noah's Ark. Such tree stands are becoming (if they have not always been) far more significant than local or regional places of interest, and should be – as the equivalent of architectural World Heritage Sites – legally protected as Green Monuments.

21.17 *For those who come after us …*

Part II Culture

CHAPTER 22

THE ART OF SURVIVAL

Ancient wooden artefacts are rare. Wood disappears quickly unless it is preserved by favourable conditions, that is, sealed off from the air by water, peat, clay or ice. In north-western Europe, many timber pieces (mostly oak but some yew as well) have been found, attesting to the flooding of woodlands, peat or moorland formation, a local rising of the groundwater table, or simply the falling of objects into water or moor, or deliberate immersion, as with votive offerings. Interestingly, most early yew finds are associated with the activities of humans,[1] and the occurrence of yew artefacts through all phases of cultural history testifies to the great utility of this wood as a material resource.

The oldest wooden artefacts in the world are two hunting weapons made of yew. The first is a yew spear found at Clacton, Essex, and dates from about 150,000 years ago. The second, also a spear, was found in Lower Saxony, Germany, between the ribs of a straight-tusked mammoth *(Hesperoloxodon antiquus)*, and is about 90,000 years old.[2] Other excavation sites across north Germany have yielded at least eight yew bows dating from the mid-sixth to the mid-third millennium BCE.[3] They all belong to the Holmegaard type with its characteristic D-shape of the arms (in cross-cut) and specially formed handle. The two specimens from the site at Dümmer, Lower Saxony, were about 127 and 146cm long before one end broke off on each of them.[4] The Barleben bow, still complete at 152cm in length, yields an interesting observation: at both ends the wood is deeply penetrated by tiny grains of sand – hence the bow must have been rammed hard into sandy soil a great number of times. Its age, however,

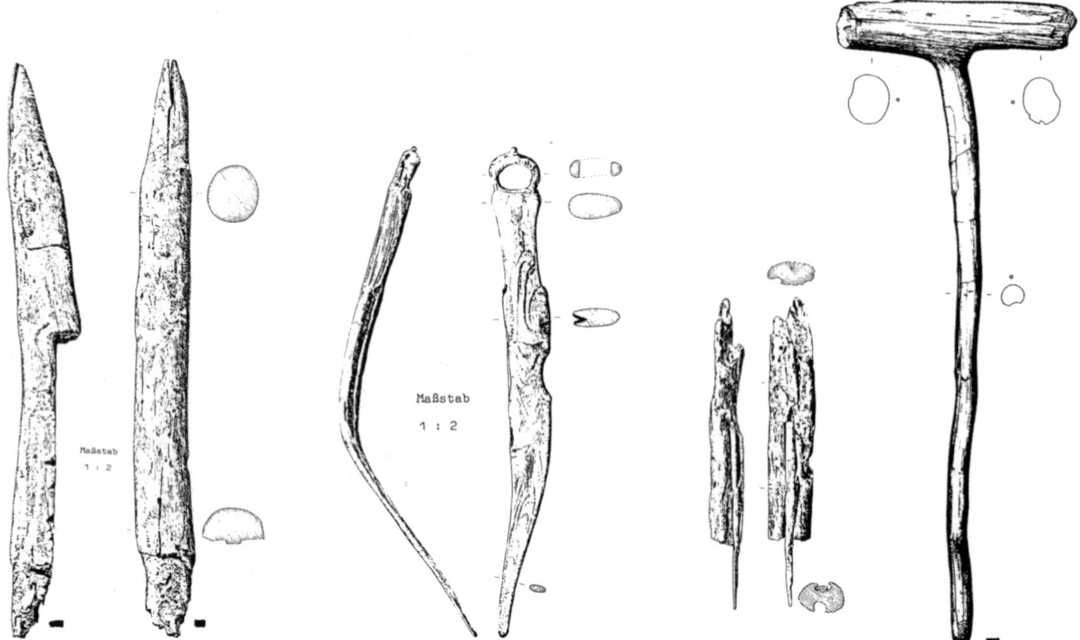

22.1 *(left) Spear head made of yew wood, Neolithic, Burgäschisee, Switzerland.*
22.2 *Neolithic yew tools: harvest knife (with flint blade) and poker from Burgäschisee, hammer from Twann, Switzerland.*

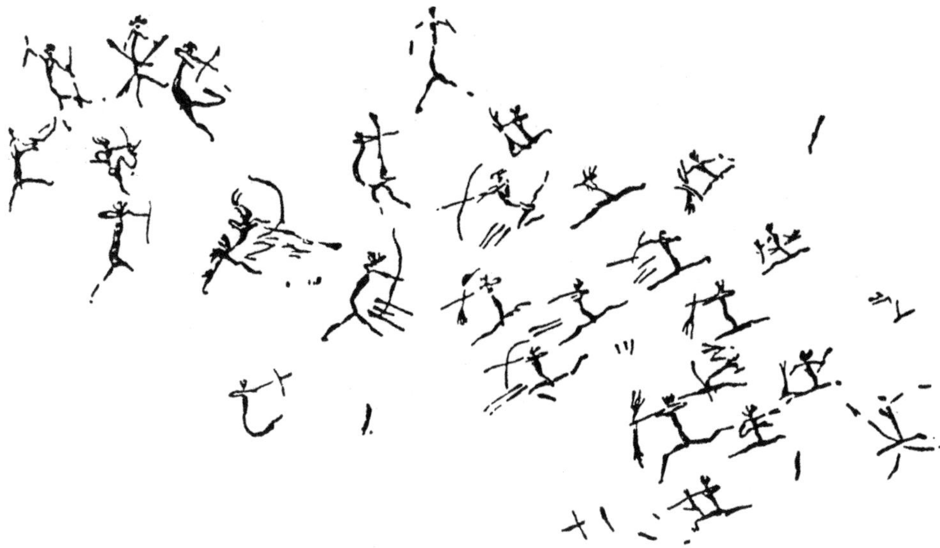

22.3 *Battle scene with bows. Mesolithic rock painting, Les Dogues, Castellón, Spain.*

cannot be established. It was found in the 1930s when the bed of the Elbe river was widened; it emerged from a load of mud, which also contained bones of Ice Age animals and objects from the Neolithic and Bronze Ages right through to the Middle Ages. Soon after discovery the bow was soaked for preservation in a wax-like substance, which now makes carbon-dating impossible.[5]

The Neolithic Age (the Younger Stone Age) saw the introduction of agricultural produce, but hunting still remained an additional source of food and yew bows[6] and longbows (see next chapter) were kept in use right through to the Middle Ages. But as agriculture spread and a settled lifestyle gradually replaced the nomadic existence, a large variety of domestic tools appeared. Yew artefacts from Denmark, Germany, Switzerland and the British Isles include plates, bowls of all sizes, spoons, ladles, buckets, boxes, chests, knives, cutting boards, needles and awls, combs; and field work implements include hoes, pokers (for sowing), harvest knives, hammers, axe handles, and saw frames or handles (for flint blades).[7] Yew twigs were found as part of the construction of the youngest of the Neolithic trackways in the Somerset Levels, England, and also at a late Neolithic site on Lake Zürich, Switzerland, where 70 per cent of the wood specimens consisted of hazel, silver fir or yew.[8] At a number of sites in south England, yew was found among charcoals from the Neolithic to the early Iron Age.[9] Wood and seeds were found at a Mesolithic layer on the Isle of Wight,[10] and in Co. Mayo, Ireland, yew stakes found beneath peat appear to have been part of a Bronze Age deer trap.[11]

This hard but flexible wood is highly water-resistant and at times has been used for fence poles, irrigation pipes, barrel taps and other waterworks, in various countries and until relatively recently. Furthermore, in some locations the foundations for lake dwellings were fashioned of this wood, for example in the Neolithic lake villages in Switzerland.[12] When 500-year-old pillars underneath some of the palaces in Venice, Italy, were replaced in the nineteenth century, the yew beams were still in a fit state to be refurbished and used elsewhere.[13]

In early medieval times, yew nails were found holding together Viking ships.[14] Among the most curious artefacts, however, are the Dover and Ferriby boats. In 1937, the remains of three Bronze Age boats were discovered in the clay of the upper Humber estuary at North Ferriby, Yorkshire. Their oak planks are sewn together with yew fibres, which is a most unusual choice of material. The boats are some 16m long and slightly over 3m wide, with room for

22.4 *The excavation of the Dover boat.* ***22.5*** *The oak planks are sewn together with yew fibres.* ***22.6*** *Close-up of the yew withies.*

nine oarsmen on each side. The boats have been carbon-dated to about 1800 BCE,[15] which pre-dates any other sea-going vessel found to date. Their unique style of architecture has no comparison anywhere in Europe, the only kinship being with the royal barge that was found in a rock-crypt in front of the Great Pyramid of Giza (this ancient vessel has been dated to *c.* 2600 BCE, and is believed to have served at the funeral ceremony of Pharaoh Khufu (Cheops) himself, built to carry his soul on its afterlife journey; it is 43m long and made of cedar wood, sewn together with grass rope).[16] In Britain, another 'sewn boat' was found in 1992 underneath the streets of Dover, Kent. Its oak planks have been sewn together with yew withies 1.4–1.8m in length and six to sixteen years old. The fibres are believed to have derived from coppiced yew.[17] The connectedness of Bronze Age Britain and ancient Egypt is an unexplained mystery in archaeology.

The ancients did not choose their wood by practical criteria alone. The usage of different woods in the Stone Age and early Bronze Age cannot be judged solely in terms of wood characteristics and availability. Magical qualities and mythological meanings were assigned to different plants and at times these determined the plant's use or non-use even more than the physical qualities.[18] Yew, throughout humankind's history, has been associated with the gates of death, birth and immortality. Hence, the killing of animals with a yew spear or bow is likely to have had a shamanic significance for the Palaeolithic hunter, such as ensuring the rebirth of the animal through the presence of the Tree of Life in the moment of its death. In this sense, the agricultural tools from the Neolithic period are no surprise either, as the yew seems to have been intimately connected with the goddesses of the earth and the underworld who also brought forth the abundance of the fields and gardens (see Chapters 32 and 34).

European yew artefacts from Roman to medieval times include spoons, spindles, needles, awls, nails and wedges, tool handles, keg tires, as well as parts of weaving looms.[19] In later centuries, the domestic nature of the items became progressively less 'essential', but nevertheless included indispensables such as egg cups, toothpicks, billiard balls and cufflinks.[20] Today, yew wood is rare and expensive and mainly used for (furniture) veneer, sculpture and woodturning.

Within its distribution area in western North America, the Pacific yew was used by most Native nations as a traditional material for bows[21] and among some tribes for arrows also.[22] In traditional fishing communities harpoons were made of yew.[23] Furthermore, Pacific yew was a traditional material for canoe paddles (Clallam, Hesquiat, Hoh, Nitinaht, Oweekeno, Quileute), combs (Southern Kwakiutl, Oweekeno, Quinault, Coast Salish), needles (Karok, Nitinaht), and snowshoe frames (Thompson). The Pomo additionally used the roots for basketry and twined them as fabric,[24] while the Quinault also made spring poles for deer traps from yew.[25]

CHAPTER 23

THE LONGBOW

EARLY HISTORY

The invention of the bow was a huge advantage for humankind. It enabled the hunting of game with success and comparative safety, and also the ability to strike enemies from a distance. The earliest signs of the use of the bow that have come down to us are arrowheads, about 50,000 years old, from Tunisia, Algeria, Morocco and the Sahara. They are made of flint or obsidian, later ones also of chalcedony or jasper. The stone arrowheads were no doubt preceded by bone, horn, tooth and thorn, or by simple sharpened sticks for arrows, but none of these could have survived.

The best bow-timber is yew, followed by wych elm. In the north, other early bows have been made of ash, hazel, fir or oak. Where wood was not available, composite bows of horn, bone, sinew and gut were crafted in various combinations. Advanced composite bows rely for their success on the fact that horn compresses (and hence was used on the belly side that faces the archer) and that sinew is elastic and lengthens (and hence is glued on the back, the surface away from the archer). The two types of bow also fused to create a composite bow with a wooden core, and in the most developed composite ones, the speed of the limbs' return to their normal position is greater than that of any kind of timber. (This is also the reason why contemporary archery has widely abandoned wooden bows for composite ones – of synthetic fibres, however).[1] Until fairly recently, various Native tribes of northern California[2] still made composite yew bows, and so did the Kalapuya and the Umpqua in Oregon. Their bows had a thin line of sapwood on the back side of the bow, which was grooved out and had deer back sinew glued into the grooves.[3] The natural material of choice for the bowstring is sinew.

The Mesolithic (*c.* 20,000–*c.* 7500 BCE) and Neolithic (*c.* 7500–*c.* 3500 BCE) rock paintings in the Sahara already show many different kinds of bows, and there were probably wooden ones among them. The gradual process of the Sahara turning into desert took about 5,000 years and was not complete until about 500 BCE. Pollen samples and remnants of campfire charcoals suggest pine, holm oak, alder and lime trees as contemporary trees at the time of the oldest arrowheads. The most common timber in historic Sahara bows, however, seems to be acacia.[4]

In Europe, more than twenty prehistoric bows or fragments of bows have been found in Denmark, Germany, Switzerland and Britain. They are all longbows (ranging from *c.* 155.5 to *c.* 175cm in length) and they are all made of yew.[5] An exceptional yew longbow was found in 1991 in the possession of the 'Ice Man' from the Tyrolean Alps near the Austrian–Italian border: it is 183.4cm long (its owner was only 160.5cm tall) and about 5,000 years old.[6] About a millennium older is the yew longbow found

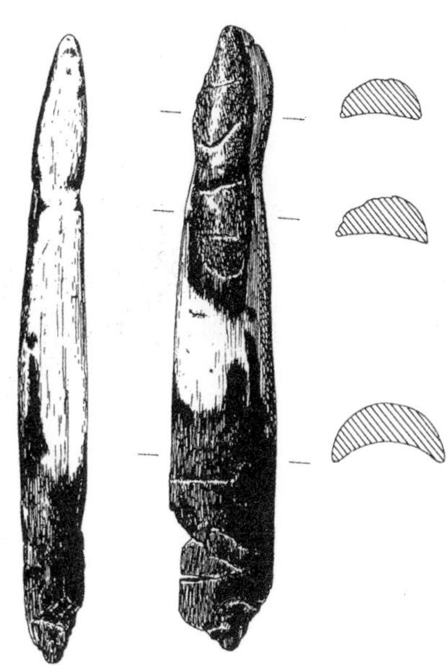

23.1 *Charred end fragment (length* c. *37cm) of a Neolithic yew bow, Niederwil, Switzerland.*

in 1990 in the peat of Rotten Bottom, a remote boggy plateau in Dumfriesshire, Scotland. A third of it is missing but its original length has been estimated at c. 174cm.[7] All these bows were fine weapons, but as Robert Hardy, the internationally acclaimed expert on longbows, points out, they were cut from the core of the yew log – their makers 'had not yet discovered the magic that lies in the use of the sapwood and the heartwood of yew together'.[8]

Archaeological traces suggest that the wooden bow (not always of yew) was widespread in Europe from the tenth millennium BCE and around this time also made its way around the globe, covering Africa, Asia, the islands in the Indian and Pacific Oceans, and the Americas. Almost everywhere, hunters and warriors were buried with their bows. In Europe the bow's importance only declined from about 1500 BCE, in the wake of the successive spreading of agriculture and the Neolithic pattern of life.[9]

How powerful is the longbow?

The draw weight of a bow is the weight that is held momentarily at a full arrow's length – between *c.* 70–80cm (28–32in), the arrows being matched to the archer's build and arm length.

The draw weights of the Stone Age and Bronze Age bows ranged from about 15kg to a maximum of about 30kg (*c.* 33–*c.* 66lb), which clearly distinguishes them as hunting bows. A replica of the Rotten Bottom longbow, for example, shows a draw weight of about 16kg (35lb) and has a maximum range of *c.* 50m. But to fell a large mammal such as a deer would then require a distance as close as 5–10m.[10] In the eighth century CE, the first bows with draw weights of up to *c.* 50kg (*c.* 110lb) appeared.[11] The weight of these Viking weapons anticipated that of the late medieval warbows, and they were not designed for hunting but for combat. But the most drastic change in bow architecture had already dawned during the first three centuries CE, although in Scandinavia: the usage of the different properties of yew sapwood and heartwood. If fashioned from the region where the two meet underneath the outer bark, a wooden yew bow becomes as powerful as a composite bow – the heartwood compresses well (on the belly side facing the archer) and the sapwood (forming a thin layer on the back of the bow) is elastic and lengthens. The strength inherent in a longbow, combined with the qualities of the 'composite' yew forged the most effective killing device known to mankind at that time. The earliest examples of this are eight bows from eighth-century southern Germany.[12] The stage was set for the devastatingly gruesome battles of the thirteenth to fifteenth centuries (see below), and for the foundation and tearing down of dynasties and empires.

23.2–23.3 *Remake of an armour-piercing arrowhead (left), and penetration of a 1mm steel sheet.*

The medieval warbows were assumed to have weighed from about 36 to 54kg (80–120lb)[13] which, in bows that make full use of the different qualities of yew heartwood and sapwood, gave them a strongly improved range and velocity, compared to any of their predecessors. In 1188, Giraldus Cambrensis reported from the siege of Abergavenny Castle six years previously that the attacking Welsh archers drove their arrows straight through the tower's solid oak door which was 7.5–10cm (3–4in) thick. A century later, the Gwent archers (South Wales) were feared for their ability to nail armed enemies to their saddles by piercing the armour, the leg, the leather saddle, and penetrating deep into the horse. The yew longbows found in the wreck of the *Mary Rose* (which sank in 1545)[14] were even heavier: they weighed up to 80kg (175lb), with the majority around 63–68kg (140–150lb) which can be taken to represent the warbow of the late Middle Ages. They could reach about 300m and certainly still do damage at about 240m. At this distance the loss of velocity was only some 16 per cent. At point blank range, almost no metal could withstand the armour-piercing tips of the arrows.

***23.4** Longbow battle scene; painted wood carving by Leslie Rendall, in the collection of Robert Hardy.*

Nevertheless, the bow remained an important weapon throughout the Neolithic, Bronze and Iron Ages. Its traces can be found amid ancient Germans, Romans, Parthians, Greeks, Cretans, Numidians, Huns, Avars, Tartars, Magyars, Ottomans and Arabs, to name but a few.[15] And in Roman Europe it was on the rise again: the Romans equipped their cavalry with short composite bows borrowed from the East (longbows are ill-suited to mounted archers), and the Germanic tribes, having forgotten or abandoned the composite bow by the end of the Bronze Age, gave the Romans considerable trouble with their longbows.[16]

THE MEDIEVAL WARBOW

In 1066, William the Conqueror won the battle of Hastings which, as tradition has it, was finally decided by an arrow piercing the right eye of King Harold of England. With this arrow from a longbow, Saxon England fell and Norman rule began. It is not known whether the Welsh archers picked up their extraordinary longbow skills from the Norman rulers, the Danish raids, or whether they (re)invented the longbow themselves. But soon after Wales was subdued by the (Norman) English monarchy, archery became gradually incorporated into the English army. The military ascent of the longbow is mainly thanks to Edward I (r. 1274–1307), who inherited unreliable, headstrong feudal armies and gradually turned them into 'cohesive, disciplined, well-paid units, with a growing proportion of archers, mostly Welsh, among them' (Hardy 1992).[17] He also decreed by law that every able-bodied man in the country, with the exception only of priests and judges, was obliged to have in his possession a bow and arrows and to practise with them and keep them in good order, ready for immediate service.[18] In 1298, Edward I defeated William Wallace's Scots at Falkirk, and this battle is generally seen as the first classic victory for the longbow. Sixteen years later at Bannockburn, Robert the Bruce defeated an English army twice the size of the Scottish, partly because he found a way to ride down their inefficiently deployed and unprotected archers.[19] Bannockburn in 1314 was a defeat for the English but the final necessary strategic lesson for a nation that was to take archery to its military peak. At the next major English–Scottish conflict, in 1332 at Dupplin Muir, only 500 English knights and men-at-arms and about 1,500 archers defeated a Scottish army perhaps 10,000 strong.

Five years later, the Hundred Years War (1337–1453) began. Edward III moved into France, and the longbow with him. At the battle of Crécy in 1346, the English were outnumbered ten to one, but still won the day. Seven thousand English archers darkened the sky with 70,000 arrows per minute, 'flying in the air as thick as snow, with a terrible noise, much like a tempestuous wind preceding the tempest', an eyewitness chronicler wrote.[20] Edward renewed his grandfather's edicts about the common practice of archery and commanded the sheriffs of London to announce that 'every one of the said city, strong in body, at leisure times on holidays, use in their recreations bows and arrows ... and learn and exercise the art of shooting' (12 June 1349).[21] With the future of a large pool of archers secured, another victory awaited them in France. In 1356 at Poitiers, forgetful of the lessons of Crécy, thousands of French perished in the arrowstorm, and in the decisive battle of the English–French war, on 25 October 1415 at Agincourt near Calais, 20,000 to 30,000 Frenchmen, many of them mounted knights in heavy armour, were defeated by only 900 men-at-arms and 5,000 archers.[22] For the time being, England seemed invincible ...

DEATH AND TAXES

From the thirteenth century onward, the yew longbow made England a major political player in Western Europe; but there was a price to be paid. War, apart from the large-scale human suffering that it brings, is also a huge national economic burden, and all over England and Wales the bowyers, fletchers, longbow-stringmakers and arrowsmiths were kept hard at work. This is, of course, in addition to the manufacturers of lances, swords and armour, and the producers and organisers of essential supplies to keep thousands of men alive and on the move. Archers, however, were especially expensive because they had to keep practising all year round, even in times of peace. In the 1270s, for example, the pay for an archer was 2*d* a day (60*s* per year), a group leader (vintenar) earned 4*d*, and the mounted company leader (centenar, constable) 1*s*. For comparison: at the time, one pound of wheat was $\frac{3}{4}d$, 28 pounds of oat flour 2*s*, a gallon (3.78 litres) of wine $2\frac{1}{2}d$, and a pig 2*s* 8*d*.[23]

***23.5** A bowyer in Cracow, Poland, meeting clients in his workshop;* Behaim Codex, *1505.*

In the 1270s, a yew longbow from the branch cost 1*s*, from the bole (trunk) 1*s* 6*d*. The standard equipment of a bow, a replacement bow, 48 arrows, a quiver and a quiver belt would have cost about 5*s* 6*d* per archer; the supply of 5,000 archers therefore amounted to about £1,375 sterling. Prices remained about the same for 200 years: in 1480, for example, bowyer John Simpson sold ten bows for 20*s*, 288 arrows for 34*s* 8*d*, and a red leather quiver for 9*d* – but in 1483, the legal price limit had to be adjusted because of increasing supply problems (see next chapter), and rose drastically to 3*s* 4*d*.[24] The wages for archers did not change, however, but still added up for the Treasury: in 1473, the House of Commons granted the king £51,117 4*s* 7*d* in full payment of wages for 14,000 archers for one year.[25] All this implied severe taxation of every town, city and county. There were also additional demands: for example, in February 1417, six feathers from every goose in twenty southern counties had to be at the Tower by the following March, and in the following year, demand becoming more precise, the sheriffs had to supply 1,190,000 goose feathers. The destination of the annual light-winged journey? The workshops of the arrowsmiths.[26]

CHAPTER 24

THE CATASTROPHE

MEDIEVAL ARMS TRADE IN BRITAIN

The medieval warbow was, at its best, made from local yew wood, at its very best, from imported yew.[1] When Edward I innovatively decreed that every man should own and use bow and arrows, the British Isles was too small to satisfy long-term demands for yew wood. Around 1350, demand began to seriously outstrip supply. During the reign of Edward II (1307–27), yew wood was imported from Ireland and Spain (and relatively inexpensively: one cargo of 180 dozen Spanish bows is recorded to have cost £36).[2] Soon, an international trade in yew timber began to develop, with shipments from Baltic, Dutch, Spanish and Mediterranean ports.[3] The continental wood was superior because for the most part it derived from slow growth at high altitudes.[4]

The oldest document relating to the trade of bow timber is a customs scroll from the Dutch city of Dordrecht, dated 10 October 1287. The first evidence of an actual import of bow timber into England dates from 1294. The next glimpse comes from Newcastle where six ships from Stralsund in the Baltic Sea brought 360 *baculi ad arcus* (bow staves) on 8 January 1295. But for decades this import trade remained inefficient because the English kings had to set price limits for bows so that all their qualifying subjects could take part in compulsory archery training. The controlled bow prices, however, were too low for the importers who had to cover their high costs for obtaining and transporting the raw bow staves across Europe.[5] In 1470, with the supply of bow wood for the army running out, and in the middle of the Wars of the Roses, Edward IV renewed the compulsory archery training and included hazel, ash and laburnum as possible bow woods for common practice. Already Henry IV (1399–1413) had ordered his royal bowyer 'to enter upon the lands of private individuals and cut down yew or any other wood for the public service',[6] but all these measures did not solve the problem. What did solve it was the Statute of Westminster of 1472 which decreed 'that for every tun-tight of merchandise [...] four bowestaffes be brought' by ships and vessels unloading in any English harbour. About ten years later, Richard III demanded even ten yew staves per casket of wine (which suggests that there were significant yew stave imports from the Mediterranean!).[7] This law ingeniously shifted the yew supply pressure from the government to the trading companies. The compulsory import of yew

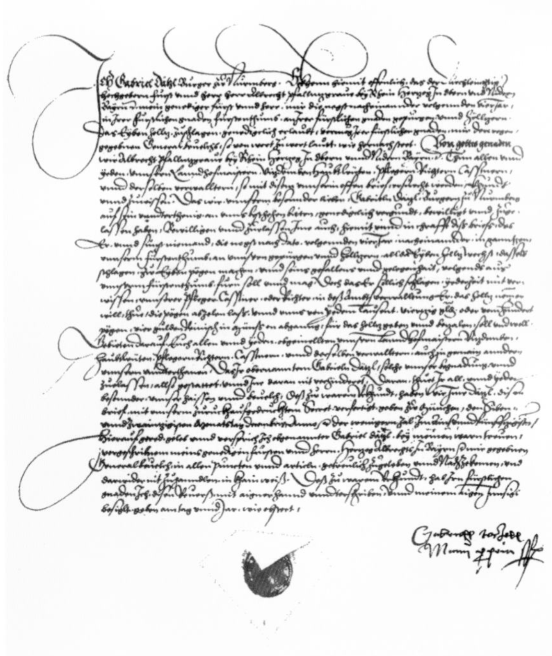

24.1 *Yew privilege manuscript, granted to Gabriel Dätzl by Albrecht, Duke of Bavaria, in 1551.*

staves, enforced by threat of heavy penalties, led to an early capitalistic monopoly in central European forestry and timber trade – and to the annihilation of the majority of Europe's yew stands.

Already by the mid-fifteenth century, most staves for English longbows were imported, mainly from southern Germany[8] and Austria, where the monarchy wanted its share of the vibrant yew trade. The trading companies had to petition the royal authorities and put in bids for the Eibenmonopol, the Yew Monopoly or Privileges, for a certain region. The respective kings of Upper and Lower Bavaria, Tyrol, and Upper and Lower Austria sold the exclusive privileges of extracting yew timber from their lands to these private trading companies. The privileges were limited to a number of years (usually three to six) and included extensive obligations: the trading company had to spare other trees and also yew trees that were too young; the woodcutters had to be experienced in bow stave-cutting (the shape and position of a 'raw' bow stave in the trunk has to be recognised precisely – a wrong cut can waste everything) and were to be monitored; and the sale of yew staves to the 'infidels', the Tartars and Turks in the East, was absolutely forbidden.[9] The privilege specified the geographical region of its validity; it also regulated the annual export volume, the payment of tolls and customs, of forest tax, and, last but not least, it stated the sum to be paid to the royal treasury. In 1521, for example, the Austrian monarchy demanded 5 rh. fl. (Rhineland guilders[10]) per thousand bow staves for the Yew privileges for Tyrol, and Balthasar Lurtsch bought a volume of 20,000 staves. In 1523, Joachim Rehle was granted twice that amount for his outlandish offer of 100 rh. fl. per thousand. Since he did not pay in the end, Lurtsch got the privileges back two years later, for 60 rh. fl. At the next bid, in 1528, the price levelled to the more realistic 32 rh. fl. per thousand.[11]

All across Europe,[12] the trade routes of the yew staves employed a mixture of carts for land transport, barges for inland navigation, and ships leaving the harbours of the Baltic and North Sea to head for London or other English ports. From Tyrol in Austria, Lurtsch used three routes (see map): on the first, the timber was taken up the Danube, carted across the Swabian Alb mountains to the Neckar river, down the Neckar to enter the Rhine at Mainz, and downriver to Antwerp in Belgium. The second route led down the Danube to Krems (Austria), by cart through the Moravian Gate (modern Czech Republic) to the Wisla river at Cracow, down the Wisla across the whole of Poland, and to the Baltic harbour of Gdansk (Danzig, where the yew stave trade of the Teutonic Order can be traced back at least to the fifteenth century[13]). The third route left the Danube at Linz (Austria) and cut over land to Prague to meet the Elbe river, where the barges took the staves past Hamburg to the English warehouses at Stade. An average cartload was thirty-two bundles of twenty staves each, or a total weight of about 1,500kg (3,300lb). During the sixteenth century, the yew privileges for large parts of Bavaria and Austria were held for over eighty years by the company of Christoph Fürer & Leonard Stockhammer.[14] Fürer shipped the staves up the Danube to Regensburg (Lower Bavaria), carted them past Nuremberg (the company's residence) to Bamberg, where they were loaded on to barges once again to go down the Main and Rhine rivers, first to the trade centre at Cologne, from there to the Dutch dockyards or the main trade centre at Antwerp.[15] The timber from the

24.2 *Yew stave transport routes in the early sixteenth century. (*Taxus *distribution after Hegi 1981, modified)*

Carpathian Mountains and western Russia either went to the Wisla barges waiting in southern Poland, or down the Tisza river (named after the yew) into the Hungarian Plain, eventually to reach Venice or, much more likely, cut across to the Danube and meet up with the Austrian timber vessels. This was the main trade network. There were also deliveries from other dealers, which joined the main timber stream from the privileged companies. There is little information on the extent of these additional exports. One piece of evidence is a letter from 1530 in which the Austrian monopoly holder Lurtsch complains to the government that a Swiss company, unburdened by monopoly fees and forest taxes, is undercutting yew prices at the main trade centre in Antwerp.[16]

The Mediterranean exports are less well documented. The Italian trade has not been researched yet;

24.3 *Deforestation battlefield.*

Hats off to the few survivors!
24.4–24.5 *The Steibis Yew is the oldest in Bavaria (girth* c. *500cm, height* c. *7m).*

all we know is that in 1483, the price for staves from Lombardy (northern Italy) increased from £2 to £8 per hundred, and that Henry VIII asked permission from the Doge of Venice in 1510 to buy 40,000 bow staves – at that time, Venice received massive quantities of bow staves from Dalmatia across the Adriatic.[17] To this the Doge only agreed after doubling the price to £16 per hundred.[18] (At this late date, incidentally, the high price might reflect growing scarcity of yew wood (see below) more than shrewd Venetian business practice.) In the far west, Spanish yew exports probably came from the north of the country. Some early English sources mention tight-grained Spanish yew wood as far superior to the knotty British bow timber, but no actual trade documents have been identified yet to tell us about the possible timespan and volume of Spanish exports. Probable trade routes would have been the near harbours of Santander and Bilbao (and possibly river barges down the Ebro – named after the yew – to Barcelona). Richard III's 1483 decree (to import ten yew staves with each casket of wine) is an important hint that Spain, Italy and southern France must have been exporting significant amounts of yew, at least for a few years. It is also possible that during the Hundred Years' War the English invaders could have removed French yews without leaving written records of this exploit (unless the French yew trees had already been taken during the Roman occupation of Gaul, or lost during the Dark Ages – in any case, what happened to the yews of France is a historical mystery).

The complicated trade routes and repeated un- and uploading from carts to barges and vice versa, plus all the fees, taxes, tolls and bribes that had to be paid, made it an expensive business. A breakdown of the expenses for 1,000 staves for the year 1549, for example, reads: timber cutting costs 21 per cent, forest tax 5 per cent, monopoly interest 20 per cent, transport 24 per cent, customs 5 per cent, provision for agent 8 per cent, sweets and marzipan for Mr Vitzthomb 2 per cent, permission letter 11 per cent, 'agent' for procuring permission letter 4 per cent.[19] And yet, if all went well, the staves could be sold with up to 32 per cent profit.

FOREST DEPLETION AND PROTECTION EFFORTS

The ecological disaster did not occur overnight, nor did it happen without resistance. A number of documents, particularly from the German-speaking Alpine countries, display the growing concern and even resistance to the wholesale destruction of yew stands. As early as 1507 – just thirty-five years after

24.6 *A forgotten giant tucked away in the Pyrenees: the ancient yew at Massane (girth* c. *6m).*
24.7 *The Solcava yew (male, girth 428cm) is the oldest tree in Slovenia and protected since 1951.*

Edward IV had first decreed compulsory yew imports – the Holy Roman Emperor, Maximilian I, negotiated with the dukes Wilhelm and Ludwig of Bavaria for a total stop on felling yew wood in both Bavaria and Austria (the country of his residence).[20] But nothing much changed: in 1518 some measures against illegal felling were taken (unauthorised yew theft occurred in both Bavaria and Austria, and particularly along the borders), and a ten-year felling stop was agreed (far too short!), but, otherwise, business went on as usual. That is until 1532, when an additional remark to the export volume appeared – the usual amount of yew staves granted was commented upon with the words, 'if there are that many'. In 1542, the Bavarian government wrote to Ferdinand, deputy of the current Holy Roman Emperor, Charles V, regarding the permitting of yew privileges. Essentially, the letter is an intense plea to leave the yews alone. It lists all possible reasons for that and also explains that yews have been severely depleted, and that only young trees and brushwood are left. However, it could not prevent the continuation of the monopoly practice.

The second paragraph of the letter is worth noting as it sheds light on felling practices of the time: 'Where a great yew tree stands amidst the forest and is felled, by which a gap is created, wind breaks occur immediately, which brings great damage and disadvantage to the high and low conifer forests and also the home forests of the subjects [of your majesty].'[21] This clearly indicates that not only young or even coppiced yews were being felled for bows, but all sizes and ages, even including old or ancient trees whose fall left critical gaps in the forest. The letter also mentions how much wood is wasted and left behind as only the area where heartwood and sapwood meet is used for bows. And Ascham's treatise on archery[22] of 1545 states that the wood from the bole (trunk) of the tree is superior to that of the branches. Additionally, the typical twisted and knotted growth patterns of *Taxus* limited the yield even more. It is an intriguing idea that perhaps someone somewhere could possibly have thought of creating yew plantations to meet the long-term demands, and some small-scale efforts might have existed. But the bow stave trade with continental yew wood was not about the quick growth from coppice stools 'shooting into the green', its *raison d'être* was exactly the opposite: the slow-grown, tight-grained, elastic timber from high and extreme mountain stands. What is more, nobody could plan a hundred years ahead – the time it takes for a yew to grow to the size yielding just four bow staves.

Inevitably, the Bavarian Duke Albrecht had to reject an application for a yew monopoly in 1568, stating that he had recently declined a request by the Duke of Saxony because of the almost complete exhaustion of yew supplies. The year 1589 saw the total end of the granting of yew privileges for Bavaria, and the Austrian lands as well, because no tree of size was left. In Württemberg (south-west Germany) yew had already completely disappeared from the timber trade documents ten years earlier. During the entire seventeenth century, *Taxus* remained absent from the forestry records of the whole of the northern Alpine countries.[23]

Some of the English kings had tried to be helpful: Richard III had ordered a general planting of yew in 1483; and Henry VIII (r. 1509–47) in 1511, while reinforcing the compulsory imports, also decreed the planting of yews everywhere in England.[24] But reality was catching up fast: Queen Mary (r. 1553–58) had to accept a doubling of the price limits: 6*s* 8*d* for a bow made of the best foreign wood and 3*s* 4*d* for an inferior sort.[25] And in the time of Elizabeth I (r. 1558–1603), the stocks of wood, both at home and abroad, were running out. Her Act of Bowyers decreed that for every bow of yew, a bowyer had to produce four of inferior wood, for example, wych elm. By then, yew had become so scarce that yew bows had to be shared around, and those under 17 were forbidden to shoot 'in a yew bow'.[26] Elizabeth also directed that yew staves be imported from the Hanse towns and other places, which left only Carpathian yews coming via Poland, and possibly timber from the north-eastern Baltic countries via Reval (modern Talinn). It was all in vain: on 26 October 1595, Queen Elizabeth I decreed that the army was to replace its longbows with shotguns.[27] All firearms were clearly inferior and would remain so for another 200 years – any longbow could outshoot a musket with six arrows to a bullet, with greater reach and greater precision – but there was no other choice to be made.

The overall exploitation during the longbow centuries was of astronomical proportions: the Fürer company alone, holding yew privileges in southern Germany and parts of Austria over eighty years (1512–92), can be assumed to have exported 1,600,000 yew staves – and there were other companies. Right across Central Europe, yew populations were exhausted. And they have never recovered, for various reasons. Human short-sightedness and subsequent wars have caused incalculable damage and fundamental loss to the European countryside that has not healed in 500 years. And even if all measures were taken today, it would still take another 500 years to have old yews back in many of the regions they once graced.

24.8–24.9 *The oldest tree in Switzerland is the Gerstler yew near Burgdorf, Kanton Bern (girth c. 4m); historical photograph (above), and today.*

24.10 *A fine specimen in Austria: the protected ancient yew at Kirchberg.*

24.11 *A famous tree in the Czech Republic: the yew at Krombach.*

24.12 *A much appreciated tree in Saxony: the yew at Schlottwitz.*

CHAPTER 25

A POTENT MEDICINE

Since early history, the yew has not only been employed for lethal weapons, but also for healing purposes. Given the fact that most parts of the tree are poisonous to human beings, the wide range of the tree's medicinal use is somewhat surprising.

ALLOPATHY[1]

Native America

In the east of the American continent, the Iroquois use a compound of the European yew to help with menstruation, rheumatism, coughs and colds, and might also include it 'in all medicines to give them strength'.[2] The native Canadian yew is used in traditional medicine for rheumatism (Abnaki, Quebec Algonquin, Chippewa, Menominee), colds (Penobscot), for gonorrhea and as a diuretic (Potawatomi). Among the Algonquin and Tete-de-Boule, women use it for irregular menses, and among Micmac women for afterbirth pain and blood clots. The Iroquois take arils and leaves, mix them with cold water and maple sap, and ferment into a 'little beer'.[3]

On the other side of the continent, the Pacific yew has a wide range of internal and external medical uses. For example, it is used for stomach pains or general internal ailments (Haihais, Karok, Kitasoo, Klallam, Tsimshian), and as a dermatological poultice to put on wounds (Cowlitz, Quinault). The Chehalis use an infusion of crushed leaves as a wash to improve general health, and the Swinomish rub smooth twigs of young yews on the body to gain strength. The Thompson take a decoction of the bark for any illness; similarly the Yurok take a bark decoction to 'purify the blood', and the Tsimshian use the plant for (skin) cancer treatment.[4]

Europe

Despite its toxicity, the yew has been used in allopathic medicine throughout history. In the Greco-Roman world, yew poison was used as an antidote to adder bites, and the Roman Emperor Claudius (r. 41–54) eventually released an edict in this regard.[5] Other early uses include stimulating menstruation, and the treatment of insect bites and worms. In old Coptic medicine, the successor of Egyptian medicine from Pharaonic times, a decoction of yew leaves was used for skin problems and particularly for persistent skin ulcers.[6]

From the early Middle Ages until the industrialisation of medicine and the advance of synthetic remedies in the twentieth century, yew preparations were used in a variety of treatments. These included, along with the uses just mentioned, aids for wounds, parasites, epilepsy, diphtheria, rheumatism, arthritis and tonsillitis.[7] Overdosing with yew preparations, however, has time and time again led to patients' deaths, particularly in irresponsible or desperate abortion attempts.

Yew preparations and smoke were widely used to treat rashes and scabies, also rabies in dogs, to expel parasites from the stables, and rodents (and ghosts) from the house.[8] The abbess Hildegard von Bingen (1148–79) used yew smoke for coughs and colds, and generally recommended a yew walking stick for health.[9]

Himalaya

Mountain dwellers in Nepal give the juice of the leaves of the Himalayan yew (*T. wallichiana*) for coughs, bronchitis and asthma.[10] In northern India, yew preparations are regarded as useful for loosening mucus and relieving cramps, and for stomach and heart problems. In India, yew decoctions are still sold.[11] Generally, villagers in Afghanistan, Pakistan, Nepal, northern India, Tibet, western China and South-east Asia eat the arils without medicinal necessity.[12]

In the early 1890s, the British intelligence officer Hamilton Bower purchased an ancient Sanskrit

25.1–25.4 *Healing mandalas painted by Nicole Melis during her cancer treatment.*

document and forwarded it to the Asiatic Society of Bengal. Written in the Brahmi alphabet it was unintelligible at first, but its eventual translation (1912) has shown it to be one of the oldest preserved texts of ancient Indian medicine (Ayurveda). The Bower manuscript, dating to the first half of the sixth century CE,[13] mentions a treatment for (abdominal) cancer containing parts of the yew tree mixed with clarified (yak) butter.[14]

It took the West only some 1,400 years to catch up with Asian prowess!

FIGHTING CANCER

The first sign that yew trees contain a compound that has anti-tumour activity was discovered in the United States in May 1964. About two years later the active compound was finally identified and named taxol, *tax-* after the genus *Taxus*, and *-ol* for alcohol (the molecule contains hydroxyl groups that signify it is an alcohol). After extensive chemical and clinical trials, Taxol® (paclitaxel) achieved the official approval as a marketable drug for human cancer chemotherapy in the United States in 1992, followed by many other countries. In 1996, Taxotere® (docetaxel, see box p. 114) was approved in the US and in the European Union. In the United Kingdom, both products – Taxotere® and Taxol® – received their licence for the treatment of early breast cancer from the MHRA (Medicines and Healthcare Products Regulatory Authority) in January and February 2005, respectively, as had Taxotere (docetaxel) for prostate cancer in November 2004.

A single chemotherapy treatment is four to six cycles, each comprised of an infusion of three hours (paclitaxel), one hour (docetaxel), or half an hour (albumin-bound paclitaxel, see box p. 114), respectively. Six cycles of docetaxel cost about £5,400 ($8,100).[15] (For more about the global market and its ecological impact on the yew populations, see Chapter 21.)

Clinical

Since its introduction, taxanes have revolutionised the treatment options for patients with advanced forms of breast and ovarian cancers and some types of leukaemia (non small-cell lung cancer). Advanced ovarian cancer has response rates of 19–36 per cent, previously treated metastatic breast cancer 27–62 per cent, and lung cancer 21–37 per cent. In single cases, paclitaxel produced a complete tumour remission.[16]

Taxanes have a mechanism of action that is different from any other class of anti-cancer drugs: they bind to the microtubules in the (cancer) cells.

By stabilising the microtubules, mitosis (see box p. 39) and therefore tumour growth is inhibited.

Chemotherapy, however, is not easy, and side effects are well known. Complete hair loss is inevitable, but hair grows back after the treatment. Numbness, tingling or burning in the hands and/or feet often occurs in patients. A lowering of red blood cells may cause anaemia, and a brief drop in white blood cells may cause bacterial infections. Allergic reactions, joint and muscle pains are quite common too. Less common are irritations at the injection site (taxanes are injected into a vein), mouth or lip sores, stomach upsets and diarrhoea. Needless to say, chemotherapy could harm a foetus and is not recommended for pregnant women.[17]

The human side

An increasing number of women who have beaten cancer with the 'yew treatment' become curious about the tree behind the cure. Some even take a detour when travelling or visiting the British Isles in order to visit one of the ancient yews because they want to thank the tree for their new lease of life.[18] A most extraordinary inner journey with cancer, however, was experienced by Nicole Melis from Belgium during her treatment. In the hope that it may inspire and help others she has agreed for an extract from her letter to be published here:

> When I first heard I had cancer I was very afraid of the chemotherapy because I was told it would be a very hard one, with many complications. The doctor said I should consider the chemo as my ally. This was the magic key for me because immediately I saw behind his back a very big *Taxus* tree. So I asked the doctor if he was thinking of a Taxol treatment and he said yes. I calmed down immediately – to his surprise because the Taxol chemo is the very toxic one.
>
> At home, I started my own process of connecting with the *Taxus*. I searched the internet and learned about the oldest yew trees in Britain; I looked for more information, but finally I connected *inside* with the yew energy. By means of intensive inner focus I received the information that what the cancer cells were missing is the new 'inside structure' to take part in a global transformation process. I had to find this inside pattern and then bring the cancerous cells back into connection with my whole system. I suddenly felt confident that *Taxus* would give me the necessary code and inner structural pattern my cells needed.
>
> During the chemo, I inwardly received a lot of images, colours and instructions about what to do with all of this. I started to make pictures of all

these images. Everyday I was drawing, even in the hospital. And in producing these images I felt good, I was not so sick as they had warned me I would be. The colours and patterns brought me into a balance and my system could tolerate the chemo without the expected complications.

How the images developed was very surprising. In the last stage I was drawing geometrical forms and finally, the platonic bodies.[19] From these I had to choose 'my geometrical pattern'. I did, and from that moment on I felt very good. The chemo stopped. I am now in recovery-time.

I am busy with remaking and working up all the drawings, and sometimes I work with them with other people and they give me feedback. The effect of the drawings as energy fields is amazing. Now I am working with connecting the paintings with sounds and tonal resonance.

So, that is, in short, something of my experience with a powerful helper, the *Taxus*. The *Taxus* chemo-material and the connection with the *Taxus* energy made it possible to go into the insights and experiences of this 'all-chemy-process'.

So, I wish you: beautiful things to see, to find and to go through.

Nicole Melis

A labyrinth of terms

All parts of the yew tree except the red aril contain, to varying degrees, toxic substances (mainly alkaloids). All yew poisons are generically called **taxoids**.

The chief active ingredient among the taxoids was discovered in 1890 and called **taxine** (empirical formula $C_{47}H_{51}O_{10}N$), but was later identified to be a complex compound of taxine I, taxine II, taxine III, and seven other bases.

Different from taxoids are the **taxanes**, which are not soluble in water, and pass the digestive tract of an organism without causing harm. In conjunction with a solvent, however, and injected, they can kill cells. The anti-tumour compounds belong to this group.

A large number of taxanes have been identified from yew trees; one of them is **taxol**, the pure substance used in cancer chemotherapy. It was given this name at the time of its discovery in 1967, but on 26 May 1992, the company Bristol-Myers Squibb obtained the patent on this name for their exclusive use as a pharmaceutical product, Taxol®. The new generic name to fill the gap in the public domain is **paclitaxel**.

Originally extracted in pure form from the bark of the Pacific yew, tumour-active taxanes can be semi-synthesised from a substance called **10-deacetyl baccatin III (DAB-III)**, which is found in the needles of all yews and is about ten times more abundant than the pure paclitaxel in the bark.[20]

Paclitaxel is a white to off-white crystalline powder with the chemical formula $C_{47}H_{51}NO_{14}$. The pharmaceutical product **Taxol®** is paclitaxel in polyethylated castor oil (Cremophor EL, produced by BASF in Germany) plus ethanol, as paclitaxel is not soluble in water or oil – a general problem with all taxanes in medicine.

Taxotere (docetaxel) is also a semi-synthetic taxane derived from DAB-III, and its vehicles are polysorbate 80 and an ethanol diluent. It is produced by the French company Sanofi-Aventis for cancer chemotherapy.

Since the 1990s, a number of other companies have begun to produce yew-products for cancer therapy. Most noteworthy is **Abraxane®**, a nanometer-sized albumin-bound paclitaxel particle, produced by American BioScience Inc., Santa Monica, California. Instead of synthetic solvents it uses albumin, a natural carrier of lipophyllic molecules in the human body. Phase III trials with women with breast cancer in 2005 have shown much higher response rates, fewer side effects, and no hypersensitive reactions – even despite the absence of pre-medication.[21]

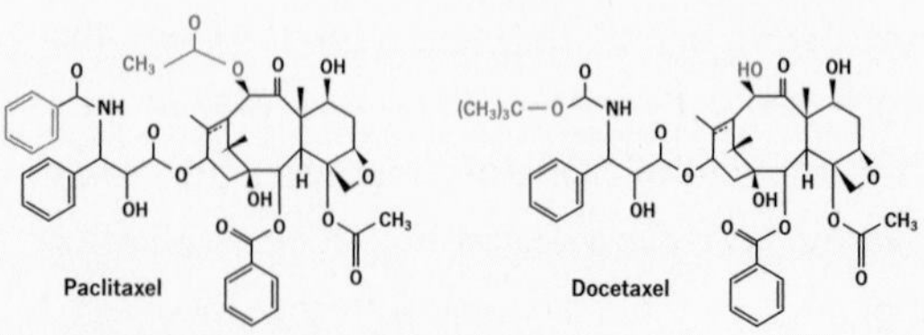

Diagram 13 Paclitaxel and docetaxel

***26.1** Cherry blossom at Kiyomizu Dera, Kyoto's 'Pure Water Temple', Japan. A grey heron among the dwarf yews –* Taxus cuspidata *var.* nana *– is hunting carp.*

CHAPTER 26

FOR THE SENSES

GARDENS

With the awe-inspiring old trees having disappeared during the longbow centuries, and their lore and mystique forgotten, the path was set for a more playful approach to trees and plants. With the beginning of the Renaissance (which started in early fifteenth-century Italy), playfulness, display and luxury entered European societies as never before, in the wake of more peaceful and prosperous times. Italian city-states began to thrive, and all over Europe private tradespeople accumulated riches and began to spend them on extensive gardens and parks. The castles of kings and the aristocracy changed from strongholds to palaces, surrounded by large gardens that served the purposes of pleasure and display. Elaborate garden and landscape designs became an expression of a new vision of art and culture. Such gardens needed evergreens and, even more important, woody plants that responded well to pruning. *Taxus baccata* was there right from the beginning, together with box *(Buxus sempervirens)*, laurel *(Laurus nobilis)* and cypress *(Cupressus sempervirens)*. In cooler climates, hornbeam *(Carpinus betulus)* was one of the favourites, and, again, yew.

***26.2**. The annual clippings of the world's largest labyrinth at Longleat, Wiltshire, amount to c. 2 tonnes and are given to taxane production.*

26.3 *Yew topiary in the palace garden at Würzburg, Bavaria.*
26.4 *The famous yew topiary at Levens Hall, Cumbria.*

Italy remained the cradle and school of European garden art for about 200 years. In the mid-sixteenth century, however, France adapted Italian garden designs and developed the 'French garden' or 'Baroque garden' with its typical large-scale ground plan and its profusion of geometrical shapes, as well as animal and mythical figures. Villandry in the Loire is an excellent example, so are Vaux-le-Vicomte, Saint-Cloud, the Tuileries and, of course, Versailles. Topiary, the 'art' of clipping yew trees and hedges into fanciful shapes, spread to other countries, and remained popular through the Baroque period. *Taxus baccata* became the most popular topiary plant in Western Europe. Famous contemporary examples are the stunning yew hedges of Levens Hall in Cumbria and Montacute in Somerset, which are also popular with film producers as a scenic backdrop in period dramas.

Topiary and the French garden style fell out of international fashion around 1760, gradually fading out by the end of the century, and was replaced by the 'English garden' with its return to natural forms.[1]

26.5 *Old postcard showing the gardens at Sedbury Park, Monmouthshire.*

One of the unsung heroes of horticulture is William Robinson (1838–1935) who is regarded by many as the inventor of gardening as we know it today. Robinson despised formal arrangements, loathed Versailles as 'indescribable ugliness and emptiness',[2] and campaigned for informal planting and the qualities of native plants. The two handsome yew trees that stood watch over him and his garden at Gravetye Manor in West Sussex are still there.[3]

26.6 *William Robinson, the 'inventor' of the modern English garden, underneath his yew trees.*
26.7 *Gravetye Manor today, seen from Robinson's yews.*

26.8 *Irish yew.*

An apt indicator for the transition from French to English gardening is the (in)famous Harlington Yew in Middlesex, England, which was allowed to return to 'normal' around 1800. Its existence, like that of the whole town of Harlington, is currently under threat by the controversial extension plans for Heathrow airport, making it, once again, an indicator of the spirit of the times.

Around 1820, an increasing passion among garden owners for rare and curious plants favoured new and rare varieties of yew. Unusual foliage (e.g. yellow needles, or light-coloured needle edges) and differing habit (e.g. columnar or weeping forms) became essential structural elements in garden design, and led to yet another comeback of *Taxus* at the end of the nineteenth century. Furthermore, the fact that yews are among the darkest conifers makes them an ideal visual background for flowerbeds or colourful perennials. Over 100 different garden varieties of the European yew have been developed, more than any other conifer; but in the early twenty-first century only some twenty of them are widely in use and available from nurseries and plant centres, together with the pure forms of the European as well as the Japanese yew, and their hybrid, *Taxus* x *media*.[4]

The most famous and characteristic variety of the European yew, however, is the Irish yew var. *fastigiata*. The original two plants were discovered in the 1760s on a rock (Carricknamaddow) in the hills near Florence Court, Co. Fermanagh, by the Irish farmer Willis. He dug them up, planted one in his garden and gave the other to his landlord, the Duke of Enniskillen. Willis's plant died some eighty years later, but the other lived on. Its cuttings were given to English nurseries who propagated and distributed them (it first appeared in an English garden catalogue in 1818, as *Taxus hibernica*). One of the oldest Irish yews now stands at Fougerelles-du-Plessis (Mayenne) in Normandy, 14m high and with an estimated age of 250 years. It is probably the tallest Irish yew of France.[5] All Irish yews are descended from the mother plant at Florence Court. As cuttings they are generally all of the same sex (female) although, curiously, males exist, too. Also, like the other garden varieties, they do not breed true but revert to the original form of *Taxus baccata*.[6] In fact, natural yew regeneration can be observed in the vicinity of cities, where the offspring of garden varieties escape the villas in the suburbs and recolonise nearby woodlands (for example, in Berlin Grunewald).[7]

26.9 *After the stump of this 350-year-old* Taxus baccata *was saved from an old house, it has taken over twelve years to create this bonsai, size: 76cm wide, 43cm high. Created by Tony Tickle, United Kingdom. Pot is Japanese Tokanoma.*

WOODEN OBJECTS

The Renaissance and Baroque periods saw furniture-making reach new and unprecedented heights. Whether as veneer or solid wood, the decorative value of *Taxus* began to shine. In the seventeenth century, gateleg tables with elaborate yew inlays and also grandfather clock cases with yew veneer were fashionable. In the eighteenth century, yew wood was widely used in Windsor chair-making. Yew veneer for all kinds of fine furniture became even more popular during the Victorian era.[8] The idea, however, was not entirely new: yew veneer was already known in ancient Egypt. Yew artefacts from Egypt include grave furniture and sarcophagi (see p. 153), and two busts of Queen Teye (see figure 39.3). Like the Egyptians, the Hellenistic Greeks, too, knew how to make the most of the precious wood by using it as veneer. Theophrastus attested to this by using it for armchairs, chests and other furniture. At Athens, archaeologists unearthed a yew comb and a yew bedpost from the Archaic period (*c*. 700–*c*. 500 BCE).[9] In early Greece, yew was also one of the principal woods for images of the gods.[10]

With its tight grain, yew wood provides excellent sound transmission and lends its qualities well to musical instruments. The oldest wooden instruments of the world are the yew pipes from Greystones, Ireland (see p. 230). Historical examples include the lutes from fifteenth- and sixteenth-century Bavaria. For the round back of the sound body of these lutes (forerunners of the guitar), their masterful makers steam-bent strips of yew wood about 50cm long, bent them into semicircular form and laminated them together. Yew wood is the only coniferous wood that does not crack in steam-bending.[11] However, the craft perished along with the destruction of the continental yew populations caused by the longbow trade. Other historical examples of musical instruments include eighteenth-century flutes and bassoons (solid yew), and two Viennese fortepianos (yew veneer) from the early nineteenth century.

26.10 *Mid- to late-nineteenth-century coopered yew wood container with copper hoops.*
26.11 *Early nineteenth-century turned yew wood tobacco jar.*
26.12 *Mid- to late-eighteenth-century yew wood 'priest', used by fishermen to administer the* coup de grâce *or last rites to salmon and other game fish.*
(All yew antiques courtesy of Richard Large)

27.1 Pennant Melangell Yew 3 *by Hans Diebschlag; 2004, watercolour on paper, 152 x 104cm.*

CHAPTER 27

POETRY

The yew tree appears in Western poetry from the seventeenth century onwards. Mainly because of its physical features and sombre impression, it is usually employed as a dark mood enhancer or even as a prophet of tragedy and death. The variations on doom and gloom begin with John Webster (1580–1625), who muses on the bones upon which the tree in the graveyard feeds, 'Like the black and melancholy Yew Tree, Dost think to root thyself in dead men's graves, And yet to prosper?'[1] Followed by Abraham Cowley's (1618–67) 'black Yew's unlucky green',[2] the dark themes continue through the centuries and are taken up even by John Keats (1795–1821): 'How a ring-dove Let fall a sprig of yew tree in his path; And how he died.'[3] The yew appears in the works of many poets and writers but rarely do the associations go beyond the graves and bones, the ghosts, the gloom, and the melancholy of the poet's own mortality. Essentially, it is graveyard poetry. The poets do not venture beyond what they see and know already, and the few who are lucky enough to encounter the tree outside the church walls, in the wild, soon think of the yew longbow – returning with the speed of an arrow to their preoccupation with death. It seems that the guardian of the gate lets no one in; the image of the ancient graveyard yew stands unshakeable as the 'tree of death' – a term coined by seventeenth- and eighteenth-century poets, not earlier.

In accord with the folklore of his time, William Shakespeare (1564–1616) lets the witches stir 'Gall of goat, and slips of yew Sliver'd in the Moon's eclipse' into their boiling cauldron (*Macbeth*).[4] But he is also one of the first writers to actually break the mould, to allude to the deeper layers of the yew's potential. In *Romeo and Juliet* the tree provokes a dream of the future: 'As I did sleep under this yew tree here, I dreamt my master and another fought,

And that my master slew him.'[5] Another sort of anticipation appears in the lines of Thomas Hardy (1840–1928),[6] (see Chapter 28, 'Commemoration'):

Thin-urned, I have burrowed away from the moss
That covers my sod, and have entered this yew,
And turned to clusters ruddy of view,
 All day cheerily,
 All night eerily!

The first poets to really break through the barrier of the conventional views of yew and ascend to new visions of transformation and transcendence are to be found among the Romantics. William Wordsworth (1770–1850), the major English Romantic poet who brought a new perception of nature in the late eighteenth century, was the first to suggest the redemptive possibilities of the natural world.[7] Samuel T. Coleridge (1772–1834), too, cast aside the graveyard gloom perception of this tree: while climbing a hillside in Somerset in May 1795, 'From the deep fissures of the naked rock The Yew tree bursts! Beneath its dark green boughs [...] I rest: – and now have gain'd the topmost site.'[8] Many years later, Coleridge – preparing for death – wrote his own epitaph, and discussed the design of his tombstone with the greatest interest. In the words of his biographer, Richard Holmes, Coleridge 'wavered long over a tender, voluptuous Muse figure; and then a broken harp; but finally decided that an old man under an ancient "yew tree" was more suitable'.[9]

***27.2** The ancient yew at Crowhurst, Sussex; nineteenth-century drawing.*

27.3 Kingley Vale Yew *by Hans Diebschlag; 2005, watercolour on paper, 104 x 152cm.*

The yew features widely in post-Romantic writing as well. In the work of T.S. Eliot (1888–1965)[10] the yew tree becomes again what it once had been in prehistoric times: the symbol not of death but of immortality. In 'Ash Wednesday', written in 1930, there appears a mysterious female in the yew's vicinity. As if transported to another, timeless dimension, the poet glimpses her as a veiled 'sister' and the world seems to stand still until the wind shakes 'a thousand whispers from the yew'. His return to our world he subsequently calls 'exile'.[11]

We will meet her again below. Alfred Lord Tennyson (1809–92), poet laureate of England for over twenty years, mentions the 'gloomy' tree frequently, but eventually bursts through to another level of consciousness:

Old Yew, which graspest at the stones
 That name the under-lying dead,
 Thy fibres net the dreamless head,
Thy roots are wrapt about the bones.

William Wordsworth
Yew-Trees

There is a Yew-tree, pride of Lorton Vale,
Which to this day stands single, in the midst
Of its own darkness, as it stood of yore,
Not loth to furnish weapons for the Bands
Of Umfraville or Percy ere they marched
To Scotland's heaths; or Those that crossed the Sea
And drew their sounding bows at Azincour,
Perhaps at earlier Crecy, or Poitiers.
Of vast circumference and gloom profound
This solitary Tree! – a living thing
Produced too slowly ever to decay;
Of form and aspect too magnificent
To be destroyed. But worthier still of note
Are those fraternal Four of Borrowdale,
Joined in one solemn and capacious grove;
Huge trunks! – and each particular trunk a growth
Of intertwisted fibres serpentine
Up-coiling, and inveterately convolved, –
Nor uninformed with Phantasy, and looks
That threaten the prophane; – a pillared shade,
Upon whose grassless floor of red-brown hue,
By sheddings from the pining umbrage tinged
Perennially – beneath whose sable roof
Of boughs, as if for festal purpose, decked
With unrejoicing berries, ghostly Shapes
May meet at noontide – Fear and trembling Hope,
Silence and Foresight – Death the Skeleton
And Time the Shadow, – there to celebrate,
As in a natural temple scattered o'er
With altars undisturbed of mossy stone,
United worship; or in mute repose
To lie, and listen to the mountain flood
Murmuring from Glaramara's inmost caves.[12]

27.4–27.5 *The 'fraternal' yews of Borrowdale in 2004. The fourth tree was lost in the great storms of 1883.*

The seasons bring the flower again,
  And bring the firstling to the flock;
  And in the dusk of thee, the clock
Beats out the little lives of men.

O not for thee the glow, the bloom,
  Who changest not in any gale,
  Nor branding summer suns avail
To touch thy thousand years of gloom:

And gazing on thee, sullen tree,
  Sick for thy stubborn hardihood,
  I seem to fail from out my blood
And grow incorporate into thee.[13]

For Sylvia Plath (1932–63), too, the yew stands for 'blackness and silence', but its path does not lead downwards into the grave but points up like a 'Gothic shape'. She enters this cathedral and embarks on a remarkable inner journey.[14] In a poem called 'Nearing the Brink',[15] the contemporary musician and poet Jehanne Mehta (born 1941) is also ready to step into a new mode of being. She respectfully pauses at the threshold and asks permission to enter – while playing on the identical sounds of the words *yew* and *you* (hence this poem gains from being read aloud):

27.6 Dark Mother *by Jan Fry; 2000, oil on canvas, 97.5 x 120cm.*

Nearing the Brink

Every path winds away into wilderness
like wet ivy clinging to old walls …
I am nearing the brink of yew.

Rath of vertiginous shadow,
teeming with metaphor,
I cannot reach yew with words …
the phrases break off and crumble.
Everywhere is edge.

For yew I have to relinquish my deaths,
truly inhabit my heart, for
with yew time has no measure:
it opens like a mouth.

I could fall into yew
and come up green and smiling
in the garden
before the fall;

But yew will not let me in before time.
I need first to awaken to the rhythm
of roots,
to the folded growth
always new, always stirring under my skin,
potent and soft as berries.

I am at the brink.
Already so many of my words have been swallowed
  by yew …
pages, without trace.

Wet winds, wilderness,
old walls …
The path disappears.

I want to be with yew,
at the heart of yew,
where I can hear yew singing,
that close.

Will yew open?
Will yew let me in?

CHAPTER 28

SYMPATHY

COMMOTION

Something strange was going on in the churchyard of Buckland-in-Dover. A number of men were digging heavily among the graves, determined to create a trench 1.2m wide and 1.5m deep, leaving a large block of earth measuring 5.4m by 4.8m around the bole of the ancient yew tree of the churchyard. The date: 24 February 1880. The goal: to move the tree 18m across the grounds, to give it a new home a safe distance away from the church, which needed an extension built. Way back in about 1770, the tree had been struck by lightning, split and shattered, demolishing the church steeple in the process. Half the trunk had been left lying on the ground but continued to grow as if upright. A hundred years later, although the form of the yew was now described as 'rude and grotesque', the village loved the ancient survivor and did not want to destroy it – hence the relocation. After cutting out the block of earth around its root, a trench to the new location was dug and – with the help of huge planks of timber, chains, rollers and windlasses – 'the whole mass of the tree, estimated at 55 tons, began to move. It arrived within a yard of its destination at dusk on the 4th March' (parish magazine, 1880).[1] The tree still thrives today.

***28.1** Moving the Buckland-in-Dover yew in 1880.*
***28.2** The same tree in 1999.*

The idea was not unique, though. The possibilities of moving a tree had already been discussed in Berlin, Germany, in the 1850s. Half a century earlier, the young prince of Prussia had played in a young yew tree in the grounds of a high official's Berlin mansion. As Theodor Fontane (1819–98) relates, 'The prince never forgot the old yew tree. Those who are grateful of character make no distinction, human or tree. Perhaps something else was stirred in the imaginative mind of the boy; perhaps he saw a stranger in the handsome, alien tree which had rooted under the pines of the March; perhaps the tree had come here with the Hohenzollern Dynasty itself, and a mysterious life thread was weaving between this tree of life and the prince's own Frankian house. Did it not rise in this very place like a tall fir among the pines?'[2] Anyway, Friedrich Wilhelm IV became the King of Prussia from 1840 to 1861, and was known as 'the Romantic on the Prussian throne'.[3]

In 1852, the Prussian Upper House by chance bought the above-mentioned estate[4] and planned an extension to the building. King Friedrich objected to the destruction of the tree and suggested moving it to his nearby palace at Sanssouci. His brother and later successor supported the King's inclination and offered his residence Babelsberg as another option. But no one had any practical experience of moving trees, and the idea (including the option of a short-distance move within the grounds only) was discarded. The King, however, had the building plans changed and the new walls had to stop short of the tree, which was protected during building works by a wooden enclosure. For years to come, the Prussian Upper House would meet while the yew was lurking just beyond the window, and it is to this that Fontane's famous short poem relates:[5]

> The yew
> Knocks on the window,
> A sparkle
> In the dark.
> Like ancient times, like pagan dream
> Glances through
> the window the yew.

At one time, the tree got even closer to the Prussian politicians: to celebrate the homecoming of the army in September 1866, a marquee was erected, with parts of its roof skilfully woven in canvas strips through the branches of the yew. The King himself – by now Friedrich's brother, Wilhelm I – was sitting to the right of the yew trunk. Fontane elaborates: 'All around the gas was burning in suns and stars [lanterns], a scenario the old tree could hardly have dreamed of in his early days.'[6]

Decades later, when the entire Upper House building was to be demolished and rebuilt, the tree was moved after all, together with its neighbouring yew. The two trees' root systems were carefully prepared for years in advance, and the trees were successfully moved within the grounds in April 1899. A photo shows the bigger tree still thriving in 1929, its girth measuring 175cm. Both trees, however, are believed to have perished in the Berlin bombings of 1944/45.[7] There is an echo of this theme – high politics with the yew outside the window – in present-day New York: the building of the United Nations has an extensive yew hedge in front of it. The windows are rather high to see through, though, and the yews are regularly clipped.

The most spectacular tree relocation happened in 1907 in Frankfurt, Germany. Here, the journey was not confined to the direct vicinity – the old Botanical Garden in the city centre – but led right across the city and to its outskirts, where the new Botanical Garden was being mapped out. Altogether 4,340 plants were moved, but only one tree: a yew with a 73cm trunk diameter, a 12m height, and an estimated age of 230–260 years. In 1905, a consultant from England was brought in whose advice was to progressively and tentatively cut the roots a certain distance and depth from the trunk so that the tree could develop a denser network of fine roots closer to the centre. In this way, the block of earth could be limited to a square of 4m at a depth of 2m, when building work began on 29 April 1907. The block of earth was framed in a wooden box, and a shaft was dug underneath to insert strong beams. Windlasses lifted the block in small steps, to allow space for ever new layers of beams until the block was propped up to transportable height above ground level. On 24 May, the journey began. The 42.5 tons were moved on a system of rollers and pulled by two steamrollers. In order to protect the roads from collapsing under this enormous weight and plummeting into the sewers beneath, a cross-hatching of wooden planks was laid out in front of the tree to distribute the weight more evenly across a wider surface. Because of the total height of 15m, the overhead tram wires had to be removed in places. To minimise traffic obstruction the transport was conducted mainly at night. Under these conditions it took seventeen days to travel the 3.5km (2.2 miles), and the tree arrived safely at its new location on 12 June. It was carefully turned to its original alignment on the compass, and then the entire earthwork operation had to be performed in reverse. Work finished on 26 June 1907, and the tree has not moved a foot since. It still stands *in situ* today. The cost of the operation was estimated at 8,000 Marks, 2,000 of which were donated by the magistrate of the city and 4,805 Marks by the public, leaving the remainder to be covered by the Senckenberg charity that ran the Botanical Gardens. The moving of the yew tree was an extremely popular event, and the entire journey was accompanied

28.3–28.4 In 1907, a yew crossed the city of Frankfurt, Germany.

by crowds of people. Only years later it transpired that the real costs for the move had been much higher than the original quotes, and had escalated to about 28,000 Marks.[8]

This new kind of care for trees had its roots in the spirit of the times. In Germany the yew was considered a rare and threatened species (see box p. 89), and also one with a significant historical tradition.[9] In 1899, in Berlin, Fontane's four-volume travelogue of the March of Brandenburg (published between 1862 and 1882) was highly popular, stirring a widespread interest in the tree's links to the royal house. But there is a bigger picture. At the same time that Fontane in Brandenburg and the Senckenberg charity in Frankfurt were investigating the natural world in this way, the Brothers Grimm combed Central Europe for its lost myths and fairytales, and Sir James Frazer in Cambridge compiled folklore and ethnological accounts from across the world. The nineteenth century witnessed important cultural changes. Industrialisation, rationalism and physical materialism in general thrived, but also triggered opposing attitudes: an emphasis on the individual, the imaginative, the emotional, the visionary, and even the transcendental. Romanticism and its deepened appreciation of the beauties of nature had been leaving its mark since the late eighteenth century. All of this was part of the spirit of the nineteenth century, and the first suggestions for the protection of nature began to appear in literature from 1825 onwards. In 1860 the first negotiations began regarding the conservation of natural areas, and also of birds. And in 1819 the German naturalist and explorer Alexander von Humboldt (1769–1859) coined the term 'tree monument' for the protection of ancient trees.[10]

INSPIRATION

Felix Mendelssohn-Bartholdy (1809–47) composed incidental music for a production of Shakespeare's *A Midsummer Night's Dream* in 1826 at the tender age of 17, and he did so beneath the yew trees in the Berlin garden mentioned above. The Mendelssohns lived in the villa in the grounds of the mansion from 1825 to 1852.[11] Fontane relates how 'the merry arts gathered under the yew tree, which looked solemn as ever, but not morose. Felix Mendelssohn, still a youth under the tree's canopy interwoven with sparkling moonlight, heard the music of dancing elves therein.'[12] Not surprising then that the young composer had the première of *A Midsummer Night's Dream* given under these very trees.[13]

28.5 Robert Louis Stevenson's yew in Edinburgh, Scotland.

The famous Scottish writer Robert Louis Stevenson (1850–94) spent part of his childhood with his grandfather at Colinton Manse in Edinburgh. He used to play in an old yew tree that today still shows the remains of Stevenson's swing, which hung from one of its branches. The tree, first recorded in 1630, is a fine specimen with a heavily fluted trunk and a girth of 3.64m. Now known as Stevenson's Yew, it is a living link with one of Scotland's great literary figures. Stevenson later said about this tree, 'A yew, which is one of the glories of the village. Under the circuit of its wide, black branches, it was always dark and cool …',[14] and in a poem:[15]

Below the yew – it still is there –
Our phantom voices haunt the air
As we were still at play,
And I can hear them call and say,
'How far is it to Babylon?'

While Stevenson wrote these lines, another famous writer, the novelist George Meredith (1828–1909),

28.6 Druids Grove: etching from c. *1840.*

lived close to the ancient yews of Druids Grove in Surrey, south England. He had moved there in 1867 and encouraged his visitors to experience these trees, telling them that 'anyone walking under them should remember that they were saplings when Jesus Christ came to earth'.[16] And some of his visitors certainly were inspired – whether their visit to Druids Grove took place before or after the creation of their greatest works we are not always able to tell, but Meredith's guest list included T.E. Lawrence (Lawrence of Arabia), E.M. Forster (*A Passage to India*), R.L. Stevenson *(Treasure Island),* Thomas Hardy (*Tess of the d'Urbervilles*), Sir James Barrie (*Peter Pan, the boy who refused to grow up*), Lewis Carroll (*Alice in Wonderland*), the poets A.C. Swinburne and W.B. Yeats, and the young G.M. Trevelyan.[17]

28.7 The churchyard yew in Downe, Kent.

COMMEMORATION

Not a writer of fiction, but far from uninspired, was the English naturalist Charles Darwin (1809–82). Darwin at first shocked religious Victorian society by suggesting that animals and humans shared a common ancestry (1859),[18] but by the time of his death his non-religious biology had spread through all of science, literature and politics; 'evolutionism' is still a fundamental paradigm of Western thinking today. In 1842 Darwin settled with his family in Downe, Kent, and for the rest of his life did not doubt that his hometown's cemetery would become his final resting place;[19] he looked upon 'Down[e] graveyard as the sweetest place on earth'.[20] To be more precise, his thought was to lie beneath the old yew tree under which he had so often sat while taking a break on his daily walk. And indeed, the day after his death on 19 April 1882, the local papers announced that he would be buried in St Mary's churchyard at Downe. In the words of his biographers, A. Desmond and J. Moore, 'He would lie under the great yew that had stood sentinel for six centuries at the lychgate – next to his infant children and beside Erasmus [his grandfather]. Darwin had expected to be placed here'.[21] But influential groups thought otherwise; they wanted a somewhat grander commemoration than a village funeral, and the Royal Society requested the family's permission for a state burial. The canon of Westminster Abbey was persuaded to bury the agnostic, and so Darwin was accorded not his own wish but the ultimate British accolade of burial in Westminster Abbey, London, on 26 April 1882.

Others were 'luckier' and their wishes were respected. These include T.S. Eliot, Lewis Carroll[22] and William Wordsworth. Wordsworth planted eight yew trees in the churchyard of St Oswald's, Grasmere, Cumbria, and later had himself and his wife Mary buried beneath one of them. Thomas Hardy was certainly least 'lucky' as his wish was partly respected, while he himself was completely disrespected: his ashes were interred at Westminster Abbey while his heart was quite literally buried beside a yew tree in Stinsford, Dorset!

To end on a somewhat lighter note, we turn briefly to an entirely different kind of memorial: postcards. Over the last 150 years or so yews have featured widely in local memorabilia of this sort. The postcards' value lies not only in the personal messages entrusted to them, but also in their worth to researchers today as the old photos hold information about the location, state and size of certain trees in days gone by. Such postcards also show, incidentally, how often the age of old yews used to be overestimated (no surprise then that early twentieth-century dendrologists turned so sharply against supposed grand ages for yew trees: see Chapter 20). This fate of yews, however, is shared by oaks and lime trees: every old village tree is obliged to be 'a thousand years old'. Apparently, there is a magic about that expression; it is not a mathematical *number* at all, but a *quality* that is emphasised. Meeting a 'thousand-year-old' tree is something that moves the heart, that makes us think about what life is all about. It does not matter at all, for example, that none of the sixty or so 'thousand-year-old' oaks of Germany is older than 600 years.[23] They will remain 'a thousand years old' even though they will never reach that age in 'real time'.

A fascinating photo has been published on a present-day postcard from the monastery of Fonte Avellana, Marche, Italy. The cloister is located next to an ancient yew and it is not known whether the tree was planted at the foundation of the cloister in 982 CE, or if it was there before. Anyhow, it is the pride of the community and its presence is incorporated into booklets and other tourist paraphernalia. The postcard shows a circle of monks demonstrating the tree's vast girth, and there is an ease and a joyfulness to the atmosphere that one would never expect to see if monks were to pose with a tree in a northern country.

ROBIN HOOD

Robin is a kind of Green Man figure with pronounced socio-political overtones. His origins have been traced back to a Robert Hode who was outlawed by the justices of York in 1225.[24] The oldest folklore material also comes from Yorkshire, from where his tales spread rapidly and the storytellers gradually incorporated many other themes. For centuries now Robin Hood has been a representative of the archetype of the native citizen who has to take refuge in the forest from the injustice of invaders or oppressors – a theme that has endured in the British Isles from the times of the Celtic invasion through to when the Romans, the Saxons, the Vikings and the Normans followed suit, and at many other times of civil and political unrest.

Robin is an archer, and an excellent one at that. He and his merry men live by the bow; a good bow is vital for them, and all sources agree on the timber: yew. Indeed, referring to Robin's men as *yeomen*, 'yew-men', was common for centuries. One of the earliest written accounts describes Robin's death wish, 'And lay my yew bow by my side'.[25] In later tales he is also said to have been buried beneath a yew tree. In a popular version of the legend he shoots an arrow from his deathbed, requesting to be buried where the arrow lands, and it is found at the foot of a yew tree.[26]

28.8 *(top) Postcard from the monastery of Fonte Avellana, Marche, Italy.*
28.9 *Early twentieth-century postcards (from top left): Iberg Cave entrance with yews, Bad Grund (Harz), North Germany; '2000-year-old yew' at Krombach, Czech Republic (see fig. 24.11); Stoke Gabriel, Devon; Painswick, Gloucestershire; chuchyard yew at Jabel, Mecklenburg, Germany (top right); Tzschochau Palace Park, Germany.*

29.1 *In all probability the oldest tree of Europe: the yew in the churchyard of Fortingall, Scotland.*

CHAPTER 29

SANCTUARIES

CHURCHYARDS

'Why are yews in churchyards?' This is a question still frequently asked by people from all walks of life. A popular answer over the last few centuries is that they were planted for longbows. This is an assumption, however, that gets nowhere near the truth. The volume of yew staves needed by the English military could never have been met by a few trees growing in churchyards. As the briefly heard but unechoed voice of G.A. Hansard had already stated in 1841, 'Every yew-tree growing within the united churchyards of England and Wales [...] would not have produced one-fiftieth part of the bows required for military supplies.'[1] And Hansard did not have all today's evidence of the sheer scale of the operations (see Chapter 24). In other words: if every churchyard in England and Wales were to have planted four yews for longbows in the fifteenth century, it would have taken them not 100 years to match the sixteenth century's volume of continental imports, but 5,000 to 10,000 years. The Crown did not consider churchyard yews anyway. For one, the tight-grained wood from continental mountains was highly sought-after, and always priced at about 60 per cent above bow wood from England. And when Henry IV ordered his royal bowyer 'to cut down yew or any other wood for the public service' he explicitly exempted the estates of the religious orders.[2] Furthermore, the opposition of faithful Christians to the military use of wood from consecrated ground would have left written traces. That is not to deny that local exceptions did occasionally occur, but on the whole it was not the longbow demand that helped yew trees to survive inside the churchyards, but rather it was the churchyards that saved the yews from the longbow mania. Today, Britain has the most old and ancient yew trees in Western Europe, and a great many of them are located in churchyards.

***29.2** Alltmawr, Powys, Wales (girth 869cm at 60cm in 1998).* ***29.3** Crowhurst, Surrey (girth 961cm).*

(Photos of churchyard yews on the British Isles are spread throughout this book, see 'churchyard yew' in the index.)

The reasons why yews are in churchyards might be of a religious nature after all, and some of these reasons reach back to pre-Christian times. The clues, however, need untangling. Until recently, it was acknowledged that the majority of Christian churchyards in Europe are located on older, pagan sacred sites, but this acknowledgement is now fading. True, there is the oft-quoted letter from Pope Gregory the Great to Abbott Mellitus, in June 607, asking him to tell his missionaries to convert the pagan sites for Christian use instead of destroying them; but so far no archaeological excavations in Britain have uncovered any traces of pre-Christian temples beneath Saxon churches. Besides, earlier papal orders had favoured the destruction strategy, and Gregory contradicted his own letter by ordering Augustine to destroy the pagan sanctuaries in his mission to Britain – which archaeology suggests is what happened.[3] On the other hand, there are a few churchyards on sites with an apparently older history, and with ancient yews, too. For example, the ancient yew (male, girth 1,158cm) in the churchyard of Ashbrittle, Somerset, grows on a prehistoric tumulus; five ancient yews grow, each within a circular stone wall, at Mynyddislwyn, Monmouthshire, where the churchyard lies adjacent to a tumulus and was probably originally circular; and another veteran (male, 615cm girth at ground level in 2004) grows at Warbleton, Sussex, where the church (thirteenth century) is built on a Neolithic mound. Richard Morris, in his excellent study of the history of British churchyards, refers to 'survivals from pre-ecclesiastical landscapes' and calls such sites 'natural pegs which join sanctuary and churchyard together through time'.[4] There are other examples, but far too few to sustain the argument of a general practice of the conversion of sites.

The archaeological report is still patchy. Also, excavations usually look for stone foundations and postholes of timber construction, but what if a sanctuary comprised only living trees? No trace would be found. Or, the ancient tree might still be in the grounds, unnoticed, silently casting its shadow over the archaeological activities …

The truth (as so often) may not just lie somewhere in the middle, but might also be somewhat independent of the exact physical locations of worship. Post-Christianisation, it was, after all, the same people with the same habits, and (most of their) old traditions and values. If a yew was the right tree to plant at the place of worship and of burial, there was

29.4 *Barfrestone, Kent (girth 523cm at the ground).* ***29.5*** *Breamore, Hampshire (girth 1,112cm at 60cm in 1999).*

no reason to change that. Of course, the Catholic authorities made clear at all times that the worship of nature and other elements of paganism was strictly forbidden;[5] in no country did Christian bureaucracy ever feel at ease with the veneration of trees. Nevertheless, many churchmen themselves preferred to find their final rest beneath *Taxus*.[6] The Cistercian Order often chose natural sites with established yew trees on them, and planted more (see Chapter 44).

The tradition of planting yew trees in consecrated grounds has a spiritual as well as a practical meaning for Christianity. In the British Isles and the maritime fringes of north-western Europe, *Taxus baccata* is one of the few native evergreens. With its dense crown it comprises a well-rooted windbreak that protects the church building during winter storms. At the same time its evergreen foliage is a comforting emblem of the continuity of life. Yew boughs (and/or those of other evergreens) were used as decorations for church services,[7] and particularly as a substitute for palm fronds in Palm Sunday processions to the extent that in some areas – for example, in Ireland, Devon and Kent – yew trees were still called 'palms' in the eighteenth and nineteenth centuries.[8] Villagers would wear a sprig of this tree in their hat or buttonhole from Palm Sunday to Easter Day. Other 'palm' branches were burnt after the procession and their ashes used on the following Ash Wednesday. In the Middle Ages, many churchyards were not merely the location for religious service but also functioned as community centres for the various gatherings and festivals of the year. Among the popular activities were mystery games or plays, which brought to life for the mostly illiterate audience biblical stories such as that of the Garden of Eden. The main prop for the play about Adam and Eve was a 'paradise tree'[9] – and in England, the female yew had its 'fruit' perfectly ready for 24 December, the religious feast day of Adam and Eve. The German winter, however, was too cold, and the people set up a paradise tree inside their homes (see below).

The great importance of the yew in medieval Wales can be glimpsed in the Laws of Hwel Da (tenth century):[10]

> A consecrated yew, its value is a pound.
> A mistletoe branch, three score pence.
> An oak, six score pence.
> Principal branch of an oak, thirty pence.
> A yew-tree (not consecrated), fifteen pence.
> A sweet apple, three score pence.
> A sour apple, thirty pence.

> A thorn-tree, sevenpence halfpenny.
> Every tree after that, four pence.

One pound was more than most people could ever earn in a lifetime. But in hindsight, this assigning of worth does not seem particularly necessary as for centuries to come the consecrated yew was to remain an important and venerated presence in the churchyard.

The yew tree has deep links with the concepts of personal transformation, rebirth and immortality, in many cultures all over the world (as will be shown in the following chapters). Christianity adopted a part of this heritage, although the idea of 'rebirth' was replaced by the concept of 'resurrection'. The positive Christian attitude towards *Taxus baccata*, however, is summed up in a sermon by Revd John Mason Neale (1818–66), warden of Sackville College, East Grinstead (West Sussex), who preached on the biblical text 'O all ye green things upon the earth, bless ye the Lord':

> The Yew ... may be accounted a fit emblem of a Christian. You see it hath little outside bark, only a small rind; to teach us not to make a great outside show of religion.[11] Then it has a very lasting timber, much harder than oak, to show the soundness and sincerity of a Christian. It has many branches, large and fair, to remind us to be plentiful in good works. It is always green and prospering, to declare unto us that a Christian should always grow and thrive in grace. Yea, green in winter and the hardest weather, to show that a Christian is best in affliction; yea, then it hath berries on it, to teach us, as then we are the best Christians, so then to bring forth most fruits of righteousness. It is a long-lived and lasting tree, to be unto us a type of immortality and lasting life ... All this we confess when we set up the Yew.[12]

The Christmas tree

In Germany, the homeground of the Christmas tree tradition, yew had been used before fir and spruce. A poet from the March of Brandenburg sang in *c*. 1800, 'The Taxus tree abounds with apples, behold! And sparkles all over with silver and gold'[13] (for red yew arils being called 'apples' see pp. 149–50). This certainly sheds some light on where the idea of the red 'apples' (made from papier-mâché, etc.) hung in the dark Christmas tree foliage might have come from. One of the earliest known records in the English language of the customs surrounding the Christmas tree are the descriptions of Samuel Taylor Coleridge, English lyrical poet, critic, philosopher and, with Wordsworth, herald of the English Romantic movement (see Chapter 27). In a letter from Germany he described a family Christmas in December 1798: 'On the Evening before Christmas Day one of the parlours is lighted up by the Children, into which the parents must not go; a great yew-bough is fastened on the Table at a little distance from the wall, a multitude of little Tapers are fastened in the bough, but not so as to burn it until they are nearly burnt out – and coloured paper etc hangs and flutters from the twigs. – Under this bough the Children lay out in great neatness the presents they mean for their parents.'[14] Coleridge wrote these lines in Ratzeburg in the north-west of Germany[15] – the area poorest in *Taxus*. In England, the Christmas tree was popularised in the mid-nineteenth century by the German Prince Albert, the husband of Queen Victoria, and Prince Albert's original tree, too, was *Taxus*.[16] The use of evergreen trees and wreaths to symbolise eternal life was a custom already known to the ancient Egyptians, Hebrews, Hittites and Chinese,[17] among others. The custom of placing gifts beneath the tree may be traced to Hittite yew-tree rituals from the second millennium BCE (see Chapter 36).

SHRINES AND TEMPLES

Shinto is the ancient and indigenous religion of Japan, and has pervaded Japanese culture for more than 2,000 years.[18] Like many of the world's major belief systems, the Shinto faith acknowledges a superior or divine realm that nourishes and guides the visible world and human existence. Shinto belief and practice are deeply entwined with the worship of supernatural beings called *kami* who oversee all aspects of nature and human life. Although often

29.6 *Torii gate and sacred yew at Hakusan Jinja, Nagano prefecture, Japan.*

translated as 'deity', *kami* designates a wide range of spirit-beings, supernatural forces and 'essences'. Mountains, trees, rivers and waterfalls have their own *kami*; a 'spirit of place' is a *kami*, and the souls of the dead transmute into yet another category of *kami*. Most *kami* are benign, some are malicious, but Shinto tradition does not believe in an absolute dichotomy of good and evil: all phenomena are thought to possess both a 'rough' and a 'gentle' side, and any entity can manifest either of these characteristics according to the circumstances.[19] Many Shinto rituals revolve around honouring the *kami*.

Many *kami* live in the sky and come down to earth periodically to visit sacred places and shrines. Natural places of extraordinary beauty are conceived of as carrying the powerful essence of the *kami*. The Japanese way of honouring sacred space is not with sumptuous pomp but with great sensitivity and minimal interference. In order to enter a *jinja*, a 'shrine', one has to pass through the ceremonial *torii*, or sacred gateway. This gate consists of a pair of posts topped with two crossbars, the upper of which extends beyond the uprights. Often the *torii* is painted in the sacred colour, red. The *torii* represents the boundary between the outer, secular world and the confines of the sanctuary. In passing through it, the visitor undergoes a symbolic purification before he or she enters the shrine. Inside the shrine, each of the sacred stones or trees which are imbued with the presence of *kami* is tied with a rope, *nawa*, made of rice straw, and hanging from the rope are *gohei*, paired strips of paper each torn in four places.[20] In Shinto, all evergreen trees are considered sacred because they maintain their 'life' during the winter and are believed therefore to partake in the divine realm in a special way. If, additionally, a tree is perceived to be the passage that a *kami* takes in coming down to earth, it is designated as a *shinboku*, a sacred tree.[21]

There is no doubt that the tree that features most in Japanese folklore and tradition is the *sugi*, a false cypress *(Cryptomeria japonica)*, often translated as 'Japanese cedar'. Adopted (in historical times) as the national tree of Japan, it is frequently found at sacred sites of both Buddhist and Shinto tradition.[22] However, the Japanese yew (*Taxus cuspidata*) is strongly connected to the emperor's house (see p. 210) and is called *Ichii*, 'of the first rank'. Some of the most awe-inspiring and beautiful ancient yews in the world are to be found at the Shinto shrines of Japan. It is a curious fact that on the far ends of the Eurasian continent, two islands of fairly similar size – Britain and Japan – have preserved in some of their small local religious sites some of the world's oldest trees.

A great number of the *jinja* yews on the main island of Japan are located in the prefectures of Gifu and Nagano. This should be no surprise as their legendary 'birthplace', Ichiimori Hachimanjinja, is located in Gifu prefecture, near the geographical centre of Japan. The region has many links to tradition and mythology. Of the many old and ancient sacred yews, a few shall be named here.[23]

The Hakusan Jinja yew tree is about 13m tall and has a girth of 6.1m. The tree was 'declared a god', that is, acknowledged to be imbued with the presence of *kami*, in 1673 and has been strongly protected ever since. Another thing that makes this tree special is its location in the valley below the active Asamayama volcano (2,568m) at just under 24km (15 miles) from the summit. The volcano last erupted in September 2004 and it is fair to assume that the yew in its estimated ten centuries of life has seen many eruptions. It is a hollow tree with many visible cavities in its trunk, and people put coins or small pebbles into them to accompany their wishes and prayers. This tree is known to bring good luck particularly with teeth problems and hence, with some humour, is also known as the 'God of the cavities'.

The Arai yew, locally known as Koyasu sama, is one of the largest-girthed *jinja* yews, and is also a good example of how little interference Shinto exerts.

29.7 The Arai yew, one of the largest-girthed jinja *yews.*

Apart from a low wooden fence to keep tourists off the central part of the root system, the tree has practically remained a woodland tree. Only a small wooden shrine to its side marks the tentative religious activity at this place. *Koyasu* means 'protection of children'. The local story is that about 700 years ago the pregnant wife of one Kanehira Imai planted the tree for the protection of her child during birth. All went well, the child and the tree thrived, and soon enough other women came to ask the tree for the same protection. The custom is still alive today, and there is also a public festival *(matsuri)* traditionally held on 15 August (changed to 11 September in 2003). *Arai* means 'new well/spring'.[24] (For the association of yew with water and childbirth, see also Chapter 34.)

The Kunimi yew is a male tree 19m tall, has a girth of *c.* 7m and an age of *c.* 700 (Environmental Agency data) to *c.* 1,000 years. According to local oral tradition, three yews were planted 700 to 800 years ago in a triangle around the village, to protect it and help to bring prosperity. One tree was lost to fire, one withered, and only this one remains. A spring has appeared in the hollow at its foot. The tree was designated a Natural Monument on 1 November 1967.

The shrine at Jirobei has three yew trees with memorial stones beneath them. The oldest tree of the group is 18.5m tall and has a girth of about 7.5m, which is the largest measured girth in a yew tree in Japan. Age estimations for this tree range from 1,000 to 2,000 years, most probably not taking into account the fact that the entire trunk consists of a dense mass of former interior roots, the old shell of the tree having completely disappeared (compare diagrams on p. 76). Many memorial shrines in Japan are not gravestones for individuals but serve in the honouring of ancestors in a more collective sense. In Japan, in the Nagano prefecture for example, many memorial and burial stones have one or two (young) *Taxus* trees above them (figure 31.4). Yew can be found in both Shinto and Buddhist burial grounds. For two weeks in August, the ancestral realm is traditionally considered to be closer to this world than

29.8 *(above) The sacred yew at Seijo is 900m above sea level.* **29.9** *Ancient (male) yew and shrine at Hirade.*

at any other time of the year. The national holidays are aligned to this season and thus allow the modern Japanese to visit the shrines, which are often far away from their dwelling places.

Yew trees appear also at Japanese Buddhist temples, for example at Gouzen Dera and Akimitsu Dera, but most of the old and ancient trees are located in Shinto shrines. The link between Buddhist sites and the yew tree can also be found in China, in particular in the mountain sanctuaries of south-western China. After visiting the area in the 1920s, the English botanist and explorer Ernest H. Wilson wrote of Mount Omei (also called Omei Shan) that it is 'one of the five ultra-sacred mountains of China, but the origin of its holy character is lost in antiquity ... Upwards of seventy Buddhist temples or monasteries (either word is applicable, since the buildings are really a combination of both) are to be found on this mountain. On the main road to the summit,

29.10 *The altar beneath the largest yew at Jirobei.*

29.11–29.12 *Impressions from the shrine at Komatsunagi.*

there is a temple every 5 li, and they become even more numerous as the ascent finally nears the end.'[25] Of the very few trees that could still grow near the higher temples between 1,800 and 3,000m Wilson mentioned the Chinese yew.[26] The current status of the remote Buddhist sites and their trees in the mountains of south-western China is unknown; there is room for future research here.

There are no reports about the role of yew in Tibetan Buddhism. It is noteworthy, however, that the refugee Tibetans in Scotland planted a yew in the centre of their newly founded Buddhistic monastery at Samye Ling (the 'Clootie tree'), and also at their sanctuary on Holy Island off the west coast of Scotland, this one being a cutting of the Fortingall yew (in November 1993) – the ancient yew that is located in the geographical centre of Scotland and in all probability is the oldest tree in all the British Isles (see figure 29.1).[27]

29.13 *Contemporary statue of Buddha, yew wood.*

29.14 *Yew at the entrance to the Buddhist temple at Gouzen Dera, Japan.*

29.15 *The 'Clootie tree' at Samye Ling, Scotland.*

SACRED SPACE

Among the woodland tribes of North America, sanctuaries are not buildings with surrounding temple gardens but often comprise the 'virtual' sacred space created by ceremonies and prayers. Perhaps one could say that here, nature in its entirety is the sacred garden in which the human being seeks to attune with the higher forces of life. Important elements in a number of Native American traditions are the sacred pipe (for smoke offerings) and also musical instruments such as praying flutes and drums. The Karok, for example, a tribe in northern California, use yew wood for their pipes. In Washington State, the Cowlitz make yew drum frames, while the Hoh and Quileute use the wood in a number of ways in ceremonies. The Clallam on the Olympic Peninsula off Washington State mix yew needles with tobacco for the sacred pipe, and the Samish and Swinomish on the north-western coast of Washington State dry and pulverise the needles and smoke them in place of tobacco[28] (not recommended, as the European yew can contain far higher concentrations of poisons than the Pacific yew).

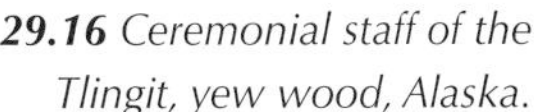

29.16 *Ceremonial staff of the Tlingit, yew wood, Alaska.*

30.1 *Hollow yew at Paterzell, Bavaria.*

CHAPTER 30

THE SECRETS OF NAMES

TREE NAMES

Researching the role of a tree in cultural and religious history requires the knowledge of the names under which the tree appears. The names for the yew tree in the present European languages can be grouped in two main categories: those that have derived from the Roman *taxus*, which was adapted during the first four centuries of the common era in wide parts of the Roman Empire, and those that are indigenous names and usually older than the Roman forms.

Some of the Roman writers mused that *taxus* must have derived from or at least was connected with *tóxon*, the Greek for bow, and *toxikon*, 'arrow poison'.[1] Virgil (70–19 BCE) and Pliny (23–79 CE) mentioned the use of yew wood for bows and poisoned arrows.[2] In Greece, Strabo (64 or 63 BCE–c. 23 CE) and Plutarch (46–after 119 CE) mentioned yew arrow poison among the Gaulish tribes.[3] Pliny also assumed that 'poisons were named *toxica*, formerly *taxica*, from the name of the tree *taxus*'.[4] It is noteworthy, however, that apart from assumptions by a few classical writers, no evidence regarding the extraction of yew toxins for the poisoning of arrowheads in war has ever been found. Nevertheless, yielding wood for bows, spears, and other military tools and constructions gave significant value to the yew's military utilisation, and, like any other empire, Rome had a keen interest in exploiting the resources of its provinces. This practical approach to naming trees remained widespread in a number of the occupied countries, and even after the Roman Empire finally crumbled, the term *taxus* for the tree was kept even though before Roman occupation there must have been indigenous names for the tree.

The possible origin of Latin *tax-* is Indo-European *tak*, 'to cut', which extended into *taks*, 'to hew, to shape', and also *teks*, 'to make, to manufacture'. *Tax* is related to *tag, tangere*, 'to touch'.[5]

Among the non-Roman names for the tree are those few from rather isolated languages, such as the Lithuanian *kukmedis Europinis*; and in the Pyrenees the Basque dialects, *agin, agina, agiña, aguiña, hagin, hagina*, and *hagintze*. The majority of the surviving pre-Roman names, however, can be found in the Celtic and Germanic language groups. The oldest Celtic forms have been found in Gaul (modern France) and seem to be variations of *eber* or *eburos*. Most of the contemporary words in the German language group are derived from Middle High German *iwe*.[6] Going further back to Old High German, the common forms are the feminine *iwa*, also *iha*. Apart from this there is the male *iwo*, and rarely the neutral *iwinboum* (*boum* = tree). Old English *iw, eow, eoh* (poetic), Old Saxon *ich, ioh* and Old Norse *yr* are masculine. Old Prussian is neutral *iuwis* (*iwis*). During the Middle Ages, the Old Irish *eo, ibhar, ibar, jubar* and Middle High German *iwe* denoted the tree as much as any weapon made from it (bow, longbow or crossbow), as did Old Norse *yr*, although it co-existed with *yrbogi* (*bogi* = bow).

Table 9 European names for *Taxus baccata*[7]

a) based on Roman *taxus*	
Swiss	taisch, tasch
Italian	taxo, tacc, tass, tasso
Spanish	tejo, teixo, teix
Portuguese	teixo, teixera, teixero
Russian	tiss
Polish	cis
Czeck, Slovak, Croatian	tis
Slovenian	tisa
Hungarian	tiszafa, tisza
Ukrainian	tys
Rumanian	tisa
Croatian	tisovina
b) based on older etymological roots	
Germanic	
English	yew, yeue, eu, u, iuu, iw, iewe
French	if, ifreteau
German	Eibe, Ibe, Iba, Ife, Ifen, Ibar
Swiss	Eibe, Iib, Iba, Ibä, iche, ige
Dutch	iep, iif, ievenboom
Danish	ibe
Swedish	id, ide
Celtic	
Breton	iuin
Welsh	ywen
Old Irish	idhadh, idho, eo
Gaelic	iúr
Scots-Gaelic	iubhar

Phonetically, *e* and *i* can interchange in the many Celtic and Germanic dialects, so can *o* and *u* and (English) *w*, and also *f, v* and *w* can shift to *b*, or *b* revert back to *v*. But the basic structure is quite clearly visible in all these forms: the core of the name is a vowel sequence, *io, iu, eo, eu*, with or without a central consonant, *f, v, w*, or *b* (sometimes followed by *r*).

It is the Old High German *iwe, iwa*, however, that reveals the ancient meaning of the name. This term for yew is closely related to the Old High German *ewi, ewa*, which means eternity. And the Saxon dialect forms, *eo, io, eoh, eow*, stem from Old High German *eo*, 'always'. It becomes apparent that humankind since time immemorial has been aware of the extraordinary regenerative abilities of this 'eternal' tree. A similar meaning to Old High German 'eternity' can be assumed for the Celtic names of the tree, as Celtic and Germanic languages share a common ancestry in so-called Indo-European. The Indo-European *ayu*, 'life force', has been suggested[8] as a root word for both Germanic *iwe* and Celtic *eber* (and Hittite *eya*, see below).

In the light of their Indo-European roots the current two categories of names of the tree seem to complement each other: to touch – or be touched – by eternity.

PLACES AND PEOPLES

The Greek merchant, geographer and explorer Pytheas (*c.* 380–*c.* 310 BCE) from the Greek colony of Massalia (modern Marseilles, France) made a sea voyage to north-western Europe sometime between 330 and 320 BCE. He was the first to describe the midnight sun, the polar lights and the polar ice sheets to the peoples of the Mediterranean, and the first to mention in writing the British Isles as

30.2 *Old yew in Ste Baume, France.*

well as some Germanic tribes. Ireland at the time was called *Ierne*, and other Greek, and later Roman,[9] writers picked up the name, although it changed over the centuries: *Iverna, Iuvern(i)a, Hibernia*, all meaning 'yew island'. The tribe inhabiting its southern coast were the *Iuverni* (later *Iverni*).

When St Columba (also known by his Irish name, St Columcille) sailed from Ireland in 563 CE to search for the right place to set up a monastery, he was given by the king of Dalriada a small and uninhabited island off the south-west coast of Mull (Inner Hebrides). Abbot Adamnan (627–704), who later wrote the *Life of Columcille*, called the island *Ioua Insula*, 'island of the yew tree place'. Probably through a misreading it became Iona, which it still is today. It is not known, however, whether there were yews on Iona before Columba. There are none left today.

Ancient settlements named after yew include Eburodunum in Switzerland (compare the Neolithic lake villages in the Alps, p. 99) and York in northern England. York was originally called Eborakon, a name committed to writing in *c*. 150 CE by the Greek geographer Ptolemy, and thought to be derived from the Old British Eboracon. When the Roman commander Agricola made it the headquarters for the Roman legion it became Latinised to Eboracum and Eburacum, and the surrounding region of Yorkshire became known as Eburach. In ninth-century Welsh manuscripts it appears as Caer Ebruac or Caer Ebrauc. The Anglo-Saxons, in calling it Eferwic (in *c*. 897) and Euorwic (*c*. 1150), added another meaning, that of *eofor*, 'wild boar'. Under the Danes and Vikings it became Iórvik (962), later Iórk, and finally York in the thirteenth century.[10] York, by the way, is the mother city of New York in the United States.[11]

Yeavering Bell, the royal residence of the Anglian kings of Bernicia (Northumberland), means 'place of the (sacred) yew tree'. The incoming Anglians usually did not replace placenames, as part of their general strategy of integrating with the indigenous people. Yeavering therefore must be a Brittonic and not a Germanic name; 'clearly a Celtic name', concludes Clive Waddington of the Milfield Archaeological Landscape Project (Dept. of Geography, University of Newcastle).[12] Around 1300 the name appears as Yeure in Scottish texts (relating to Scottish *ure*, 'yew'), and Yverne under Franco-Norman influence. During the centuries it changes to Yevere (1359), Yevern (1404), Yvern (1442), Yeverin (1637), Yeverington (1663), and finally Yeavering (1796).[13] Bell may have come from Gaelic *bile*, 'sacred tree'. In Saxon, *bell* denotes a bell-shaped mound, which also applies perfectly to the site, except that it is *not* a Saxon one. Bede's (Ad) Gefrin, which translates as 'goat hill', might just be the Welsh rendering of the name, although goat sacrifices (my guess) would suit the context, given the appearance of the goat in various World Tree traditions (see pp. 162, 229).

Some other old yew placenames have changed beyond recognition, for example, Avrolles (from Eburobriga to Hebrola to Hevrora) in the Yonne *département* (also a yew name). Most noteworthy, however, are the many occurrences of Eburodunum, which stretch in a broad belt from the British Isles across France and deep into Germany and Italy: Eborakon (York) in north-east England; Din-Evwr in medieval Wales; Embrum, Château d'If[14] and Yèvre-le-Château in France; Eburodunum and Yverdon in Switzerland; Iburg in Germany; Eburini, Eboli, Inveruno and Invorio in Italy; and possibly Brünn in Moravia.[15] It was discussed during the early twentieth century whether these names indicated places founded by individual Celtic leaders named Eburos, or whether these places were strongholds in the literal sense – *eburo-dunum* means 'yew stronghold' or 'castle'. However, with no archaeological evidence at these places their names most probably denoted natural features, just like the many names that translate as yew pasture, well, valley, wood, hill, mountain and so on.[16] Since yew is a species that in most cases is to be found dispersed among the other species of the mixed forest, yew placenames would have been applied where this species stands out more than average, for example, colonising a pasture, surrounding a well, dominating the shore of a lake or covering a hilltop. However, one phenomenon is not represented in the placename lists:

30.3–30.4 *Yew grove at Sueve Otono, Spain.*

a cluster of yew trees that are ancient, or gigantic, or both, places such as Newlands Corner or Kingley Vale in southern England, or the towering yew stands of Alapli in Turkey and those of the Caucasus. The name 'yew stronghold' would be rather appropriate for such places. (Similarly, the Celtic *sali-duno*, 'willow castle', can be found in the name of a lake village in Ireland, Dún Salach, and in Solduno in the Alps.) Indeed, already in 1928 the etymologist Vittorio Bertoldi mentioned in this context classical reports about monumental (yew) trees around the Gulf of Lyon and in the maritime Alps between Nice and Genoa.[17] It seems that Europe once had monumental yews such as the ones still in existence in Turkey and the Caucasus. When did they disappear, and why? There is room for future research here.

Apart from the above-mentioned I(u)verni, other names of Celtic tribes have possibly derived from yew: the Eburones between the Maas and Rhein rivers,[18] the Esuvii in Calvados, the Eburovices, the Eburobrigen and the Eburomagus in Gaul (unless they referred to wild boars). The only two other Celtic tribes with a tree association in their name are the Averni ('People of the Alder', from *verna* for alder) and the Lemovices ('Warriors of the Elm' or 'Warriors of the Lemos')[19] in Gaul. Once every year the druids of Gaul, according to Caesar,[20] met in the forest of the Carnutes, which was deemed the (sacred) centre of all Gaul. Little more is known about these events or their location but it has been suggested that one of the most important cathedrals in Christendom, Chartres, was built in the area.[21] The forest of the Carnutes is gone, however, and the only remaining sign of yews in the vicinity of Chartres is the name of the river that runs through the town: the Eure.

During the Roman invasion of Gaul, Catuvoleus, the king of the Gaulic Eburones, in 53 BCE took his own life with yew poison to avoid the shame of Roman captivity.[22] And when in 19 BCE, the last part of Spain, the Cantabrian Mountains, fell to the Romans, the local Iberian tribe, the Cantabri ('People of the Mountain'), committed collective suicide with the aid of the yew tree (wooden weapons being more likely than eating handfuls of needles). And again, when the last revolt against Rome in northern Spain was crushed, the Cantabrian and Asturian defenders of Mons Medullus employed yew to end their lives.[23] As Iberians and Eburones gain their very names from the yew (*ibe, eber*), these suicides suggest a ritualistic context.

The name of the Iberian Peninsula derives from its ancient inhabitants whom the Greeks called

Iberians, most probably after the River Ebro (Iberus), the peninsula's second longest river after the Tajo (Spanish) or Tejo (Portuguese), which is also named after yew. Waves of migrating Celtic peoples from the eighth to the sixth century BCE onwards settled heavily in northern and central Spain, but left the indigenous Bronze Age Iberian people of the south and east in peace. In the borderlands (north-eastern Meseta Central, in Catalonia and Aragon), these two ethnic groups merged into what are called the Celtiberian tribes. The Iberians traded with Carthage, Greece and Phoenicia.[24]

Beyond the far end of Europe, in modern Georgia and its borderlands with south-western Russia, flourished a medieval kingdom of the same name: Iberia. Without doubt it was named after the important yew stands of the Caucasian Mountains. The old Georgian names of the yew are *Chvaebis che*, 'God's tree' or 'Tree of God', *chvyturi che*, 'divine tree', and *tciminda che*, 'sacred tree'. In the twelfth century, Queen Tamara of Iberia adamantly guarded the yew against severe economic pressures. Since then, the yew in Iberia has been protected, and also acquired two more names: *utchtovari*, 'not to be permitted', and 'Queen Tamara's tree'.[25]

30.5 *Queen Tamara, who ruled in twelfth-century Iberia (modern Georgia).*

About ten degrees longitude further east begins the distribution area of the Himalayan yew, and in some parts of the north-western Himalayas, the tree is known as *deodar*, 'Tree of God'. Sages and ascetics are said to have served beneath it in ancient times.[26] And, as reported by John Lowe in 1897, 'the wood is burnt for incense, branches are carried in religious processions in Kamaon, and in Nepal the houses are decorated with the green twigs at religious festivals'.[27]

In present-day China the tree is called *hong dou sha*,[28] 'red bean fir', and also *zi shan shu*, 'dark-red fir tree' – a bland and simple name, suitable for a society 'reformed' by a regime that has taken great care to extinguish old, and particularly religious, traditions. Of interest would be the ancient and current names of the tree in the many languages and dialects of this vast country. At least one of them was detailed by two British botanists, H.J. Elwes and A. Henry, in 1906. They describe the Chinese yew as rare in the mountains of Hupeh and Szechuan (*c.* 1,800–2,400m/6,000–8,000ft altitude) and added: 'The Chinese mountaineers ... called the tree *Kuan-yin-sha*, "the fir of the Goddess of Mercy"'[29] (see p. 254).

The Japanese name, *Ichii*, means 'number one rank'[30] and is a prestigious name referring to the Imperial Court (see p. 210). In the traditional inauguration ceremony each new emperor is given a sceptre made of yew wood as a symbol of sovereignty over the land and its people. In the Japanese writing system,[31] however, the written characters for *Ichii* can also be read as *araragi* or *kunungi*. Indeed, *araragi* is another name for the tree in the middle and south of Japan, while *kunungi* occurs towards the north. A separate Japanese name for yew is 'water pine', and the written characters for this can be pronounced either as *suimatsu* or *mizumatsu*.

In the old language of the Ainu (the indigenous people of Hokkaido, the northernmost island of Japan), the yew is called *onco* (the same as *onko*), which means 'Tree of God'. Another name that is common in Hokkaido is *kuneni*, 'Tree of the bow'. Curiously, *onco* in Latin means tumour (Greek *onkos*), and a modern Japanese biotechnology company that develops cancer treatments from natural products claims that it 'fights against cancer like the tree onko which continues to grow as a tree of God'[32] (compare Chapter 25).

THE ANCIENT TONGUES

The oldest existing word for yew, *eya(n)*, appears in texts of the Old Hittite period, *c.* 1750–1500 BCE,[33] preserved on cuneiform tablets found at Boghazköy (modern Turkey). It took most of the twentieth century, however, to establish the botanical identity of *eya*. Another ancient word, although doubtful, is Semitic *elammaku*. In a document from about the eighteenth century BCE, the king of Mari, a Mesopotamian city-state situated on the Euphrates river, boasts of his expedition to the sea and his conquest of the mountains of cedar where he also cut wood of box and *elammaku*.[34] From the ninth century BCE or earlier,[35] this word also occurs in the Babylonian-Assyrian versions of the Sumerian mythical hero Gilgamesh (see p. 200). It has not been verified by etymologists whether the Old Hebrew word *almug* generally accepted to mean 'yew', is related to the Semitic form *elammaku*.[36]

King Solomon would have been aware of the meaning of *almug*, and this might possibly explain why he particularly requested some wood of this tree for his temple, and why it is mentioned not in league with other timber, but in the same breath as gemstones. When Solomon negotiated with Hiram, the king of Tyre (and hence ruler over the vast forests of Lebanon) about the supply of materials and craftsmen for the construction of his temple at Jerusalem, he writes,[37] 'Send me also cedar, pine and algum timber[38] from Lebanon,[39] for I know that your men are expert at felling the trees of Lebanon ...' And after the completion of the temple – made of stone and timber of cedar, cypress and wild olive[40] – Solomon received 'huge cargoes of almug wood and precious stones. The King used the wood to make stools for the house of the Lord [the Temple] and for the palace, as well as lyres and lutes for the singers. No such quantity of almug wood has ever been imported or even seen since that time.'[41] (Compare figure 42.6.) The Jewish priest and historian, Josephus Flavius (37/38–100 CE), however, also mentions the use of almug wood for 'the pillars for supporting the Temple and the palace'.[42] As for the furniture inside the Temple and the palace, the combination of light boxwood with dark yew wood for inlay was familiar all over the Near East,[43] and used in tables, chests and various small furniture. Another hint as to the natural distribution of yew in Lebanon comes from Nineveh. A palace inscription mentions 'cedar wood' from Lebanon as the material for the beams, but a microscopic examination of a sample from one beam proved it to be yew wood.[44]

30.7 *Jar shard with stamp impressions on the shoulder, Nineveh.*

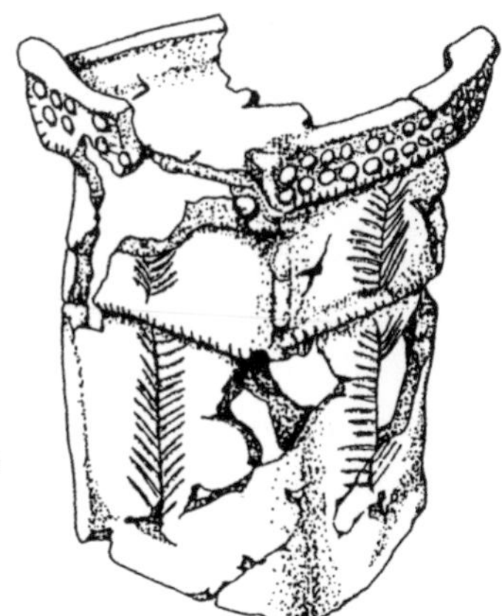

30.6 *Bronze Age domestic altar, Syria-Canaan.*

It seems that the term *almug* was not exclusively Hebrew, but was also used by the other Semitic peoples of the ancient Near East. The Assyrian king Tiglath-Pileser I (eleventh century BCE), for example, reports to have planted *almug* in his palace garden at Nineveh, together with cedar and the other trees of which he had collected specimens during his expeditions.[45] Some 650km (400 miles) west of Nineveh, two tablets found at Ugarit (modern Ra's Shamrah, on the Mediterranean coast of northern Syria) contain a shorter form of *almug*: *l-m-g*.[46]

In the Old Testament, an additional word for yew – *smilax* – is used once (Jer. 46: 14). However, Jeremiah in the original text *c.* 605 BCE does not speak of yew but of *svivcha*, 'surroundings'.[47] Most translations into European languages retained the Greek word *smilax* which probably goes back to the Septuagint (the first translation of Hebrew scripture into Greek by seventy-two Jewish translators in Alexandria, from the middle of the third and during the second centuries BCE). It is a mystery how this change of meaning occurred. Prof. G. Henslow in 1906, however, noted other possible mistranslations in the early Greek biblical texts: occurrences of 'hewn timber' might actually have been mistaken for yew wood since a similar Greek word, *smilentos*, means 'carved or shaped timber'.[48]

The first to mention yew in Greek writing[49] was the philosopher and botanist Theophrastus (*c.* 372–*c.* 287 BCE; for his full account see Appendix IV). The progression from Semitic *almug* to Greek *smilax* and *milos*[50] is not obvious at first and has been suggested by Henslow as follows: the three consonants of *almug*, *l-m-g*, by transposition of *m* and *l* changed to *m-l-g*, and, as *k* is equivalent to *g*, *m-l-k*. This became *milaks, smilax* in Greek, because the addition of an *s* at the end or as a prefix used to be common in this language (for example, the Sanskrit word for emerald, *marahata*, became *smaragdos* in Greek). In the old Attic dialect, *s* before *m* is added or left out, according to convenience, hence the form *milos*.[51] There is a possible relation to the Greek word *smao*, 'to rub or smear', which derived from an ancient root word *mao*, allied to *masso*, 'to touch, to handle' (hence our word *massage*),[52] which presents a striking parallel to the Latin *taxus* and its origin (see above). In Greece today, another name for yew, *elate*, translates as 'soft fir', and as we have no knowledge of how old this way of referring to yew could be, yew research in Greek sources should look out for forms of this word as well.[53]

30.8 *A pottery 'lid' – possibly from a vessel for offerings of incense or liquids (oil, milk, honey) from Lagash, Babylonia, with yew depictions on the front and early Hebrew lettering on the back.*

CONFUSION

Throughout history, the identification of yew – both in nature and in text documents – has been perplexing. The discussion about the meaning of the Hittite tree *eya* is a good example. The Czech archaeologist and language scholar Bedrich Hrozny (1879–1952), who opened the door to the translation of the Hittite hieroglyphs in 1915, came upon a ritual text that describes a sacred tree as an 'apple tree' standing over a spring and bleeding. To comply with the contemporary knowledge of ancient ritual, he attempted to interpret the bleeding as a reference to animal sacrifice but could not ignore that what was really being said was that the actual tree was bleeding (compare figure 30.10).[54] Hrozny, however, had to leave it at that. In 1939, the definitive discussion about the meaning of *eya* began with the suggestion that the word might have denoted a valuable fruit tree because some Hittite texts connect *eya* with the epithet 'mountain apple tree'.[55] This was soon rejected because *eya* clearly is an evergreen, and during the 1960s all eyes turned to 'fir'.[56] This, in turn, had to give way when in the 1970s it was

30.9 Ancient monumental yew in the Batzara Reserve, Caucasus.

discovered that 'fir' is the Hittite *tanau-*.[57] At about the same time, the German orientalist Volkert Haas[58] tried to make a case for 'evergreen oak' for a number of good reasons: one text[59] mentions *hurpastanus*, 'leaves', and can therefore hardly refer to conifer needles; other texts call it a 'large tree' or a 'great tree', and one reference mentions the use of its wood for spears, which is archaeologically verified for oak; and the Greek myth of the Golden Fleece (see p. 206) centres on a sacred 'oak' in Colchis (in the Caucasus, part of modern Russia and Georgia). However, since Haas, 'oak' has been verified as Hittite *allantaru*, confirming the views of those scholars[60] who had already linked *eya* to the etymology of *yew*. Yews are large (particularly the monumental trees of Turkey); yew wood, too, has been attested for a Hittite spear, and the broad, soft needles may qualify for 'leaves'. As for the argument about the Greek tree of the Golden Fleece, Greek texts do speak of an oak but the myth and the corresponding rituals are definitely Anatolian in origin. Hence the famous Greek version of the Golden Fleece story needs re-evaluation (see Chapter 39).

Greece did import not only the ritual context of the sacred tree but also its name, *eya*, though it seems that during the large-scale population movements of the late Bronze Age (*c.* 1300–*c.* 1100 BCE, the so-called 'Dorian invasion') and the successive 'dark ages' (twelfth century to *c.* 750 BCE) it got lost. In Greek literature the only trace is the term *oa, oie* for the rowan tree; the 'new' word for the yew, *milos (smilax)*, was taken from an entirely different source (see above). The Greek language lost the primary meanings of both inherited yew terms *(oa, toxon)*, and this dislocation may account for the appearance of oak rather than the yew in ancient Greek literature.[61] In other words, one might say that the language of their time failed the first writers on ancient Greek religion (Homer, Hesiod) in adequately reporting the yew's role in myth, ritual and folklore.

The study of the tree myths of the northern hemisphere reveals a number of trees that overlap with, or rather got confused with, yew. Among them are rowan and hazel in north-western Europe and fir, plane and olive, and sometimes beech and willow, in the eastern Mediterranean. There are four trees, however, where the confusion with yew has created significant distortion for historians and mythologists.

Cedar

In the Old World, people did not follow botanical criteria for the differentiation of trees, but whatever naming strategies they used suited the needs of the time just as well. In the Mediterranean timber trade from the third millennium BCE onwards,[62] the wood of best quality was cedar, closely followed by cypress.[63] The main supplies came from Lebanon and the Amanus Mountains (modern southern Turkey). The Greek label *kedros*, however, also included the three tall species of juniper,[64] which are indigenous at the same altitudes as cedar. Like the Greek term, the Roman *cedrus*, too, used to include cedar and juniper trees, and the fact that both languages additionally have a separate word for juniper – *arkeuthos* in Greek and *juniperus* in Latin – has led to some confusion among scholars.[65]

Theophrastus says that the wood of yew *(milos)* was occasionally traded as cedar.[66]

Ancient Egypt had to import most of its timber from Lebanon, the Taurus and the Amanus Mountains (both in southern Turkey),[67] and occasionally small amounts of yew wood were included in a 'cedar' cargo. No hieroglyph for 'yew wood' has yet been identified and it is possible (but unlikely) that the Egyptians may not even have had a name for it.

In the context of the sacred scriptures of Sumer and Babylonia it remains open whether the term 'cedar' really specified *Cedrus libani* as the Tree of Life,[68] or whether it might be better translated along the lines of 'most venerated needle tree'. For neither *Taxus baccata* nor, indeed, *Cedrus libani* could grow as far south as Sumer; and anyway, neither rises from the salty ocean every morning – the Sumerian Tree of Life was a *symbol* from a mythical past (see Chapters 32 and 43).

Oak

Another major confusion, and the most significant one, concerns the (evergreen) oak. It seems the Greek words *drus*, 'sacred tree',[69] and *drys*, 'oak',[70] were confused in classical times and that many an evergreen sacred tree became an evergreen 'oak' in the historical accounts. Of course, this is by no means to say that *all* references to oak are faulty. The confusion is partly caused by the fact that young yew arils look as if they are sitting in a 'cup', just like the acorns of the oak family, and in some old European texts there are allusions to the 'acorns' of yew, and of other trees. Theophrastus called sweet chestnuts the acorns of Zeus;[71] the Romans, who inherited the Greek's open approach to 'acorns', added the walnut as *iuglans* (from *Jovis glans*), 'acorn of Jupiter',[72] although this is probably the result of a translation error by Pliny (see below). This suggests at least *three* trees – oak, yew and sweet chestnut – that could have been intended in the Greek vernacular to say, 'having partaken in the myrtle of Aphrodite and the acorns of Zeus' (to do so being a synonym for initiation into the mysteries, see Chapter 35).

Furthermore, the notion that some of the sacred trees of ancient Greece were bleeding when they were injured might be more than superstitious 'myth'. There is the rare – and scientifically still unexplained – phenomenon of the 'bleeding yews'. Such trees keep oozing a thick, deep red liquid from an opening in the bark or a sawing wound. The 'Bleeding Yew of Nevern' in Wales (figure 34.12) has done so for years and has become a Christian pilgrimage site (because the 'bleeding' is taken as a symbol of Christ's wounds). Ovid in his *Metamorphoses* describes a sacrilege in the sacred grove of Ceres (the Roman Demeter). In the centre of the ancient grove stood a huge 'oak' which in itself was 'a grove' (*una nemus*; compare 'Branch layering' in Chapter 19). The centre of the tree of the goddess *(Deoia quercus)* was surrounded by priestly garlands, votive tablets and flower wreaths. As soon as the assailant's axe hit the trunk, blood streamed forth from the wound.[73] No European tree other than *Taxus* can react like this.

What added to the confusion in early literature is the different appearance of male and female yews, one being laden with bright scarlet fruits in autumn, the other remaining 'barren'. According to their cosmology, the Hellenistic Greeks understood all tree

***30.10** 'Bleeding' yew at Chillingham, Northumberland.*

30.11–30.13 *'The acorns of Zeus': (from left to right) sweet chestnut, prickly oak* (Q. coccifera) *and yew.*

and plant species to be of different sex, like humans and animals. Even Theophrastus describes 'male' and 'female' trees for monoecious (with male and female flowers on the same individual) genera such as oak, fir, pine and cedar. A truly dioecious tree (male and female flowers on different plants) like yew, then, has at times been conceived of as two distinct species – a mistake that occurred not only in the distant past but until fairly recently. A botany book from 1912 states the following about *Taxus*: 'The older botanists, not realizing the dioecious character of the tree, made two distinct species of the barren and fertile plants respectively. Thus Gérard [*Herball*, 1597], for instance, describes a "*Taxus glandifera baccifera-que*, the Yew bearing acorns and berries;" and a "*Taxus tantum florens*, the Yew which only floures."'[74] (And is the modern *baccata*, 'berry-bearing', not a remnant of this as a name for a conifer that botanically does not even have fruits, let alone berries?) It remains peculiar, however, that Pliny when describing the monoecious holm oak says that the Greeks call it *milax* – which is Greek for yew.[75]

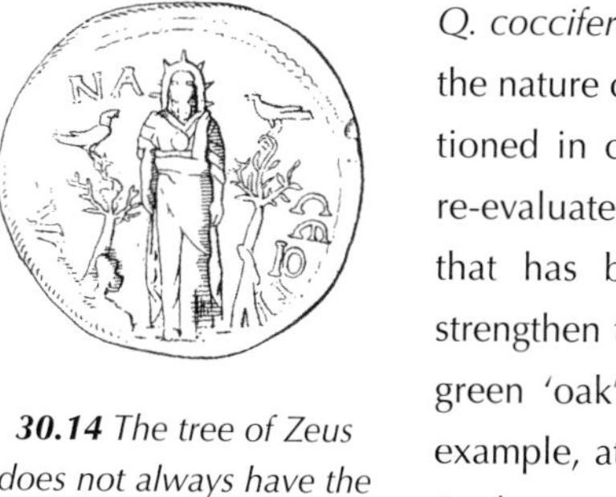

30.14 *The tree of Zeus does not always have the appearance of an oak; coin from Caria.*

With quite a large number of different species, broadleaf as well as evergreen, the oak genus was a dominant presence in the early Mediterranean landscape.[76] In the search for an evergreen 'acorn-bearing' tree to be identified as the sacred tree in Mediterranean myth, the holm oak *(Quercus ilex)* attracted much attention in the twentieth century, particularly from Robert Graves. This tree inhabits the coastal and middle zones of the Balearics (Graves had a home in Mallorca), Corsica and Italy. In the upper hills of Sardinia, for example, it is a natural companion in the yew–holly woods. But holm oak did not survive the last glaciation in Spain and Portugal where the native evergreen oak is live oak *(Q. rotundifolia)*, and, most importantly, it is rare in Greece and Crete where the dominant evergreen oak is prickly-oak (also kermes oak, *Q. coccifera*).[77] With all this in mind, the nature of many a sacred tree mentioned in classical texts needs to be re-evaluated. Much of the evidence that has been brought forward to strengthen the case for a sacred evergreen 'oak' in ancient Greece – for example, at the famous tree oracle of Dodona (see box next page) – relates to holm oak when other, northern Greek oak species should have been targeted.

In general, oak is the tree of the Indo-European thunder-god and as such can be found across (mostly northern) Europe as a sacred tree – that is, beginning with the spreading of the Indo-European peoples who suppressed or gradually assimilated the older, native religions with their tree cults. This is probably the main reason why yew lore went 'underground' during the third and second millennia BCE. What was left became further obscured by language problems. The real confusion, however, began with Pliny the Elder (23 CE –79 CE). His celebrated *Natural History* is 'an encyclopaedic work of uneven accuracy' *(Encyclopaedia Britannica)* that nevertheless was used as an authority on scientific matters up to the Middle Ages, and is still often

Dodona

The sanctuary of Dodona was the oldest religious site in ancient Greece, and functioned for about 2,000 years.[78] In classical times, its fame was second only to the Oracle of Delphi (the latter was geographically closer to the Greek city-states and hence visited more often by politicians and other famous figures), and a whole town developed nearby to cater for all aspects of Old-World tourism. However remote from Athens, Dodona buzzed with life.[79]

The earliest deity worshipped at Dodona probably was an ancient indigenous fertility goddess who in the early second millennium BCE was called Da (De) or Do by incoming Indo-Europeans, that is, Greek, Thracian and Illyrian tribes. Do developed to Dodo(ne) (related to Damater, Domater, Demeter), the mistress of the fertile earth. She had a male consort, the Illyrian god Dodon, and evidence suggests the prevalence of *hieros gamos* rites ('sacred marriage', see p. 202) of the fertile earth and the rain-bringing god. In about the eighth century BCE (when the Selloi, a branch of the Thesprotian tribe, took over the sanctuary), the role of the divine couple was given to Zeus Naios and Dione Naia.[80]

According to Herodotus (fifth century BCE), Dodona had been founded by a priestess from Egypt, metaphorically called a 'black dove',[81] and doves remained the sacred birds of the priestesses of the local goddess. The priestesses were even called *pleiades*, 'doves', but also referred to as Muses (see p. 181).[82] A ritual cauldron of renewal and consecration *(apotheosis)* was part of the ancient rituals at Dodona. Dione is often associated with the goddesses Rhea, Artemis/Diana, and occasionally Europa. Her name, like that of Zeus *(Dion)*, means the 'bright one', or 'divine'.[83]

Dodona is located *c.* 22km (*c.* 14 miles) south of Ioannina in Epirus (north-western Greece), a mountainous region that used to be densely wooded and even exported timber,[84] but after millennia of over-grazing it looks different today. Its natural mixed woodland on limestone slopes enjoys high annual rainfall and is very suitable for yew. Regarding the botanical identity of the sacred tree of Dodona, however, confusion reigns. There are Roman references[85] to a 'holm oak', which Pliny says the Greeks call *milax*, 'yew', a dioecious tree the female of which bears fruits (oaks are monoecious); while the Greeks themselves called the Dodonean tree *drus*, 'sacred tree', or *phagos*, a term for a tree with edible fruits,[86] nevertheless by many contemporary authors interpreted as Valonia oak *(Q. macrolepis)*, and which was already mistaken by a number of Latin translators for 'beech' (Latin *fagus*).[87] *Phagos* could have denoted any tree with sweet 'acorns', even the sweet chestnut, but since Dodona's limestone soil is unhospitable to Valonia oak (and even less to sweet chestnut), the most likely contender among the oaks is Macedonian oak *(Quercus trojana)*.[88] Some Dodonean votive offerings from the fifth century onwards depict oak leaves and acorns, hence it seems certain that the tree of Dodona was an oak in classical times. But there is the question of age: none of these tree species can exceed an age of 600 years maximum; given that Dodona's history spans two millennia, the sacred tree would have had to have been replanted two or three times[89] – unless it was a yew in the first place. Some time in the first half of the first millennium BCE, the Selloi probably transferred the sacred status to an oak. The true identity of the original sacred tree of Dodona, however, must remain a mystery until more evidence surfaces.

30.15 *Rock pigeons.*

quoted today. For his section on trees, Pliny mainly copied from older works (he got a huge amount of reading done with the help of well-educated slaves) or gathered stories from informants. Pliny never saw the places he wrote about, nor would he have recognised many trees. He was not a botanist like Theophrastus – his main source on trees – and despite the high standard of his predecessor, Pliny got many of his trees wrong.[90] He repeatedly mistook Theophrastus' description of the sweet chestnut for walnut,[91] translated Theophrastus' Cretan coastal pine and mountain pine on Mt Ida as 'mountain' and 'coastal' larch (which is only one species, and confined to the Alps); he even classified the deciduous larch as an evergreen; and, most noteworthy, he took Theophrastus' entire passage on *milos*, the yew (see Appendix IV), and assigned it to the deciduous ash (Greek *melia*).[92]

With the yew thus obscured, a misconception of even greater consequence for Western Europe was started by Pliny: by copying an account from Gaul he authored the cliché of the Celtic druid cutting mistletoe from an oak tree.[93] Despite the fact that mistletoe grows very rarely on oak trees, this passage became part of the foundations of the druid revivals in the sixteenth and seventeenth centuries, and the image of the white-robed druid among the oak leaves has been around ever since.[94] It also reinforced Sir James Frazer's case for mistletoe being the 'golden bough' (see p. 209 and Appendix V), and has spurred generations of mythologists and folklorists to compare the 'sacred oak' in Celtic and biblical traditions. Probably because of a confusion of Greek *drys* and *drus*, almost all 'sacred trees' in the five books of Moses have become 'oaks' in the Latin translation, and later in the other European languages. English and also German Bible texts have at least maintained some of the sacred trees of ancient Israel as the turpentine tree or terebinth *(Pistacia terebinthus)*,[95] but there is more to consider: the Hebrew texts speak of either *elah (alah)*, generally translated as 'terebinth', or *allon (elon)*, generally translated as 'oak'. Both terms contain the word for 'God', *el*,[96] and still today *elah* also means generally 'sacred tree'. In this context, the old language does not necessarily differentiate between two species but may denote a 'male sacred tree' *(allon)* and a 'female sacred tree' *(elah)*, without intention of specifying a species. *Elah* still in modern Hebrew also means 'goddess' (for further discussion see 'Asherah' in Chapter 37).

Ash

The poets of medieval north-western Europe, the Celtic bards and the Nordic skalds used metaphors, riddles and imaginative description rather than calling an object by its name – a technique somewhat similar to the epithets used by the composers of the ritual texts of the ancient Near East over 2,000 years before them. The medieval Welsh bards even pitted their wits and set challenges to see who could keep the others guessing the longest as to the meaning of their words. In Skaldic poetry there was even a technical term, the *kenning*, which is a concise compound or a figurative phrase that replaces the common name of an object. A kenning is commonly a simple stock compound, for example, 'wave-horse' for 'ship', but often requires a deep knowledge of Norse mythology to be understood. Sometimes kennings are extremely indirect; hence, many have become unintelligible to later generations. Nevertheless, Icelandic texts describing the Nordic World Tree as *ask*, 'ash', were taken literally in post-medieval times, which has led to a major misconception that has persisted in mythology ever since. The World Tree is, however, described as an 'evergreen needle-ash' and this shows we are dealing with a kenning, as the ash proper is neither evergreen nor does it have needles. (For further discussion see Chapter 41.)

Apple

Another medieval misunderstanding involves the apple. 'Apple' seems to have been a common word in history for '(red) fruit'. Already the Hittite text mentioned above describes *eya* as a 'mountain-apple'; the Icelandic *Volsung saga* mentions an 'oak' that carries 'apples';[97] and right across Asia Minor and Europe the mythological 'apples of immortality' are connected to traditions relating to kingship, ancestors and the underworld, death and rebirth (see Chapters 37, 39 and 41). Traditions of the 'real' apple *(Malus)*, on the other hand, are mostly associated with more domestic themes: love, courtship, marriage and fertility.[98] Furthermore, most of Europe did not know *red* domestic apples until the Romans began to introduce them; the native wild apple or crab apple *(Malus sylvestris)* is mostly yellow-green.

Around 425 CE, however, a Christian theologian, Cyprianus Gallus, published his epic on the creation in which he was the first to interpret the forbidden 'fruit' (Hebrew *peri*) that Eve hands to Adam as a

A hard nut to crack

In old texts of the British Isles, mention is made of a 'tree of three fruits';[99] this alludes to the female yew with its 'apples' (the mature, red arils), 'acorns' (the green unripe ones), and 'nuts' (the seeds themselves, without the fleshy aril). The 'acorns' and 'apples' we have already encountered in south-eastern Europe in much earlier times, the 'nuts', however, are endemic to Ireland – and have inevitably contributed even more to the confusion in tree lore. In the Irish *Dindshenchas* (eleventh to twelfth century), the mythical Nine Hazels of Wisdom, for example, grow over Connla's Well, which is the true source of all wisdom. Their fruits drop into the pool and feed the 'salmon of wisdom', and the hero Fionn who eats such a salmon attains the gift of prophecy. But why is it said that the salmon's pink flesh is stained by the 'juices' of the fruits? And why is the Irish word for the fish *eo*, which means 'yew', and not *coll*, 'hazel'? Furthermore, the mythical salmon shares a personal name with a human Irish hero: Fintan, which sounds rather like a reference to a 'tree-god', can be translated as 'red tree'. It is this Fintan who, in *The Settling of the Manor of Tara*, receives the sacred branch from the otherworldly messenger (see p. 238) and plants the five sacred (yew) trees of Ireland. Later, when the ancient yews disappear, he dies too.[100]

30.16 *'Acorn' and 'apple'.*

30.17 *This Celtic altar from the Comminges, Pyrenees, shows a conifer and below what could be three different stages of aril development: 'nut', 'apple' and 'acorn'.*

domestic 'apple' (Latin *malum*). This idea is repeated in a poem about Genesis by Bishop Avitus of Vienne at the end of that century.[101] The reason for this change might have been eco-geographical: fifth-century Gaul (modern France), the home of both writers, had been abundant with apple orchards since the Roman occupation. The apple might have found such smooth acceptance as the biblical fruit of sin because of its occurrence as a love gift in pagan Graeco-Roman texts, as well as in European folklore. The apple certainly is a marvellous fruit, and in the wake of the spread of its cultivation all over Europe 'new' folk customs, such as wassailing or beating the apple tree, arose to express people's joy and gratitude. In some regions, elements of such rituals are likely to have been transferred[102] from other native trees such as yew or rowan. The consequence of the preoccupation of Christian Europe with the domestic apple was on one hand the demonisation of the apple tree among certain Christian groups – Osbern of Gloucester in 1150 explained the Latin word for apple with the (Latin) word for 'evil'[103] – and on the other hand the (post-medieval) willingness to unquestioningly accept other fruits in legend as apples; for example, those of the Hesperides. The twelfth-century historian Geoffrey of Monmouth referred to the Celtic Avalon as *Insula Pomonum*, 'apple island', which has limited our understanding of Celtic afterlife beliefs ever since. However, islands that were sacred in Celtic tradition have distinct yew connections, for example the Scottish islands of Iona ('yew island') and Arran (which was known as Eamhain Ablach, 'place of yews [with] apples'), and possibly Glastonbury in south England: Celtic *glas* means '(ever)green', *tan* is 'red, fire', and *tann* is 'sacred tree'. In the old British tongue, Glastonbury would have been named after an evergreen, red tree – *if* the name were Celtic, but as it stands it is Anglo-Saxon and translates perfectly well as the *bury* (protected settlement) of the *glastan*, 'oak trees'[104] – which might bring us back to the 'apple'-bearing oak mentioned in the Volsung saga (see above), anyway. After all, in the early 1960s archaeologists found the root system of a yew at a depth of 3.6m at the sacred spring at Glastonbury Tor, where the tree had flourished in about 300 CE.[105]

31.1 *Yews line the path to the underworld.*

CHAPTER 31

THE GREAT PASSAGE

In the ethnobotany of trees, related and often similar customs and traditions regarding one tree species can be found cross-culturally, varying, of course, according to the cultural context.[1] This applies to yew as well, and among the tree-related customs of the world, material relating to the yew stands out not only as being exceptionally voluminous but also as being remarkably homogenous and coherent throughout entirely different cultural spheres and vast stretches of time. The following chapters explore the full range of the timeless, global, pan-cultural subjects that link the yew tree with almost every area of human existence. However, birth and death, transformation and the human quest for immortality are the most frequently recurring themes.

The most obvious place to start is with the well-known association of *Taxus* with the graveyard and with burial practices. The dominant role of yew in Christian graveyard design and burial practice is well documented and can easily be traced back to the Middle Ages. Giraldus Cambrensis, for example, observed yew in many cemeteries and holy places in Ireland in 1184 CE.[2] Still in the twentieth century, an old custom in England and Wales was for mourners to tuck a sprig of yew into the shroud, or carry it over the deceased and lay it beneath the coffin in the grave.[3] But more impressive than old manuscripts or folklore are the veteran trees themselves, which stand as living proof of their close and time-honoured association with human faith. The Ancient Yew Group (AYG) database presently holds over 800 graveyard sites with old yews in Britain.[4] Because of different climate and different histories, other European countries have fewer yews in churchyards and graveyards than Britain, at least fewer old and truly ancient ones. And yet, the ecologically protective aspect of the consecrated ground has preserved yews in Central Europe too. In Slovenia, for example, which lost most of its natural yew stands during the medieval longbow trade, the oldest trees are located in church-/graveyards.[5]

Most British writers never doubted that the presence of yew in the graveyard was of Celtic origin. Already in 1658, Sir Thomas Browne suggested that the druid's 'funeral pyre consisted of sweet fuel, cypress, fir, larix [larch], yew, and trees perpetually verdant'.[6] However, with an age of 4,000 years the oldest yew remains from graves in Britain are older than Celtic culture.[7] The Saxons also preferred this tree in their burial grounds (as many veteran trees in the Saxon counties of south-east England attest); the Germanic tribes in general, according to Tacitus (56 CE–c. 120 CE), avoided splendour at their funerals but took great care to be burnt with special kinds of wood.[8] And this applies to burials as well: a number of Germanic graves from the times of the Roman Empire onwards have yielded a somewhat surprising type of grave goods, that of wooden buckets. These buckets are usually coopered from a number of straight wooden pieces held together by metal hoops (mostly bronze, sometimes iron, rarely silver). About 200 buckets from this period have been found (not only in graves), but in three-quarters of the cases only the metal parts have survived. In the remainder, the type of wood often has not been determined. Where it has, the most frequent wood used is yew. One or more yew buckets have been found in aristocratic graves in Denmark, Saxonia, Thuringia, Frankia, also in a Merovingian grave, in two graves in Slovakia, and two Anglo-Saxon graves in England.[9]

If the yew trees at the graveside had not existed before the Roman invasion of Britain, the invaders would probably have introduced them, substituting their traditional southern burial trees, cypress and pine, with yew. Charles Coote *(The Romans in Britain)* wrote in 1878 of the funeral yew: 'For as of old it was connected with the passage of the soul to its new abode, so ever since the introduction of Christianity into this country it has continued to adorn the last resting-place of the body, which the soul has left.'[10] Coote's statement can be understood

31.2 *Linear yew shoots in Marche, Italy.*
31.3 *Urn shard from the third century BCE, Etruria, Marche, Italy. The decoration is reminiscent of linear yew shoots.*

in greater depth in the light of a verse by the Roman poet Statius (45 CE–96 CE), 'Amphiarus had descended into Hades [the underworld] so abruptly that there was no time to purify him by a touch of the yew branch of the Eumenis [a 'Kindly One'; see 'Furies', p. 182].'[11] *Taxus baccata*, without any doubt, played a significant role in the burial rites of Mediterranean cultures. The Romans adopted theirs from the Etruscans and from the Greeks to whom, in classical times, cypress and yew were the trees of mourning.[12] Classical writers such as Lucan and Silius Italicus describe yew as sacred to the deities of the underworld, and Silius Italicus as well as Seneca[13] mention a huge yew tree standing by the underworld river.[14] Ovid (43 BCE–17 CE) neatly expresses the essence in his *Metamorphoses*, where the yew tree is the guardian of the threshold between this world and the next:

> The path is declining, shaded by mourning yews:
> towards the infernal seats, through muted silence.[15]

The battle and funeral customs of Iron Age Germanic tribes are, somewhat surprisingly, mirrored in Georgia, where the yew is called *Chvaebis che*, 'God's tree'. The night before going into battle, the dukes of medieval Georgia traditionally slept on a bed of yew boughs – not taking the tree into battle like the Germanic warriors but spending on it what could be their last night on earth. This very custom is also reported in relation to a blue-blooded Russian, Igor Svyatoslavovych (1150–1202), better known (in Russia at least) as Knyaz Igor or Ygory, Duke of Novgorod-Seversky in the Chenigov region (modern Ukraine).[16] In 1185, before fighting the Polovtsy army over the town of Kiev, Ygory slept on his yew bed, according to the national epic, *The Song of Igor's Campaign*.[17]

Yew wood has also been found in the royal tomb at Gordion, the capital of the ancient kingdom of Phrygia (in modern eastern Turkey). The mausoleum is also called Midas Mound, after the legendary king whose touch turned everything into gold. The architecture of his burial chamber, however, carbon-dated to about 590 BCE (±60 years),[18] was wooden. The floor of the room was carefully made of cedar and yew, the inner walls of pine,[19] as was the ceiling, apart from two yew beams and one of cedar. The outer walls were made of round logs, mostly of juniper but with some yew. Inside the chamber were two wooden screens (195 x 80cm) decorated with intricate inlay work, the dark yew wood inlays providing a striking contrast to the white boxwood into which they were fitted.[20]

In 1894, Georges Beauvisage, Head of Botanics at the Faculty of Medicine in Lyons, France, was given a number of planks of wood sampled from Egyptian coffins.[21] In the first microscopic investigation of wooden grave goods from ancient Egypt to that date, Beauvisage discovered – to everyone's surprise – that the sarcophagi had not been made of native wood or any other timber famously used in the ancient Orient, but of yew. Since the group of tombs in question had suffered badly from grave robbers the dating of the finds proved difficult. However, after careful study Beauvisage concluded that the coffins in question probably belonged to the Twelfth Dynasty (*c.* 2400 BCE), but admitted a possibly greater age.[22] One of the tombs had been dated through a cartouche of Pepi II, while the main tomb is dedicated to Nefer-tum-hotep. One of the planks themselves, with fire damage on one end, bears a label 'Coffin of Ur-s-nefer' (also called S-Ur-nefer). He was the son of a man called Hotep who is pre-titled 'chief of the house' or 'chief of the administration of labour', which is interpreted as a kind of foreman for a grand feudal landlord. He was buried in three coffins inside each other, the yew plank coming from the innermost one. Some 1,900km (1,200 miles) east of Egypt, another ancient yew sarcophagus (of non-Egyptian origin) has been found at Bou Hadjar near Tunis.[23]

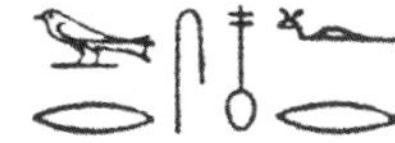

31.5 *Yew coffin inscription of Ur-s-nefer, Egypt, Twelfth Dynasty.*

31.4 *Yew graves at Minamiura, Japan.*

In North America, the northern limits of the Pacific yew stretch along the south-eastern part of the Pacific coast of the Alaska Peninsula. Here, the indigenous Tlingit people have used yew wood in a number of traditional ways, such as carving death masks and spirit whistles.[24]

The religious significance of the yew tree can be traced back for millennia (see the following chapters). As the cultural history of humankind developed, and religions too changed in due course, the veneration of yew became generally more marginalised (in some regions even suppressed and finally extinct); but with a history so old, some yew customs that are still alive today – such as certain burial customs – are among the oldest surviving religious traditions of humankind.

***32.1** Goats as guardians of the Tree of Life, ivory carving, Nimrud, Assyria, 870–860 BCE.*

CHAPTER 32

THE TREE OF LIFE

Among the peoples of the world, the Tree of Life is a recurrent religious theme[1] which has persisted over millennia. Sacred trees and vegetation rites can be found in the history of every religion, and some of these aspects are still alive in a number of tribal and/or shamanic cultures, for example in the Americas and Asia, while others have come down to us from the ancient cultures of Europe and the Orient. The Tree of Life exists as the centre of paradise in Judaic, Christian and Islamic mythology. It is the

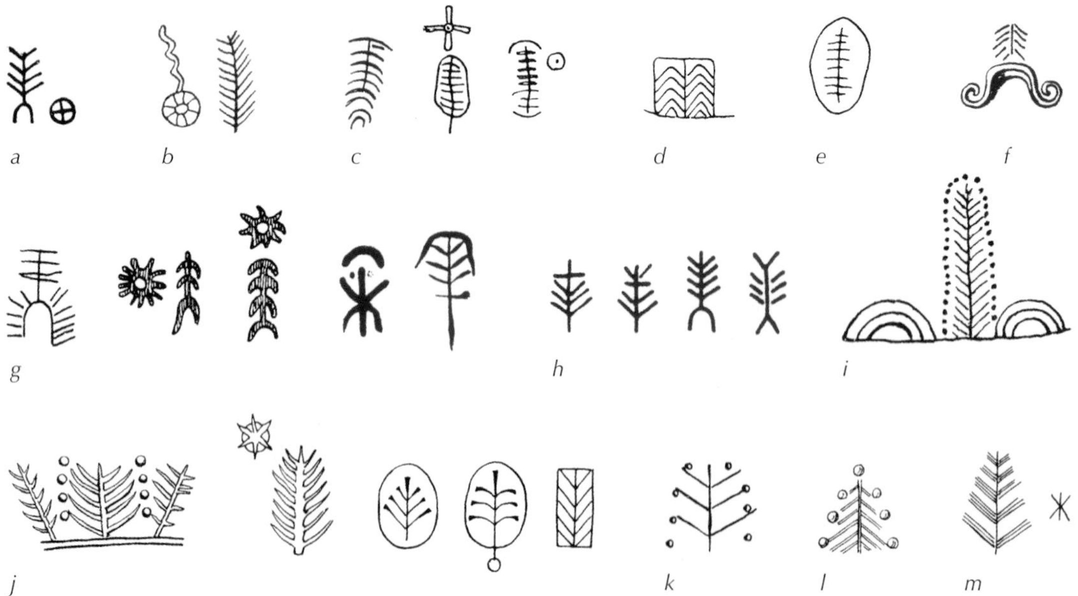

***32.2** In the art of the late Stone Age and the Bronze Age, many of the tree depictions appear rather coniferous, even yew-like. Examples from a) Owens Valley, b) Blarisden, both USA; c) Slial na Calliaghe, Ireland; d) eastern Gotland, Sweden; e) Tordos, Transylvania; f) Troy; g) La Pileta, Le Zarzalon, Las Palomas, all Spain; h) Gezer, Israel; i) Lake Geneva, Switzerland; j) Susa, pre-Elamitic; k) Poland; l) Armenia; m) Dnjepr Valley, Russia.*

centre of the world in Hindu lore and most shamanic traditions across Eurasia, reaching from ancient Celtic myth and the well-known World Tree Yggdrasil in Norse myth in Western Europe to the ancient tales of China and Japan. In Buddhism it became the 'Tree of Enlightenment'. Its history, however, is far older than any of these cultures and can be traced back well into prehistory.[2] Because of its prevalence in preliterate cultures, there are no 'sacred texts' telling us what exactly the Tree of Life meant to the different peoples. A summary of what sense contemporary writers[3] make of it follows here:

The Tree of Life, in its essence, symbolises unity – the unity of all opposites (sun and moon, night and day, summer and winter, male and female, etc.), and it connects the different levels of existence – in ancient worldview the underworld, the earth where humans live, and the heavens. The Tree of Life denotes the entire world as being one organism; each single life form, whether human, animal or plant, every manifestation of life (in animistic religions also stones, rivers and other natural phenomena have a soul or spirit), is but one leaf or seed-bearing fruit on this tree. The focus is not on the individual but the collective: no single life form can exist on its own, all life is interdependent.

The Tree of Life as the source of life, youth and immortality is also the centre of the world (and hence is also called the Cosmic Tree or World Tree). Its 'representatives' on earth, actual sacred trees, are also a 'centre' where the sacred, the eternal, breaks through into the temporary world of appearances. There can be many such 'centres' in the geographical landscape, they do not exclude each other at all.[4] Depending on the distribution of tree species across the surface of the earth, different peoples in different (climatic) regions have at times worshipped different tree species as *the* Tree of Life (although ultimately, all species are part of the transcendental tree symbol).

There is ample evidence to suggest that the yew has been a sacred tree in many traditions of the world, and, moreover, that it might have been regarded as the Tree of Life. And indeed, what other species could express more clearly what the Cosmic Tree represents? The yew is evergreen and throughout winter it maintains the highest energy levels of all conifers (see Chapter 17); its power of regeneration is second to none;[5] it has fruits in the 'colour of

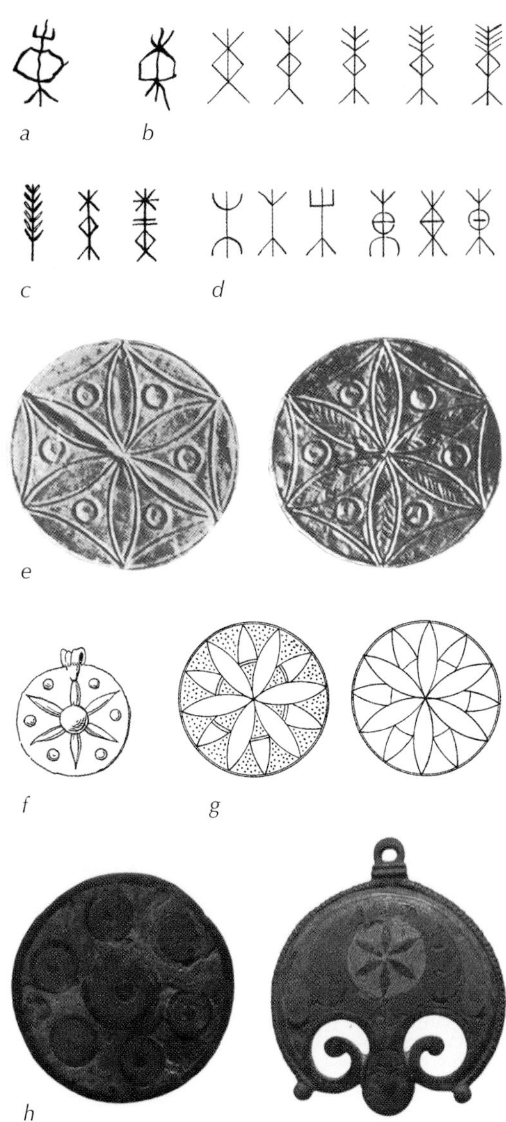

32.3 *Transition to a 'six-pointed' star and a 'rosette': a) La Pilata, Spain; b) Siberia; c) Sumer/Babylonia; d) China; e) gold sheet designs from Mykanae, Greece; f) votive offering from ancient Persian temple at Cuchinak; g) engravings on early seventh-century-BCE bronze cups, Idaean cave, Crete; h) Celtic enamelled brooches: second century CE, Chepstow, Gwent (left), and second to third century CE, Castor, Cambridgeshire. (Compare the 'arils' in figure 37.5.)*

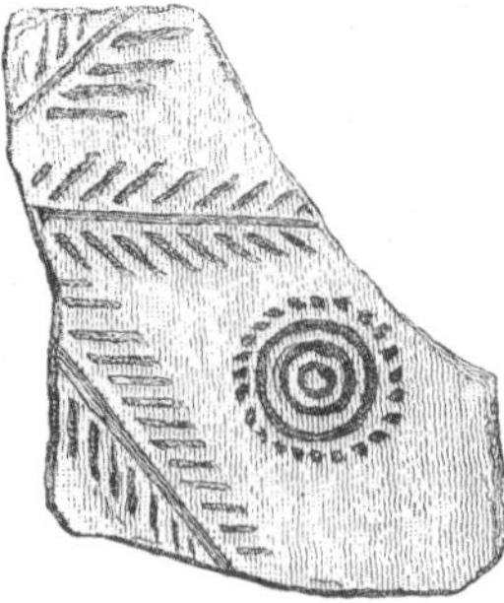

32.4 *'Sacred centre' and branch, bright-red terracotta fragment, Troy I, c. 3000–2300* BCE.

life' (blood); from its overripe arils it oozes sticky, sweet liquid like *soma* (see p. 219); and, at least from a human point of view, it is quite immortal. But there is more to it: evidence suggests the possibility that the very idea of a Tree of Life might have occurred in very early times in contemplation of the extraordinary, evergreen, long-lived, self-renewing, poisonous yew tree. It would have then spread with the human migrations and the cultural diffusion of the early Neolithic period to the many regions across the entire width of Eurasia (see the following chapters for further discussion). In areas where yew trees grow, the yew cult or parts of it could survive relatively long term (e.g. in Norse and Saxon Iron Age culture) unless they became marginalised during early history (e.g. in Greece during the late second millennium BCE). In other cases, migration led into areas completely unsuitable for the survival of *Taxus baccata*, such as the settlement of the proto-Sumerians in the southern floodplains of Mesopotamia (modern Iraq) in the late fourth millennium BCE, and forced the religious to find another venerable tree species to substitute for the sacred yew (the subject of migration is continued in Chapter 43).

Female divinity

In many ancient cultures, images of goddesses appear most intimately associated with the Tree of Life. Of course, there is a comparable number of male deities linked with sacred trees, being the supernatural guardians of certain tree species and having groves or individual trees dedicated to them. But where a divinity appears as the spirit of the World Tree itself, it is usually a goddess. For example, Hathor in Egypt (Memphis area) was worshipped as the Queen of Heaven and also called Mistress of the Sacred Tree; or the old white-haired goddess inside the World Tree of Yakut myth (Siberia).[6] This type of goddess usually emerges from the crown or trunk of the tree as far as her waist, and, more importantly, bestows a life-giving drink, whether that be the Water of Life from the spring at the base of the tree, milk from her heavy breasts, the honey-like sap of the tree, or the juice of its fruit. Universally, this liquid gives the recipient life, health, poetic inspiration, and in some cases even immortality.

Symbolically, anthropomorphic images of the divine forces within nature enhance specific aspects of human consciousness for the devotee to contemplate, be it compassion, willpower, strength, persistence, etc. In this case, the female divinity bestowing a life-giving drink accentuates the elements of (maternal) care and of nourishment as part of the same eternal life-giving principle of nature of which the Tree of Life itself is a reflection. The image of the tree, however, suggests that all of this reaches far beyond the boundaries of human existence; while the human shape changes and is not constant (even 'gods' die), the Tree symbolises stability and perpetuity. This intimate association of the tree and the female power of giving birth can be illustrated with an example from pre-dynastic Egypt: the word and sign for 'to give birth' derived from that for 'tree'; and both show the Neolithic six-pointed star in the centre.[7]

32.5 *The glyph for 'tree' (left), when combined with the female figure, becomes 'to give birth', pre-dynastic Egyptian.*

However, this archetype of a goddess inside or as the Tree of Life does not exhaust the religious complexity we are dealing with. The goddesses whom we encounter during the excavation of the yew tree's religious past display the four key functions that can be identified among the goddesses of the Bronze Age:

***32.6** Evening yews at Elmsted, Kent.*

The Mistress of the Underworld is the goddess of death, of the rites of passage, and also of initiation into the mysteries. As the fate and nemesis of all creatures, who takes their form and life away from them when their time is up, death goddesses may bring fear because death implies farewell, the letting go of all and everything one knows. At the other end of the dark tunnel of the unknown, however, waits a new life, a new youth. Maiden or 'virgin' goddesses symbolise the never-wearying regenerative power of nature. These divinities, radiant with eternal youth and rapture, are patronesses of fertility, pregnancy and birth. It is noteworthy, however, that the term 'virgin', unlike today, used to signify the complete independence of a woman or goddess from a male partner. In Hellenistic Greece, for example, certain goddesses carried the epithets 'virgin' or 'maid' (e.g. *Pallas* Athene), to acknowledge their independent female strength in life and in battle. This designation was entirely independent of their sex life. The patroness of birth, not surprisingly, blends seamlessly into the patroness of motherhood, representing nourishment and protection. This last aspect, in turn, may develop into a fierce defender: the protective mother, with the fury of the she-bear – or in fact, the fury of any mammalian mother protecting her young – defends her children (or devotees) from enemies and invaders. The fearsome goddess of the battlefield, then, takes us back to the mistress of death and the underworld.

In antiquity, these diverse functions did not occur neatly divided into different goddesses; often more than one function can be found in a single divinity. For example, the Greek Artemis is a beautiful, young and free huntress (a 'virgin') who nevertheless is a strong protector of mothers and their offspring (both humans and wild beasts); and when she takes revenge upon those who inflicted injury on those under her guardianship, she becomes the aggressor's nemesis, which has overtones of a death goddess. In ancient Mesopotamia, Inanna, and later Ishtar, combine various different functions, particularly love and war.

A deeper investigation into some of the Bronze (and Iron) Age goddesses (Chapters 34 and 37) will enable us to look beyond the 'doom and gloom' picture of the yew that prevailed in late Greco-Roman culture as well as in the Christian era. Before that, yew was not only a symbol of death but of birth too, just like the Tree of Life.

33.1 *Serpentine roots.*

CHAPTER 33

TIMELESS SYMBOLS

Only fragments of ancient yew lore have survived, and for a deeper understanding and appreciation of the substantial role that the yew has played in the cultural history of the human race we have to reconstruct the mental and spiritual context in which the tree might have been perceived. The first step in this 'archaeology of religion' – the sharpening of our tools, as it were – is the examination of a small number of symbols[1] that are intimately related and interwoven with the imagery of the Tree of Life, the yew, and various goddesses. Because the following symbols occur frequently in connection with religious yew artefacts from all over the world, the entire complex may be termed the yew–tree–goddess pattern.[2] Fragmented parts of it, in varying combinations, occur and recur across a great number of cultures and ages. Once we begin to fathom their meaning, doors can open for us and reveal glimpses of the ancient relationship of human and tree.

THE SERPENT

The flowing motion of the sap of the Tree of Life is represented by the slithering serpent. But most importantly, in sloughing its skin the serpent, like the tree, participates in the cosmic ability to self-renew. The snake, like the yew and many goddesses, is associated with water because its gliding on the ground has a rippling, water-like motion and it is able to 'insinuate itself in and out of minute cracks or holes in rocks'.[3] Also, the dual nature of the tree's sap, which can heal or destroy, is equivalent to the snake's venom. Hence the image of the snake

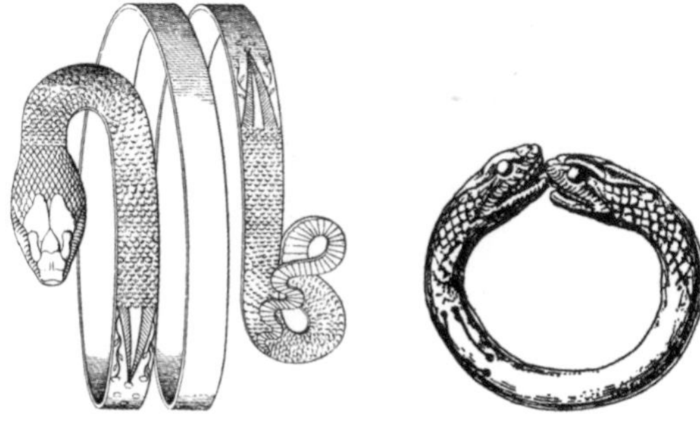

33.2 In the Old World, goddesses as well as mortal women enjoyed snake jewellery.

combines notions of life and death, the underworld, fertility,[4] healing and renewal. Particularly the themes of death and renewal connect it closely with the yew as the latter actually is a poisonous tree. And when an interior root takes over the crown of an ancient yew, and the old shell falls away (see Chapter 19), the tree is 'sloughing its old skin'.[5]

In myth and iconography, the serpent, sometimes as dragon, has appeared closely associated with the Tree of Life in the early Neolithic (in the Near East about the sixth millennium BCE).[6] In some traditions the snake or dragon gnaws at the roots of the World Tree (e.g. Yggdrasil, see Chapter 41) – an activity that seemingly threatens the tree but ultimately only confirms its powers of regeneration and indestructibility.

As guardian of the World Tree the serpent is also guardian of the entire world, protecting it from the forces of chaos. In Norse myth, for example, the world-serpent, Midgardsormr, encircles 'Middle Earth' (the plane where humans live) and completes the protective circle by biting its own tail.[7] The symbol of the snake biting its own tail, also known as *oroubos*, reccurs in mystery cults throughout history. Like the geometrical circle that has no beginning and no end, it signifies eternity. The snakes and dragons of myth are the guardians of the source of all life force, and of all the paths to eternity. They guard every 'centre' where the sacred is concentrated, and every symbol that embodies it.[8] The Greek god Hermes, who – like the yew – performs the function of guiding souls to the underworld,[9] has a staff with two serpents coiling around it (the *caduceus*).

For all these reasons serpents and dragons are a leitmotif in the history of Eurasian religion.[10] Snakes can frequently be found associated with ancient healing temples, for example those of Apollo and Asklepios (both in Epidauros, Greece), where sacred snakes were kept in an enclosure.[11] Serpent symbology also appears at many healing spring sanctuaries where curative Celtic goddesses were venerated.[12] The legendary founder and patron god of medicine, Asklepios, is usually depicted with a staff that has a serpent coiling around it (*one* serpent, while Hermes' staff has two) – an ancient symbol that is still used by medical organisations around the world. The deepest understanding of the serpent's

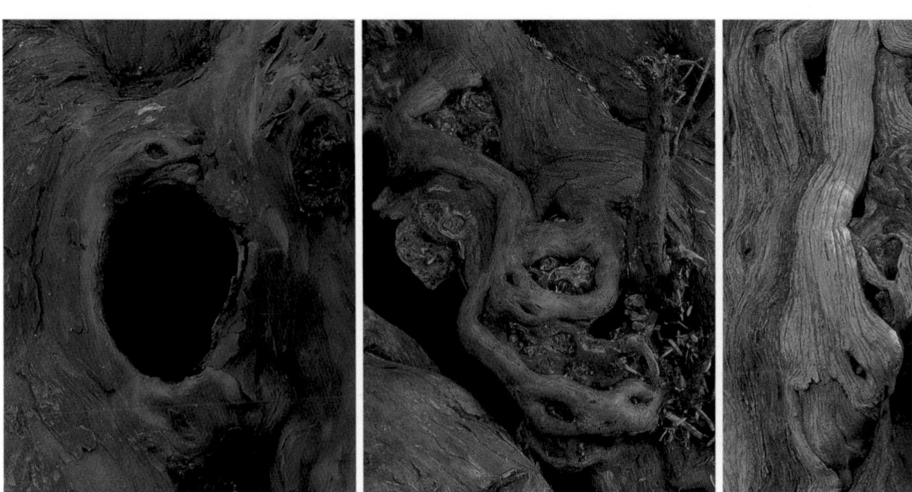

33.3–33.5 More snake impressions from yew bark. The left motif is reminiscent of the oroubos.

role in human health can perhaps be gained from the ancient Indian teachings of *kundalini yoga* (see box p. 196).

In the natural world, some snakes are attracted to yew wood for its seasonal abundance of rodents and rabbits.

THE EAGLE

For good reasons the eagle never ceased to be a dominant animal in heraldry, coats of arms and national symbolism. Like the lion, it conjures up feelings of strength, invincibility, mastership and power. The lion is the king of animals, the eagle is the king of the skies. For the ancients, however, one aspect was more important than power: the unlimited heights of the spirit. All birds have been seen as messengers of the heavens, and none flies as high as the eagle; hence it usually represented the predominant sky-god.

In the imagery of the Tree of Life, the eagle is usually perched at the tree's crown, symbolising the presence of the highest heavens. The serpent at the base of the tree represents the underworld, and hence the two animals form a dualistic couple. In some forms of the myth they fight each other, denoting the dance of opposites in nature that creates the rhythms of day (sky, eagle) and night (earth, snake[13]), summer and winter, and so on. To readers of mythology, this couple is probably best known from the description of Yggdrasil, the World Tree in Norse tradition. The serpent–eagle motif is global, however, and

33.6 *Eagle and serpent – fighting or dancing? Pottery shard from Sumer.*

33.7 *'Easter egg' or the fruit of the Tree of Life? Painted pottery motif, early second millennium* BCE, *Asine near Argos, Greece.*

occurred, for example, in Sumer about 4,000 years before the Norse tales were written down. In the Etana epic,[14] serpent and eagle swear an oath of friendship in the renewed world after the Flood; they hunt together and find food for each other and each other's offspring.

In the natural world, eagles and yews share high altitudes. Currently, one of the very few eagle's eyries in Britain is situated on a rocky precipice in the shelter of a mountain yew; the location, however, cannot be disclosed because of the threat of egg robbers.

A curious legend relating the eagle to the yew tree as the Tree of Life comes from Celtic Ireland. In *The Voyage of Maeldún,* a number of warrior heroes are taken to thirty-four mystical islands. The Island of the Eagle is forested with oaks and yews. A gigantic but exhausted eagle appears and begins to eat the red fruits from a branch that he had brought with him. Then he dives into a magical lake and emerges renewed. He flies off again with shining feathers and vigorous strength.[15] This is obviously a classical renewal myth in which the heavenly force (eagle, sky-god, sun-god, male energy) has to descend periodically to receive the life-giving strength from the elements of water (the lake) and earth (the fruits). While the water is cleansing and rejuvenating, it is the arils that bestow immortality, as in the permanent ability to regenerate. They are *soma*, the nectar of immortality (see p. 219). In this context, a motif from Greek painted pottery[16] showing an eagle-like black bird with a red 'egg' inside might be noteworthy. Is the immediate assumption always correct that every oval shape inside a symbolic bird or beast image has to be an egg? Or did this bird, too, eat an aril from the Tree of Life? Or, if it is an egg, did the idea of painting eggs red derive from a reflection on the red aril? Either way, fruit and egg represent the same principle, renewal and regeneration, only their origin (plant or animal) differs.

THE DOVE

No bird is so intimately connected with ancient female divinity as the dove.[17] In his study of the Tree of Life in relation to ascetic and trance techniques, E.A.S. Butterworth remarks on the dove symbology in the Cretan mountain cults (see Chapter 37):[18] 'The dove was originally the spirit of the tree, and this spirit could enter into the appointed priestess so that her spirit and that of the tree became one and the same.' Mircea Eliade agrees that 'birds can represent the soul as well as the epiphany of the goddess'.[19] Indeed, an impressive number of female divinities are associated with the dove, for example, Athena, Hera, Aphrodite and Eurynome in Greece, and Holla and Freyja in Germanic tradition.[20] In Celtic (Romano-Gaulish) goddess cults, doves appear as symbols of peace and healing. Their voices (like that of ravens in a different context) are linked with prophecy, and furthermore they are associated with Venus,[21] the Roman goddess of love. This might be linked to the behaviour of 'love doves', but certainly has historical roots in the connection of the bird with Greek Aphrodite on whom the Romans modelled their Venus.[22]

***33.8** Dove and sacred branches on a pre-Persian sacred vessel, Iran.*

Today, the white dove is an international emblem of peace, a tradition that goes back to the biblical tale of Noah and the Ark (see p. 201). In Christianity the dove also represents the Holy Spirit. The Christian trinity (Father, Son and Holy Ghost), however, derives from the ancient mythological mother-father-child pattern, and is a result of patriarchalisation: the female side of God (the Mother) was neutralised (to Ghost), but retained one of its symbols, the dove. 'The dove seems indeed to have become a symbol of divine spirit, as it is in the New Testament,' says Butterworth,[23] 'but instead of representing the female spirit of the tree, it has become the spirit of the Son, who hangs upon it' (he refers to a tradition that sees the Cross of Christ as a new form or developmental phase of the Tree of Life).

In the natural world, wood pigeons *(Columba palumbus)* occasionally nest in yews but unlike many other birds have no special interest in the arils. Rock pigeons[24] *(Columba livia)*, mountain yews and humans have been close neighbours as early as the Palaeolithic period (see p. 176). At times rock pigeons make their home in a crack in the rock close to a yew or in a cavity of the tree itself.

BULL AND COW

In some traditions, the bull is the emblem of the fertilising weather or sky god in its full (testosterone) splendour. In the Sumerian 'sacred marriage' of earth and sky, it embodies all the power of the skies in the shape of the mythical Heavenly Bull. Various interpreters of Neolithic symbology suggest the bull as an image of the horned moon who, by sending rain that fertilises the fields, was conceived of as male and in its rhythm of waxing and waning might have given rise to the myth of the dying and resurrected vegetation-god,[25] such as Dumuzi, Tammuz or Osiris. In later times, the male principle was identified rather with the sun,[26] and the moon 'became' female.

The peaceful, patient cow usually symbolises motherhood. The similarity of cow's milk and mother's milk undoubtedly made its impression on religious iconography; many ancient goddesses were identified with the cow, for example, Egyptian

Hathor, Greek Hera, Celtic Damona ('divine cow').[27] Both bull and cow may represent wealth and fertility. A pair of (bull- or cow-shaped) horns is an important religious icon found in Minoan Crete, and thousands of years before that in ancient Anatolia.[28]

GOAT AND RAM

Domestic goats and sheep originate from the mountain varieties that would have been encountered by humans in close vicinity to their Palaeolithic cave shelters. Horns in general were believed to act as channels for heavenly forces, which might explain why both goat and ram were sacred to the winged messenger of the gods, Mercury (the Roman equivalent to Hermes).[29] The goat was sacred to mountain goddesses like Aphrodite,[30] and later to hunting divinities like Dali in the Caucasus, who cared especially for the horned and hoofed animals.[31] The goat stimulates the (re)awakening of nature and protects young life, in some icons even the Tree of Life itself. In Greek myth, one of the horns of Amaltheia, the divine goat who had suckled Cretan Zeus when he was a child, became the cornucopia, the horn of plenty.[32] Like the bull, the billy goat and the ram represent the vigour and vitality of the male force, which often is personified as the vegetation-god (e.g. goat-legged Pan in Greece).

The ram represents spring and fertility. It became an essential symbol of various powerful sky-gods (Zeus, Ammon, Yahweh), or, in north-eastern Gaul, of the divine lord of animals, nature and plenty, the horned Cernunnos.[33] Its spiral horns, however, have always linked it to the sphere of the serpent, the moon, and of goddesses (a popular symbol of the ancient world was the ram-horned snake,[34] which combines the celestial power of the ram with the chtonic, subterranean forces of the serpent). The ram has a special association with early kingship rituals (see Chapter 39), and, because of his wool, with the spinning and weaving of the web of life (Chapter 41).

THE DOG

Dogs were associated with self-healing. There were sacred dogs at the great healing sanctuary of Asklepios at Epidaurus, Greece, and images of dogs occur at the healing shrine of the Celtic god Nodens at Lydney in Britain.[35] The principal symbols that accompany Celtic mother-goddesses in Gaul are serpent, dog, fruit basket and a baby;[36] this relates particularly to small lapdogs.[37]

Dogs, in all their shapes, can be both protective and destructive, guardians of the house and the herds, defenders of women and children, and at the same time fierce attackers on the hunt. Furthermore, the dog is a guardian of boundaries, not only physical ones but also of the otherworld (a common term for the Celtic metaphysical realm) or the underworld, the realm of the dead. This latter function is epitomised in Greek myth by Cerberus, the ravening many-headed dog of Pluto that guards the portals of Hades.[38]

***33.9** Goats and branch on a miniature pottery plaque from an offering pit in Transylvania, c. 5200–5000 BCE.*

***33.10** Motifs of needle tree, goat and an unidentified symbol on a four-sided Minoan seal (black stearite). Tholos B, Platanos, Crete, c. 2000 BCE.*

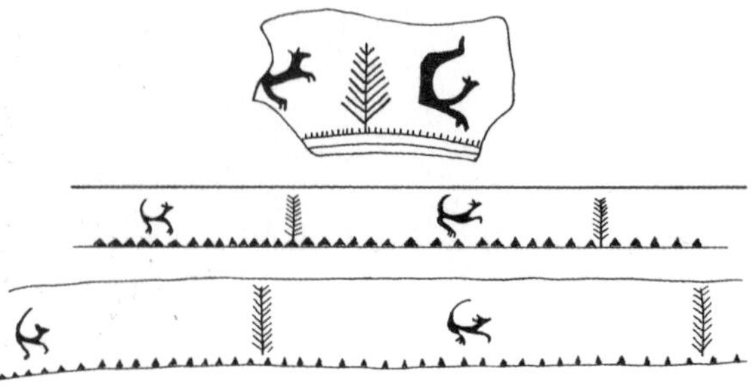

***33.11** Leaping dogs guard yew-like trees; painted pottery from Sipenitsi, western Ukraine, 3900–3700 BCE.*

33.12 The fox kami *in Shinto tradition is a white beast with red ears (just like the dogs of the underworld in Welsh tradition); he guards the yew shrine at Seijo, Japan.*

THE BOAT

A widespread symbol in burial traditions – from the Egyptian *Book of the Dead* to the Viking ship burials – is the boat. In this world, it carries its passengers over water, and it can also be conceived as ferrying the soul of the deceased across the mythical river that separates the afterworld from this one. Even more so in this last journey, the boat needs to guarantee a safe passage, and hence the different boats of all cultures used to bear symbols of power and protection. The miniature votive bronze boats from Bronze Age Sardinia, for example, have a bull's head at the prow, and the mast (central pillar) is crowned by the dove.[39] Dragon-head prows empowered the Viking voyagers, in this world and the next. In the cosmology of the Elysian Way described below – the Pythagoreans spoke of a cosmic 'ship'[40] – the boat not only delivers the soul to the realm of the dead, but, when the time comes, also bears it back, via the Milky Way, to the womb of a mortal mother, or,

33.13 The sacred branch becomes a boat to ferry the dead: Stone Age drawings from Hjørring, Denmark; Skjeberg, Norway; and two from Kalleby, Sweden.

33.14 Two of the figures of the Roos Carr ensemble, yew wood, 600–500 BCE, northern England.

for a brief interlude, to a beehive or the chrysalis of a butterfly.

In 1836, a group of figures carved from yew wood was discovered at Roos Carr, near Withernsea, in East Yorkshire. Five figures were recovered, together with a piece of wood resembling a boat with a prow in the form of a serpent's head. The figures stand upright because of holes in the 'boat'. The ensemble has been radiocarbon-dated to 600–500 BCE.

THE BEE

In the Old World, there was a widespread belief that the souls of the dead and the souls of the unborn took the form of bees. Various Greek philosophers, for example, connect the bee with the belief in the transmigration of souls; Porphyrios states that souls were thought to come down from the moon-goddess (Artemis, by his time) in the form of bees, and honey was emblematic to him of the sweet pleasure that draws souls down to be born.[41] And bees appear not only at the entry but also at the exit point of earthly existence: 'So that Sophocles was not far wrong in saying of the souls, "The swarm of the dead hums and rises upwards."'[42] The belief in 'bee-souls' and their origin in paradise is also found outside the Mediterranean, for instance in the British Isles.[43] What suggests their heavenly origin is that not only do bees produce this sweet and healthy substance called honey, but they do so without harm to any other life form in the process: carnivores kill to survive, and herbivores destroy plants, but bees live on pollen alone, or, in the words of the ancients, simply collect the honey-dew that descends every morning from the skies[44] (or from the World Tree, see p. 219).

Honey had been known to humankind since very early times,[45] and written evidence for its ritual use is as old as writing itself. When Gilgamesh, for example, prepares the (yew-wood?) altar (see p. 200) to begin the death rites for his friend Enkidu, the first thing he places on it is a cornelian bowl, which he fills with honey.[46] Theophrastus says that the first sacrifices of the ancients were made with water; afterwards, libations were made with honey, then with oil, and last with wine.[47] The fermentation of honey into an intoxicating drink, mead, is much older than the cultivation of the vine[48] and even older than beer. Mead played an extraordinary role in ancient ritual. In the Old World, libations to the dead were generally threefold and usually consisted of three of the following ingredients: water, milk, honey or mead, (olive) oil and wine, but honey (or mead) was always one of them.[49] The gods in Homer's Olympus drink mead, as do the warriors in Odin's hall, and certainly those in the Celtic paradise.[50]

Most rituals incorporating honey are connected with female divinities, with rites of life and death, the underworld and its guardian beasts, particularly the snake.[51] The complex of symbols here called the yew–tree–goddess pattern first appeared in early Neolithic art together with the beehive tomb.[52] Much later, in Greece, Persephone, the Queen of the Underworld, received libations of honey and was called *Melitodes*, 'honeyed', by the initiates into

33.15 *Ancient gem showing a bee bearing a caduceus, the symbolic staff of Hermes, the shepherd of departing souls; red cornelian, Scardona, Dalmatia (n. d.).*

the mysteries of Demeter (see Chapter 35).[53] The bee was also a symbol of the Ephesian Artemis; her priestesses were called *Melissae*, 'bees', as were the priestesses of Rhea and Demeter.[54] The oracular priestess at Delphi, the Pythia, was also known as the Delphic Bee.[55] Furthermore, some of Artemis' priestly officials at Ephesus were called *Essenes*, 'king bees'.[56] The bee, together with the lion, appears as an emblem of the Phrygian mother-goddess Kybele, the predecessor of Artemis at Ephesus.[57] Definite upperworld deities to whom honey was offered[58] include the Muses, Mnemosyne (goddess of memory),[59] Helios (sun-god), Selene (moon-goddess, later Artemis), Eos ('Dawn', Helios' daughter) and Aphrodite *Urania* (Queen of the Mountain). The Fates received offerings of flower wreaths and honey mixed with water.[60] The Muses (see Chapter 37) were said to bestow the gift of eloquence and poetry, and even of prophecy as well as the healing arts, by sending bees to the lips of a human infant or youth.[61] Hence Rome's greatest scholar, Varro (116 BCE–27 BCE), called bees the 'Birds of the Muses'.[62]

In the natural world, wild bees often build their homes in trees, and hollow ones are ideal. How close the connection of yews with bees might once have been can be illustrated with the *Bieneneiben*, 'bee yews', in Bavaria, Germany. They have a rectangular hole of *c.* 40 x 15cm sawn into the hollow trunk at a height of 3–4m, and a second, much smaller hole (*c.* 5cm) close to the ground. A beekeeper or honey-collector would have used the lower hole to smoke out the bees and once they have left, access the honeycombs through the bigger opening at the top. In the famous yew wood at Paterzell, two such trees are still standing; a third one, now gone, was photographed in 1910 and still had the rectangular 'door' for the top hole, used to close the opening and invite the bees back for the next season.[63] In the foundation legend of the church of St Baglan in Glamorgan, Wales, the saint of that name is instructed by his protégé, St Illtyd, to build the church where he finds a female yew tree, and the tree eventually chosen has 'a hive of bees in the body'.[64]

In the first century BCE, however, Virgil warns beekeepers to keep their hives away from yew trees;

33.16–33.17 *(left and centre) Until the early twentieth century, rectangular openings in hollow yews assisted the removal of the honeycombs: so-called 'bee yews' in Paterzell, Bavaria.*
33.18–33.19 *Entrances to wild bees' nests in the sacred yew at Seijo, Japan, and in an English country garden.*

Virgil seems convinced of the toxicity of the tree (see Chapter 12).[65] It is noteworthy that his warning also indicates that yew trees in his day must have been quite widespread in the Roman world (far more than today) in order to potentially spoil enough honey to justify a public warning of this kind. But Virgil must also have known of the ancient sacred status of bees and of yews, and in the last analysis we cannot know whether his true aim was to protect the honey (which is unrealistic anyway, as yew pollen is not poisonous), or the bees and trees. But with both the bees and the yews being so closely associated with rites of passage, his remark would have caused only a little surprise to his contemporary readers.

Both bees and yews, then, are to be found at the thresholds to the infinite. And the yews standing at those gates to the netherworld are at times buzzing with the sound of bees (or souls). Virgil's warning *not* to use their honey might have meant to leave them be(e), and it is by all means fair to assume that it represents a distant echo not of a bee-keeper but rather of a religious taboo: yew honey belonged to Persephone – a perfect offering to the queen of Elysium.

33.20 *'Butterfly souls' on golden disks from a shaft-grave at Mycenae.*

BUTTERFLIES AND CICADAS

Other ancient religious fragments refer to butterflies as temporary homes for souls.[66] Butterflies do not produce honey like bees, but the strength of their symbolism is the transformation from chrysalis to a beautiful winged adult, hence they represent regeneration and immortality. Similar to the butterfly, the cicada with its pupa is a symbol of the soul's potential and its journey through form and time. The cicada also appears in association with the moon symbolism of the interchange of darkness and light (new moon and full moon), signifying that after every dark period (the pupa stage) follows a light one (the adult insect).[67] In ancient Greece, cicadas were popularly believed to subsist on dew, and hence shared some of the purity and innocence of the honeybee who does not harm other life forms.

Particularly in the ancient Aegean, butterfly symbolism blended seamlessly with that of the cicada. The promise of both – the butterfly chrysalis and the cicada pupa, images of which were worn as pendants and talismanic gems[68] – goes beyond the notion that the physical body will disintegrate for the sake of other biological life forms; it implies that the individual soul – once the conscious mind has overcome its fear of transformation – will be transformed to the core and finally enter an intensified and renewed individual existence.[69] After all, *psyche*, our modern word for the 'soul', comes from the ancient Greek word for butterfly.

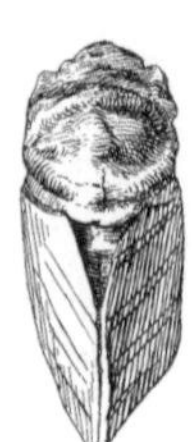

33.21 *Butterfly/moth on ceremonial bronze double axe (labrys), Phaistos, Crete.*
33.22 *Gold pendant of chrysalis from a chamber-tomb at Mycenae.*
33.23 *Gold brooch of cicada from the Artemis temple at Ephesus.*

34.1 *Sunset on the internal trunk of the ancient yew at Linton, Herefordshire.*

CHAPTER 34

BIRTH

The in-depth study of the yew's connection with the burial customs and underworld deities of antiquity reveals an equally strong link with the themes of birth and rebirth. *Taxus* as a tree displays all the characteristics of the Tree of Life symbology, and particularly emphasises the aspect of (re)birth: when an interior root (see p. 79) matures inside the opening old trunk, the yew tree perfectly illustrates the principle of birthing out of oneself, literally a 'virgin birth', and is reminiscent of a child inside the womb. Hence, from early times, one finds those divinities that are essentially associated with the theme of birth also connected with the yew tree – for example, the Greek goddesses Artemis and Hekate.

ARTEMIS

Artemis is a goddess of wild animals, hunting and vegetation, but also of childbirth. As the divinity of wild nature, who dances (usually accompanied by nymphs) in the mountains and forests, she was the most popular goddess in rural Greece. Her worship in Crete and on the Greek mainland is clearly pre-Hellenistic, and many of the local cults preserved traces of the even older nature deities with whom she merged when her worship was adopted from Anatolia (see Chapter 37). Theophrastus[1] reported around 300 BCE that her main sanctuary in the Peloponnese, the Artemision, was located in a yew grove, a hilly region in Arcadia.[2] As goddess of the tree cult, Artemis was commonly worshipped by maidens dancing to represent tree nymphs. Another yew grove was consecrated to Artemis by the inhabitants of the Greek colony of Massalia (modern Marseilles, southern France, see Ste Baume, p. 252).

Artemis was best known as Mistress of the Animals, and was accompanied by the stag or the hunting dog.[3] She was patroness of the fair hunt but avenged unjust killing. Her principal weapon was the bow, and while it is not known whether it was of yew wood she certainly (according to Homer) used

***34.2–34.4** Three females with big hips: Neolithic figurines from southern (left) and northern (right) Sardinia, ancient yew bole at Lande Patry, Orne* département*, France.*

yew poison for her punishing strikes on humans.[4] And as she 'helped' with death, so she did with life: she was one of the principal patronesses of childbirth. This aspect was particularly emphasised in Crete where she was known as Artemis *Eileithyia*, 'the birthing one'. (Artemis' Roman counterpart, Diana, was also a patroness of birth and bore the epithet 'opener of the womb'.[5]) Artemis presided over the birth water as much as over the life-giving water element in general; throughout the Peloponnese,[6] Artemis was known as Lady of the Lake *(Limnaea, Limnatis),* and as such supervised waters and lush wild growth. In myth she is not married, a 'virgin' goddess of untouched nature, and the wild cult dances in her honour reflected the wildness of the virgin forests.

***34.5** Artemis with sacred branch, bow and deer; tetradrachms from northern Greece, early fourth century BCE.*

***34.6** The tree at Gwenlais, South Wales, is one of the four female ancient yews situated at a spring.*

***34.7** A place rich in (female) legend: Pennant Melangell, Powys, Wales, with its four ancient yew trees.*

34.8 Flowing movements in (secondary) yew wood are reminiscent of the liquid element.

HEKATE

The difference between Artemis and Hekate is that Artemis is a goddess of life and death and Hekate is a goddess of death and life. The polarisation, however, into clear-cut images of the hunting, dancing, bathing, young and generally good-looking Artemis on one hand, and the dark, frightening, dangerous crone Hekate as mistress of night, death and the underworld on the other, was mainly a product of the rationalising mind of the post-Homerian poets. Hekate used to be 'a popular goddess of many roles, with a share in the earth, sea and the starry heavens, who could give luck in hunting, victory in battle, and help with crops'.[7] Often she was depicted as being surrounded or accompanied by dogs (representing healing and protection), and she could even bay like a dog, appear with a dog's head, or in the shape of a dog. In Hellenistic Greece, her function distinctly shifted to that of the death goddess;[8] as Queen of the Underworld, Hekate was sometimes called the mistress of Cerberus, the monstrous dog that guards the entrance to that realm.[9]

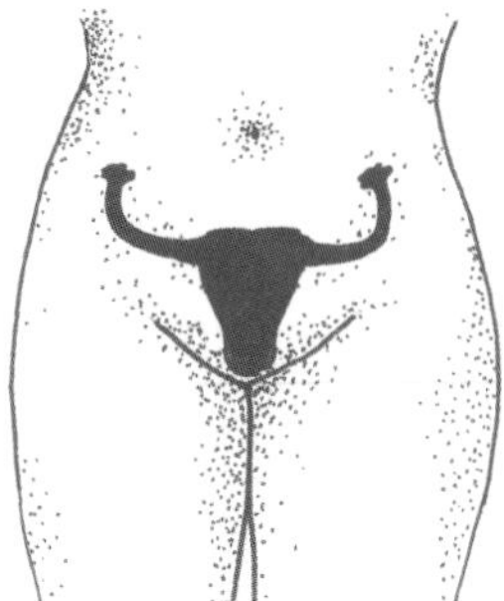

34.9 The shape of the female reproductive organs resembles a bull's head (compare figure 42.6).

In the days under Roman rule, black bulls sacrificed to Hekate were adorned with yew wreaths around their necks.[10] With the bull one of the most powerful emblems of the restoration of life in the Old World, this ritual seems to signify more than the mere appeasement of the fearsome deity of the underworld. The bull in conjunction with a goddess is an ancient life-confirming pattern that may date back to the Neolithic period, and rather than to death it points to renewal and rebirth, even more so when we consider that the shape of the bull's (or cow's) head – the religious icon in question – can be likened to the shape of the female reproductive organs (compare figures 34.9 to 34.11).[11] The ceremonial adornment of the horned head of the beast, then, would have been a substitute for adorning the underbelly of women. This persisting association of Hekate with birth was not at all alien to her cult: birth was her original domain, she had come from Egypt where she had been, under the name of Hekabe, Heket or Heqit, the goddess of birth and midwifery.[12]

REBIRTH

The intimate link between death and birth as expressed in the traditions of Artemis and Hekate, points to an ancient belief in the rebirth of the soul.

The early evolution of the functions of Hekate shows that death was once conceived as belonging to the domain of the divinities of birth: the mistress of the underworld (the realm of the dead) is the mistress of the womb who transforms and prepares the souls for the next step; thus, death is a prerequisite for birth. For the druids of ancient Gaul, for example, death was 'but the mid-point of long life',[13] a theme also reflected in the abundant Celtic myths of the cauldron of rebirth.[14] No less than Socrates (*c.* 470 BCE–399 BCE) – one of the three Greek philosophers[15] who are credited with laying the philosophical foundations of Western culture – lifts the veil (in a work by Plato):[16] 'Now this is the story: when a man has breathed his last, the spirit to whom each was allotted in life proceeds to conduct him to a certain place, and all they that are there gathered must abide their judgement, and thereafter journey to Hades in company with that guide ... There that befalls which must befall; and having there abided for the due span of time they are brought back hither by another guide; and so they continue for many circuits of time.'

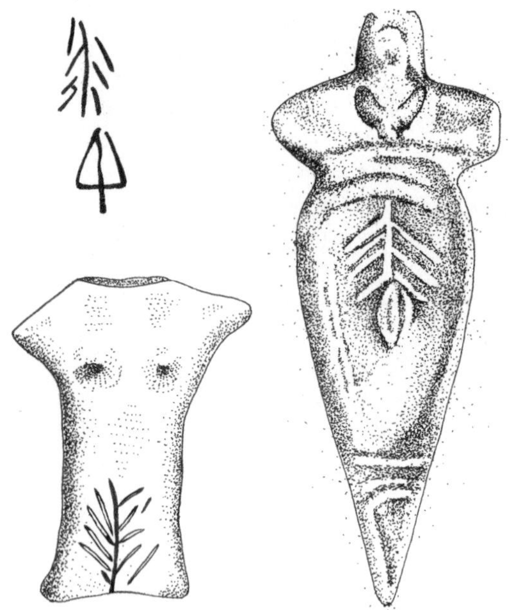

34.11 *Needle-bearing branch and vulva: cave painting in La Mouthe, France, late Palaeolithic (top left); terracotta figurine from Jela, Serbia,* c. *5200 BCE (left); bone figurine from Neolithic Italy (right).*

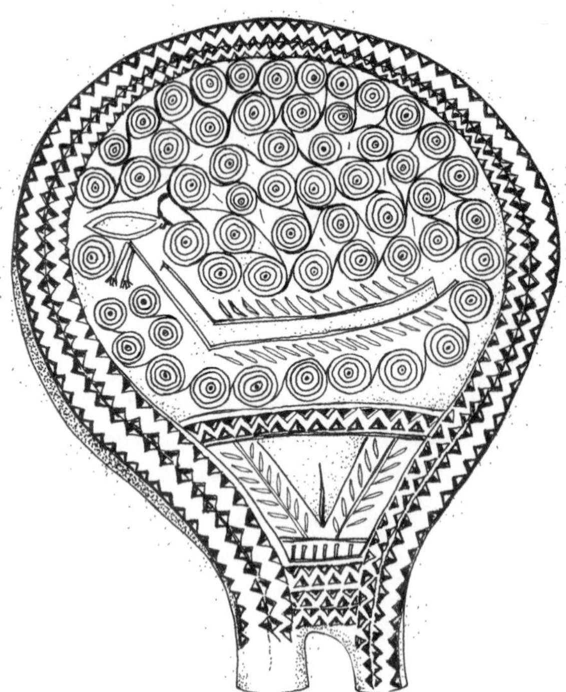

34.10 *Pubic triangle adorned with needle-bearing twigs, the birth water in the womb surrounds a ship with a bird at the prow. Painted terracotta object, Chalandriani (Syros), Cyclades, Greece, mid-third millennium BCE.*

The common conception of the afterlife in the Greco-Roman world was that the majority of people descended to the shadowy land of ghosts (Hades), and a few of the worthy (such as the heroes) might ascend to the bright and pleasant Elysian Fields (such inherent moral judgement, however, is a relatively late 'symptom' in the history of religion; it seems that once, eternal happiness could have been the birthright of everyone). The term *elysion* for the abode of the dead probably derived from *elysíe*, 'way'. In Pindar's *Olympian Odes*, for example, the just soul goes 'by God's road' to the Island of the Blessed,[17] and Ovid speaks of the soul path as a road leading to the palace of Jupiter: 'There is a road aloft in the clear heaven, milk-white and therefore named the Milky Way ...'[18] According to the Neoplatonist philosopher Porphyrios (*c.* 234–*c.* 305 CE), Pythagoras asserted that the souls 'are gathered together in the Milky Way, so called from those that are nurtured on milk, when they fall into birth'.[19]

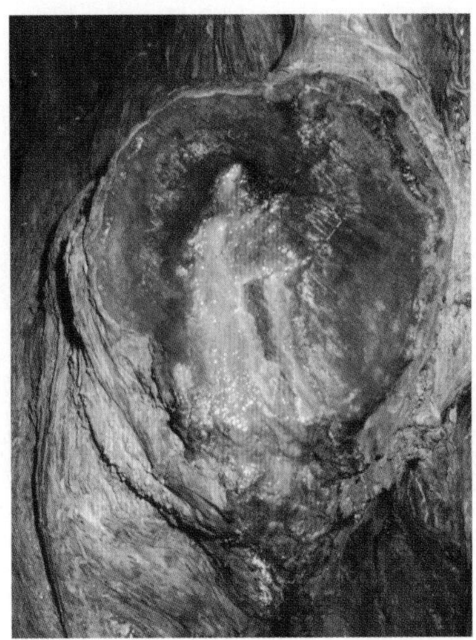

34.12 *'Bleeding' yews (scientifically unexplained as yet) might have reinforced the association with birth and menstruation in ancient times.*
Bleeding yew at Nevern, West Wales.

And Plato tells how immortal souls follow the gods round the great arch of Heaven and from its summit behold sights of unspeakable splendour: 'they are borne round by its revolution, and gaze at the eternal scene. Now of that region beyond the sky no earthly bard has ever yet sung or ever will sing in worthy strains'.[20] Some 750 years after Plato, Macrobius still speaks of the 'gates of the sun' (Cancer and Capricorn as the stations of the zodiac where it intersects with the galaxy) as the entry and exit points for the souls to the Milky Way: 'Cancer is the gate of men, because through it they descend to the lower regions; Capricornus, the gate of the gods, because through it souls return to the seat of their own proper immortality'. And he also refers to Pythagoras about milk as the 'first food offered to the new-born, because their first movement downwards in the direction of earthly bodies begins at the Milky Way'.[21]

34.13 *Falling into birth via the Milky Way: the heavenly gate as* circulus lacteus *('milk circle'), held by a half-draped female who bears aloft another female representing a departing soul; twelfth-century Latin prose work on astronomy.*

The notion of the galaxy as the Elysian Way lasted well into the Christian age. Paulinus (353–431 CE), for example, Bishop of Nola, makes Enoch, Elijah, and other pious souls ascend to heaven via the galaxy (Paulin. *Nolan. carm.* 5. 37 ff, in Cook 1925, p. 43), and the theme also occurs in Dracontius of Carthage at the end of the fifth century (*Drac. Romul.* 5. 323 ff, in Cook, *ibid.*). The heavenly gate persisted for quite some time. The illustration is from the *Vienna manuscript cod. Vindob.* 2352 (in Cook, *ibid.*).

In Elysium, the tree of our enquiry appears yet again, in a different guise. As Pindar continues: 'Where the Airs, daughters of Ocean, Blow round the Island of the Blest, And the flowers are of gold, Some on land flaming from bright trees, Others the water feeds; They bind their hands with them and make garlands, In the straight rule of Rhadamanthys [the personification of justice], Whom the great Father keeps at his side in counsel, The husband of Rhea on her all-brightest throne.'[22] The flowers of gold, of course, could denote any plant with yellowish blossom, or gold could symbolically stand for most refined, paradisiacal or everlasting. But the explanation of these flowers as opening on flaming trees[23] suggests one botanical identification in particular: *Taxus baccata* (see figure 9.1). This is further strengthened by Pindar bringing it all back home to the goddess archetype, here in the guise of Cretan Rhea (for Rhea's close connection with yew see p. 184), whose throne outshines even that of Zeus.[24]

The presence of the tree (as a religious pattern) on earth as well as in the metaphysical realms of existence – flaming and bright in Elysium; white and dead in the central plain of Hades (see *leuke*, p. 202); and dark and sombre along the paths to the respective gates – creates a truly grand image of the World Tree that is worthy of its name. And could it be that Elysium and Hades are but one place anyway,[25] and – as in our life on earth – it is our perception of it that makes all the difference?

35.1 Moonlight over a yew woodland in Wiltshire.

CHAPTER 35

THE MYSTERIES

In the second century CE, Pausanias in his *Guide to Greece* talks about a legendary tin scroll that was 'inscribed with the mystery of the Great goddess' and buried in the mountains where a yew and a myrtle tree grow.

MESSENE

Pausanias elaborates upon the origins of Messene,[1] an ancient Greek city in the south-western Peloponnese. The Theban statesman and leader Epaminondas (c. 410 BCE–362 BCE) was 'in difficulties about founding a city that would be a match for the fighting strength of Lakonia, and unable to discover whereabouts in the territory he ought to build it'.[2] Pausanias relates the local legend that when Epaminondas was in despair, 'an old man in the likeness of a hierophant came to him' in a dream, and that this same figure also made a 'revelation' to Epiteles, the chosen commander to rebuild the city. Pausanias continues:

> The dream commanded him to find where a yew and a myrtle grow together on Mount Ithome, and to dig between them and rescue the old woman, because she was tired out and fainting away, shut in her brazen [bronze] chamber. Next morning Epiteles came to the place that was described, and when he dug there he found a bronze jar, which he took to Epaminondas at once, explained his vision and told him to take off the lid and see what was inside. Epaminondas offered sacrifice and prayed to the vision he had dreamed and then he opened the jar. Inside he found a leaf of tin beaten to extreme fineness and rolled up like a scroll, and inscribed with the mystery of the Great goddess.[3]

Modern archaeology has confirmed that the summit of Mount Ithómi, 798m in altitude, indeed served as the local acropolis, with the foundations of a Temple of Artemis *Laphria,* sitting on its shoulder.[4] The passage from Pausanias confirms the religious connection of yew and myrtle with ancient goddess worship.[5] Another revealing part of this legend is what the finders actually do with the newly acquired information. First, the scroll was given to priests who wrote the mysteries into books; next an enquiry by the prophets was ordered, reading omens 'whether the divine powers would wish to come'[6] to the chosen site; then, prayers and sacrifices were offered to Dionysos, Apollo, Hera and Zeus, and to the 'Great goddesses' (Demeter and Kore/Persephone).

Finally, Epaminondas and the various groups involved with the ceremonies 'called out together to the divine heroes to return and live with them'.[7] The 'mysteries of the goddess' obviously provided the correct rituals to call upon the spirits of the dead and invite the (deceased) greatest sons and daughters of the people to come to bless their culture once again. It was this that made all the difference for Epaminondas and his co-founders: now they could trust that the souls of these famous personalities could be reached and convinced to return, so that their strength would benefit the future city.[8]

35.2 Coins showing Demeter on her serpent-drawn chariot were widespread in the Aegean of the first century BCE.

The foundation legend of Messene shows that yew was at the time firmly associated with something ultimately higher in value than gold: the rites of passage and the secrets of immortality.

But what else do we know about the 'mystery of the Great goddess'?

ELEUSIS

For almost 2,000 years,[9] the Eleusinian Mysteries[10] were an essential part of the religious life of ancient Greece, and can rightfully be regarded as one of the most remarkable religious phenomena in the history of humankind. The Lesser Mysteries were held annually in early spring on the fringes of Athens, and included ceremonies of cleansing and purification, fasting, sacrificing, dancing and the singing of hymns. The Greater Mysteries, celebrated annually at Eleusis near Athens, lasted nine days and nights in the early autumn; they were more than a reprise of the spring festivities: for one they happened on a much grander scale; and, most significantly, they incorporated the initiation of applicants into the secret rites.

Before the time of the celebration arrived, special messengers were sent to the other Greek city-states, proclaiming a holy truce and asking for the official delegations to be sent to the goddess. Such delegations would also arrive from many of the islands, and at times from as far as Egypt, Syria and Antioch.[11] When the festival, on its fifth day, culminated in a huge procession from Athens to the temple precinct of Eleusis (a distance of 22km, or 14 miles), these delegations would join the Athens city officials, the priesthood and priestesses of Demeter, the initiates and their mystagogues, and the many thousands of people from Athens and elsewhere. After a day spent walking the prescribed route, the ecstatic peak of the ceremonies was the night-long singing and dancing in honour of the goddess. The following day the crowds would depart and only the aspirants for initiation would stay, awaiting their preparation for initiation into the mysteries the following night. The initiates were sworn to absolute secrecy, and, as George Mylonas in his excellent study of Eleusis remarks, it is 'amazing indeed that the basic and important substance of the secret rites was never disclosed, when these Mysteries were held at Eleusis for some two thousand years, when a multitude of people from all over the civilized world was initiated'.[12]

These rites were secret but not exclusive; everyone – men, women, children, even foreigners[13] or slaves – had the right to take part, provided they had no blood on their hands (had not committed murder), spoke Greek (necessary to understand the words spoken during the initiation), and had gone through the preliminary rites in spring. Furthermore, Eleusis exhibited an extraordinary combination of religious teaching and religious freedom: the Mysteries bestowed some secrets about death and the afterlife onto the initiates that, as all sources seem to agree, kept them happy for the rest of their lives. Yet they were not obliged to return to the sanctuary periodically (to worship and to spend more money), or to follow a certain pattern of life or rules of conduct; nor did anyone ever found a band or club or a church of Eleusis initiates. The initiates were free to return to whatever life and religion they chose, enriched by the Eleusinian experience, which per-

The myth of Demeter and Kore

Kore, the 'maiden', plays in a meadow picking flowers when the earth gapes and Hades, the King of the Underworld, appears in his golden chariot and carries her down into the abyss. Demeter, her bereaved mother and, as it happens, goddess of the fertile soil, roams the earth for nine days, holding a torch in each hand, looking for her daughter. Eventually she meets Hekate, and the two approach the sun-god who tells them the whereabouts of Kore and that she is to marry Hades. Demeter, in wrath and grief, quits the world of the gods and in the guise of an old woman arrives at Eleusis, mourning her daughter being lost to the 'land of no return'. Next, she meets the local queen and accepts the invitation to bring up the infant prince, Demophoön. She does not nurse him in the usual way but rubs him with divine ambrosia, and one night as she holds him over the fire 'to burn away his mortality'[14] – a process destined to change the human child into an immortal god – his mother interrupts the scene and thus breaks the spell. Demeter now reveals her full splendour and leaves the palace. She retires into the sanctuary newly built for her at Eleusis but, still grieving for her daughter, curses the earth to bear no fruit. As humans and gods begin to starve, Zeus intervenes and (with some aid by his and Hades' mother, the goddess Rhea) brings about the release of Kore. But because she has eaten seven seeds of a pomegranate offered her by cunning Hades, she now has to spend three months every year with Hades, and only the remaining nine with her mother.

haps helped them to become 'more pious, more just, and better in everything' as Diodorus phrases it.[15]

The story of Demeter and Kore is usually interpreted as an agricultural myth, that is, the divine creation of a food plant by the death and resurrection of a divinity. Kore is the seed that has to spend a part of the year in the earth to germinate, her symbolic 'death'. She has sacrificed her life in the heavens and goes to the dust, to the underworld. But she is reborn when the plants sprout in spring. By her cyclic death and rebirth she provides humankind with the sustenance of grain. This mirrors the ancient belief that our food is part of the body of a deity.[16] However, there is more to it: once she has descended into the underworld, Kore does not remain the innocent girl she once was. She is never described as a displaced victim of abduction, a tender damsel distressed by a cruel dark lord, but, rather, appears to be fulfilling her destiny. At first guarded (or tutored?) by Hekate herself, she adapts astonishingly quickly and soon changes into mighty Persephone, the mature mistress of the lower world, the worthy queen at the side of Hades. She has annulled – like the Sumerian goddess Inanna with her visit to the infernal regions some 2,000 years before her – the unbridgeable gap between the upper and the lower worlds. As the mediatrix between the two divine realms, Persephone can now intervene in the destiny of mortals.[17] As Persephone, she plays an integral part in classical Greek myth and cult. And the King of the Underworld, also called *Plouton*, 'wealthy', is no mere baddie either. On the contrary, he doubtless was the third chief deity at Eleusis, and was honoured at the Ploutonion, a sacred natural cave with its opening appearing immediately to the right where the Sacred Way (the road from Athens) enters the sacred precincts. Sometimes he and Persephone are simply referred to as *Theos* and *Thea*, God and Goddess.[18]

Generations of scholars have marvelled at the possible content of the secret rites of Demeter, and at what could possibly be hidden in the underworld journey of the wheat grain, the mythical abduction of Kore, and her happy reunion with her mother. Kore and Demeter are often interpreted as mere personifications of wheat, but at least Persephone cannot be reduced as easily.[19] Strangely, it seems to have escaped the discussion almost entirely[20] that the most remarkable aspect of the myth actually is the transformation of Kore, not only from girl to woman, but into the divinity that holds the key to immortality. The statement that the Eleusinian Mysteries consisted of rites taught by Demeter (as opposed to Persephone) does not contradict this argument, for there is the easily overlooked episode with the infant Demophoön. Demeter too was in charge of granting immortality;[21] she taught the art of living and dying. One thing is for certain: the initiates were generally known to have lost their fear of dying. This is reflected in a local grave inscription: 'Death is no evil but a blessing.'[22]

As far as ethnobotany is concerned, the main sacred plant at Eleusis seems to have been the myrtle tree *(Myrtus communis)*,[23] at least in the public cult. Those to be initiated carried the *bacchos* (a staff made up of branches of myrtle tied with strands of wool), and wore a myrtle wreath at the procession and during the first night at Eleusis (the dancing). For the initiation into the secret rites, however, their wreaths of myrtle were replaced by 'wreaths with ribbons'[24] – but from which plant?

The evergreen myrtle is a convenient symbol for the Demeter–Kore myth: at the time of the Lesser Mysteries (spring) the myrtle wreaths bore white blossoms – fair and pretty for the girl Kore – while the wreaths at the Greater Mysteries in autumn would

***35.3** Light at the end of the tunnel (hollow yew at Ste Baume, France).*

***35.4** Persephone with snake; silver coin from Sicily, fifth century BCE.*

have carried purple–black berries – fruits of the earth as mother. Furthermore, in herbal lore myrtle was regarded as preventing premature delivery by closing the uterus,[25] which relates it to midwifery, a domain of mother-goddesses. It is no surprise, though, that the myrtle wreath should have been replaced before entering the Hall of Initiation. For all we can tell, the secret rites would have required another plant, one that has strong associations with death, rebirth, and the soul's journey through eternity. It is possible, however, that the wreath or branch of the tree of the secret rites would not have been worn or held by the mystagogues at all, but simply shown to them in the revelation of the sacred objects, called the Hiera. This tree remained as secret as the mysteries. The myrtle cannot possibly have been the well-guarded secret of Eleusis because it was the well-known public emblem.

There is no direct evidence of yew at Eleusis, however, and any sacred trees at the sanctuary can be expected to have died in the fires when the Goths destroyed the sanctuary in 395 CE, or perished any time thereafter. Dead wood does not last as long as stone, and living trees are not considered at all in archaeological studies. Demeter and Persephone, however, appear to have inherited a number of symbols of the yew–tree–goddess pattern, in particular the serpent: both goddesses drive snake-drawn chariots, and Persephone is not only the bride of the chtonic serpent lord, Hades, but is able to take on a serpent shape herself.[26] Furthermore, Demeter is usually depicted holding two torches, otherwise a symbol of the underworld goddess Hekate.[27]

The proposition that yew secretly complemented myrtle at Eleusis is strengthened by the foundation legend of Messene (as related by Pausanias, see above), and the notion of the yew tree as the tree of the underworld as well as its connection with Persephone (even though this connection is not named in the surviving classical texts).

CHAPTER 36

ORIGINS

For more than a million years, early humans and their hominid ancestors used cave dwellings for shelter from the elements, protection from predators and scavengers, and for the availability of drinking water as well as of exposed stone materials (for toolmaking) that mountainous locations usually provide.[1] Only between 20,000 and 10,000 years ago (differing with location) did humans leave the caves and begin to *construct* dwellings along the rivers.[2] Of the areas with well-developed caves in the world, over 70 per cent are formed of limestone,[3] the very rock that provides the foundation for the prime stands of *Taxus* worldwide. Hence yew can be assumed to have been a continuous 'neighbour' of human dwellings from a very early stage.[4] The encounter of early human with yew would have taken place not long after the first human populations left Africa and began to colonise Eurasia.[5] This is further attested by the oldest wooden artefacts being made of yew wood. It is also here, in the higher altitudes, that early human expressions of religious awe have left their mark, for example the famous cave paintings in Spain and southern France that constitute a part of the oldest religious art on earth,[6] and, from somewhat later periods, a great number of stone settings such as altar stones, standing stones and stone circles. And again, yew can be regarded as having been close to early religious sites at all times: in the French Pyrenees, for example, all religious Mesolithic stone settings are to be found at elevations between 600 and 1,300m – exactly the altitudinal range of *Taxus baccata* in the region.

__36.1__ (above) Yew leaves. ***36.2*** *The branch motif in Palaeolithic cave paintings at Castillo, Puente Viesgo (a), Lascaux (b–e), Marsoulas (f) and Niaux (g–i).*

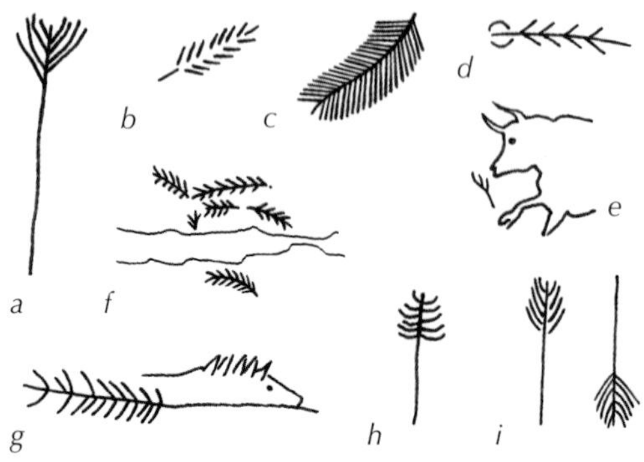

The religious pattern incorporating depictions of a (divine?) female figure as well as tree and serpent motifs seems to have originated in western Cilicia and the Taurus Mountains (modern southern Turkey and northern Syria). Overlapping with this region occurs the earliest use of painted ornament on pottery vessels, which also coincides with the first traces of settled communities in northern Mesopotamia. These early cultural stages – named after the sites of archaeological excavations Hassuna, (Hassuna-)Samarra, and Halaf – are generally thought of as successive stages but have considerable overlaps: radiocarbon-dating for an early Halafian level is as early as those for the Hassunan – about 5750 BCE. While the Halaf period is associated with the Syro-Turkish-Iraqi borderlands, the Hassuna period appears to have its focus in the Syro-Cilician corner of the eastern Mediterranean[7] (see map p. 214) – the very same region where *Taxus* in Asia Minor survived the last Ice Age and from where it spread again.

In the art of the Hassuna period (c. 5750 BCE–c. 5350 BCE), stylised and abstract forms are the rule. The only attempt at using a representational or naturalistic design is a so-called 'sprig pattern', which might represent a branch or tree. All other patterns

36.3 *Yew trees framing a cave at Humphrey Head, Cumbria.*

are exclusively geometric.[8] One possible interpretation of the relationship of these two patterns is that the geometric motifs – cross-hatching, herringbone pattern, (overlapping) lozenges and triangles – could have developed from the representation of the apparently coniferous branch, especially so if the Hassunan branch pattern were a symbolic reference to the Tree of Life, then a simple geometric pattern

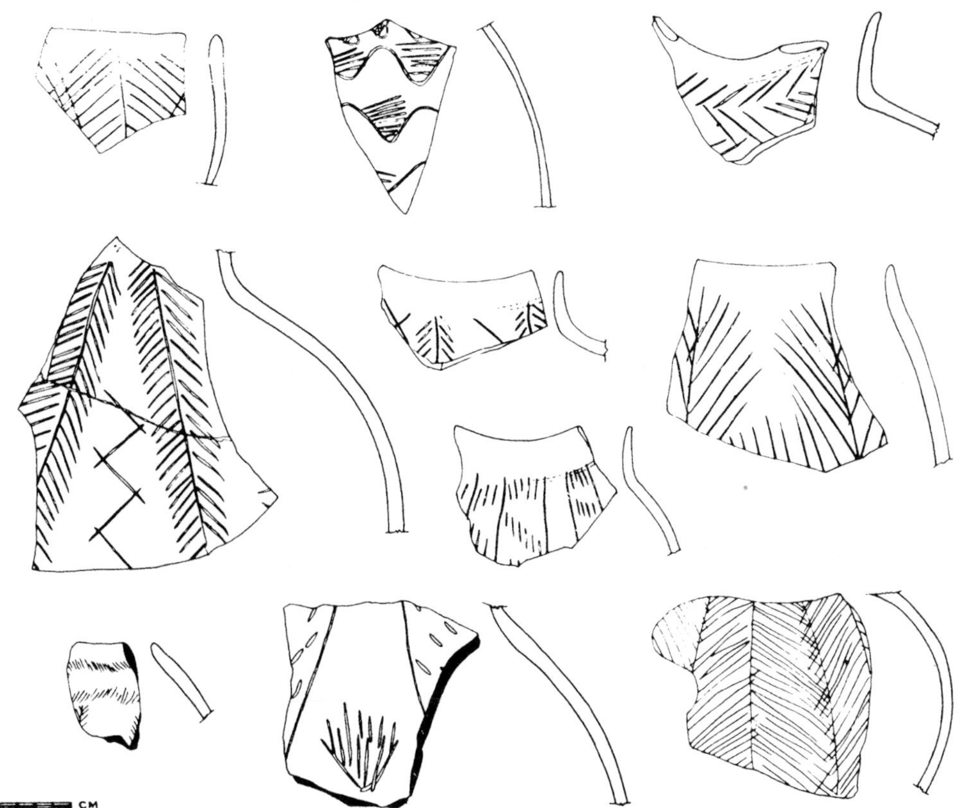

36.4 *The 'sprig pattern' on early Hassuna standard incised ware, sixth millennium BCE.*

like cross-hatching could have been a reference to the symbolism of weaving (see pp. 182–3, 226–8). Be this as it may, Hassunan art certainly shows the potential of coniferous branches for abstract art. Hassunan and Samarran pottery were used domestically but have also been found in graves, particularly those of infants.[9]

During the Halaf period, the importance of geometric designs was maintained and yet enriched by depictions of various animals (e.g. bull, serpent, the dove and other birds) in later cultures associated with goddess cults. A variety of female figurines as well as pendants and amulets of birds or double-axes also appear during this period.[10] Alongside the 'Neolithic revolution',[11] this complex of images (and possibly their symbolic meaning) spread southward onto the mudflats of riverine Mesopotamia where certain peasant villages began to grow into market towns and eventually, in about 5000 BCE, the settlement of the south of the Euphrates and Tigris valley

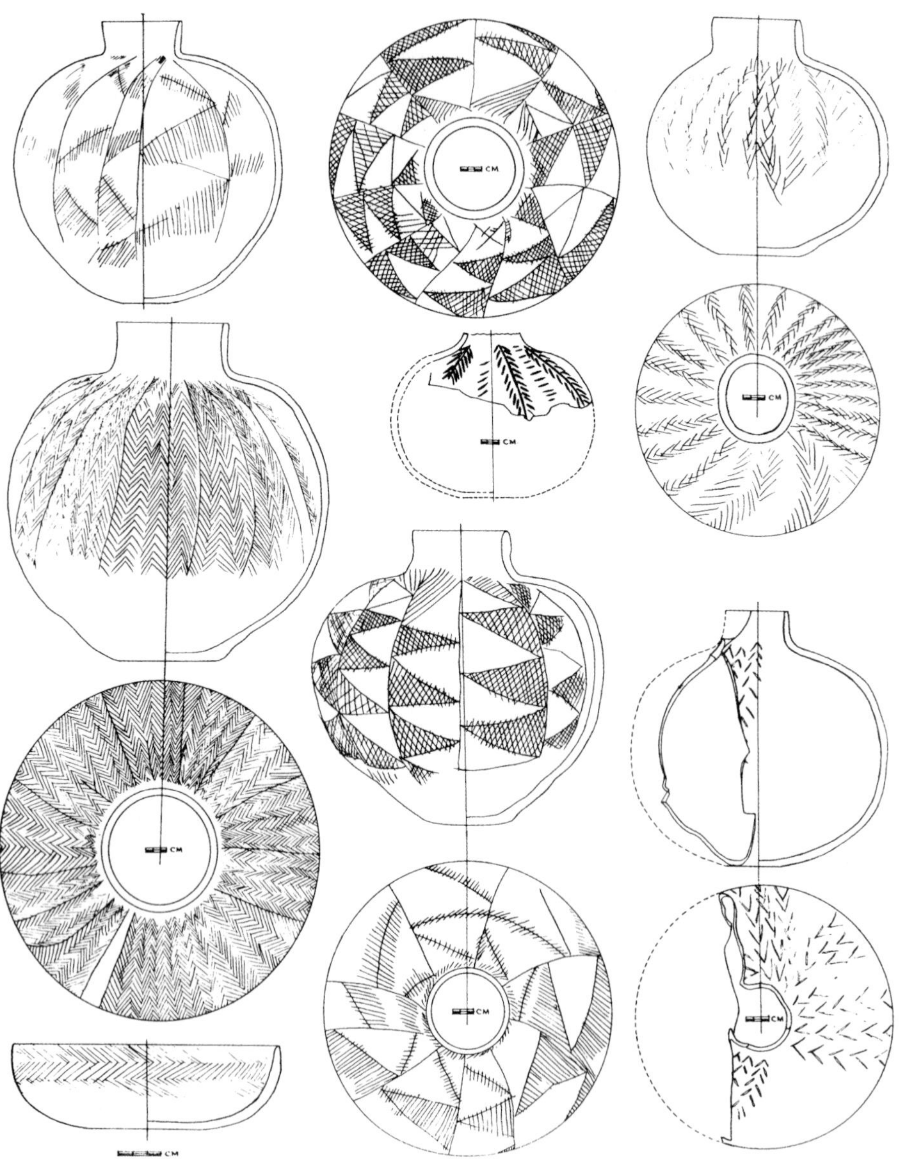

36.5 *The 'sprig pattern' on Hassuna pottery soon revealed its geometric potential: cross-hatching, herringbone patterns, overlapping lozenges and triangles.*

36.6 *The sacred branch: a) in Hittite seals; b) on pottery from Troy (compare figure 37.6); c) on a Punic votive stone to the goddess Tanit, Atlas Mountains, North Africa.*

took place at Eridu.[12] In Sumerian religion, the idea of the goddess and the dying and resurrected god lived on in the figures of Inanna and her spouse Dumuzi. Along the Nile it developed into the religion of Isis and Osiris. At least 1,000 years after its appearance in the southern Taurus region, the yew–tree–goddess pattern appears in Crete (according to ancient legends via Egypt as well as directly from western Anatolia). From Crete it spread further to Greece via the Mycenaean adaptation of Minoan culture, and was carried by sea through the Gates of Gibraltar and northward to the British Isles. On land, radiations from the Tigris–Euphrates region crossed the Caucasus (as well as the Black Sea), and via the Aegean penetrated the Balkans in the fourth millennium BCE.[13]

The earliest written documents from Anatolia stem from the Old Hittite period, *c.* 1750–1500 BCE.[14] The Hittites were an Indo-European group that entered Anatolia about 2000 BCE. They subjugated the native Hattians, and also assimilated the culture of the Hurrians, another non-Indo-European population located in Cilicia and the Taurus Mountains.[15] Their cultural and religious symbiosis created a pantheon in which divinities of Mesopotamia stood side by side with those of Hattian and Hurrian origin, with the Indo-European heritage even becoming the least significant. In fact, historians of religion[16] have remarked on the surprising religious continuity that existed in Anatolia, spanning thousands of years from the late Neolithic period to the introduction of Christianity. Presiding over the Hittite pantheon were two chief divinities, the storm-god and an equally powerful goddess. In Hittite times, both were known principally by their Hurrian names: the storm-god Teshub, and the goddess Hebat (known later, in Phrygia, as Kybele).[17] This Anatolian goddess was thought to have received a great part of her magical powers from her close relationship with underground (chtonian) forces – like a tree. And indeed trees played a significant role in the myths and rites of the Hittites,[18] especially the yew, which is mentioned in a great number of Hittite cuneiform tablets. The *eya* or *eyan* tree featured widely in their ceremonial and even in their legal texts: for example, when planted in a certain way beside a house, it served as a marker for tax-exemption. It was of particular significance in the kingship rituals (see Chapter 39) but also in those of the common people. Excerpts from the tablets referring to rituals read: 'the yew tree which had been placed apart on the altars, the priest of Telipinu sets [it] up';[19] 'on the altar above a yew tree they slaughter';[20] 'before them shall stand a yew tree, as token of their being free';[21] 'set up the yew tree for me and make me free!'[22] Yew wood was also used in ceremonial fires: '[he] kindles on the hearth; one [piece of] yew tree he holds with his right hand and one with his left hand'.[23]

An ancient custom, which can still be found today all over the Middle East[24] and also across Asia as well as Europe, consists of tying a piece of cloth or a ribbon to a sacred tree in order to accompany and amplify a prayer (see also figure 29.17). At the house of the Virgin Mary in Ephesus, Turkey, for example, Moslems and Christians alike pay tribute in this way. In Iraq, people tie a white ribbon to a tree for the soul of a child who has been killed in violence.[25] In the large rural districts of Turkey, too, a wide array of folk customs and shamanic traditions is still alive, and in different parts of Anatolia, conifers can be found as prayer trees adorned with ribbons. To make the prayer last, the most long-lived tree is chosen, a yew if available, otherwise another evergreen tree.[26]

37.1 *The shrouded peaks of Mt Olympos, seen from an old yew in the Enipéas (Mavrólongos) canyon, Greece.*
37.2 *Old yew at the old Dionysos Monastery, Enipéas canyon.*

CHAPTER 37

THE MOUNTAIN-MOTHERS

Since *Taxus* is a sacred tree that grows in the hills and mountains from the Iberian Peninsula across the Mediterranean to Asia Minor and the Caucasus (and across the entire width of the Himalaya range, South-east Asia, and to Japan), we need to look at the religious activities in those elevated sites, the associated deities, and the (human) qualities these deities represent. This chapter focuses on the female divinities whose worship originated in the 'high places' (the male ones are discussed in the next chapter). In some cases the connection with the yew tree seems wanting at first, but will become clearer in the following chapters.

In the Neolithic and Bronze Age Near East, although human activity focused increasingly on the fertile river plains, mountains retained their sanctity, and they would do so for millennia to come, right into the era of Christianity and Islam. As shown in Chapters 6 and 36, *Taxus baccata* in the Mediterranean was originally a widespread mountain tree, and thus would never have been far away from early religious sites at high altitudes (not to mention those cases where a tree constituted the sanctuary itself). And even where the cultural diffusion entered the lowlands, we find the divinities with yew associations maintained their connection to the mountains for a long time. Hence, the first Sumerians in the lowlands of Mesopotamia built a temple to the goddess Ninhursag whose name means 'lady of the mountain'[1] (and who was also known as Ninmah, 'lady of the holy grove').[2] Further east, the important Indian goddess Parvati is the 'daughter of the mountain'; in Japan, the Artemis-like mistress of the animals, Yamanokami, resides in the mountains, just like her relative,[3] Dali, the hunting goddess of the Caucasus. And when Greek Artemis is offered by Zeus (her father in Olympic myth) the granting of her wishes, she answers at once: 'Pray give me eternal virginity; ... a bow and arrows like his [Apollo]; the office of bringing light; ... [and] all the mountains in the world; ... because I intend to live on mountains most of the time.'[4]

One aspect of the ancient mountain cults are the so-called mountain-thrones, which have been found scattered up and down Asia Minor, Greece and western Asia.[5] Throughout these regions, various mountains and hilltops have been adorned with seats or thrones cut into the living rock, and sometimes with a few steps leading up to them. Their creators are unknown, as is the time of their creation, but they are thought to date back at least to the fourteenth century BCE.[6] Usually there are no inscriptions or dedications. Two inscriptions, however, have been found: one at a double throne on a hilltop on the small island of Chalke off the west coast of Rhodes, dedicating the throne to Zeus and Hekate; the other at a throne not far from Lartos on Rhodes itself, mentioning Hekate. The script of the latter is not later than the third century BCE, which does not necessarily determine the creation of the rock throne itself and could have been a rededication. The connection with an underworld deity is noteworthy though, because the Hellenistic Greeks generally rededicated the mountain thrones to their sky-god, Zeus.[7] However, in pre-Hellenistic religions, goddesses of the heavens, of the earth, and also of the underworld can be found in association with mountains and mountaintops. In Ephesus, on the Aegean coast of modern Turkey, the Phrygian goddess Kybele was known as Artemis *Protothroníe*, 'She of the First Throne'.[8] In mountain-less Egypt, the goddess Isis, Artemis' counterpart, was depicted as wearing a throne-shaped crown – possibly a relic of her pre-Sumerian mountainous birth (in the Taurus? See Chapter 43). However, before these goddesses developed their full individuality, they may have appeared rather nameless, in groups of three or of nine, as the Fates, the Muses and (less often) the Furies.

37.4 *Mountain-throne to Zeus and Hekate, Chalke, near Rhodes.*

37.3 *Old yew 'on the rocks', Ardasai, Sardinia.*

THE MUSES

Early Greek literature refers to the archaic female deities of the mountains as the Muses, 'Mountain-mothers' (from Greek *Moúsa*).[9] In Hellenistic times, there were temples and altars to and statues of the Muses all over the Greek world,[10] but the Greek centre of the cult of the Muses was located at Mount Helicon in Boeotia. The name of this mountain is reminiscent of *Helíke*, the Greek name of the constellation of the Great Bear (Ursa Major), which refers to the constellation's turning around Polaris, the Pole Star,[11] and suggests its meaning as a sacred 'centre', or an access point to the cosmic axis, often symbolised by the mythical world mountain on which is placed the Tree of Life.[12] In early times, however, the cult of the Muses most probably involved a ritual of sacred marriage (see Chapter 39) in which the sky-god would meet and fertilise the mountain-mother (representing the earth). Their followers celebrated with song and dance, pipes and drums and string instruments (see Chapter 42).[13]

The Academy

On the outskirts of Athens, Plato founded the famous Academy, a school that originally taught mathematics, dialectics, natural sciences and preparation for statesmanship. The property that Plato acquired about 387 BCE included a sacred grove[14] comprised of olive, yew, white poplar, elm and plane trees.[15] Legally, the school was a corporate body organised for worship of the Muses. It was closed in 529 CE by the Byzantine emperor Justinian; the trees, however, had already been destroyed by the Roman commander Sulla to build war machinery for his siege of Athens in 87 BCE.[16]

At Alexandreia in Egypt, the temple of the Muses was called the *Mouseion*, 'seat of the Muses', from which the word museum is derived.

THE FATES AND THE FURIES

The Fates *(Moirai)*, also called the Graces, weave the fates of man and all beings into the great tapestry of destiny, or creation; hence they have been envisaged as spinning and weaving. The origin and the original number of these female immortals are unknown. Plato describes them as sitting on thrones next to 'Lady Necessity' who turns the spindle of the entire solar system in her lap (see box p. 186). The Fates are daughters of Necessity and, with the planets moving in their orbits, join the music of the spheres: Lachesis, the 'measurer', sings of the past; Clotho, the 'spinner', sings of the present; Atropos, 'she who cannot be turned, or avoided', sings of the future.[17] Being located at the cosmic axis, they also govern the gates between the heavens, the earth and the underworld. For these reasons, spinning and weaving were of immense symbolic significance from the Neolithic period onwards (see figures 37.6–7).[18]

The Furies (*Erinyes*) are the agents who ensured the fulfilment of the decrees of the Fates. They were engaged to protect burial sites and also to witness oaths – because oaths created an assignment, a *moira* ('fate'). Early in their history the Erinyes were morally neutral – hence they were also called Eumenides, 'the Kindly Ones'[19] – but the public sense of divine justice changed and eventually the 'growing sense of guilt'[20] of a later age transformed the understanding of their activity into one of punishment, and the Kindly Ones themselves into ministers of vengeance. By Hellenistic times they were much feared, and imagined as fearsome wild women swaying torches of yew wood and punishing human breach of (divine) law with yew poison – the yew was sacred to them.[21] The torch-bearing motif is reminiscent of Demeter and Persephone, who show the path through darkness – but where does the light of the Erinyes lead? It is noteworthy that they particularly seem to avenge injuries inflicted on a mother,[22] which suggests a pre-patriarchal origin. The sacred yew branch gives them ultimate legislation for judgement and execution (compare Chapter 39).

Some 1,600 years after Plato's description, the Icelandic *Poetic Edda* tells of the Norns, three immortal women who live by the primeval spring of life, the 'Well of Wyrd', at the foot of the World Tree, where every morning they host the council of the gods.[23] Their names are Urd (another form of *Wyrd*),[24] Skuld, 'fate', and Wer-

***37.5** The World Tree on top of the 'world mountain' which holds aril-like circular patterns; seal from Mesopotamia, 3000–2900 BCE.*

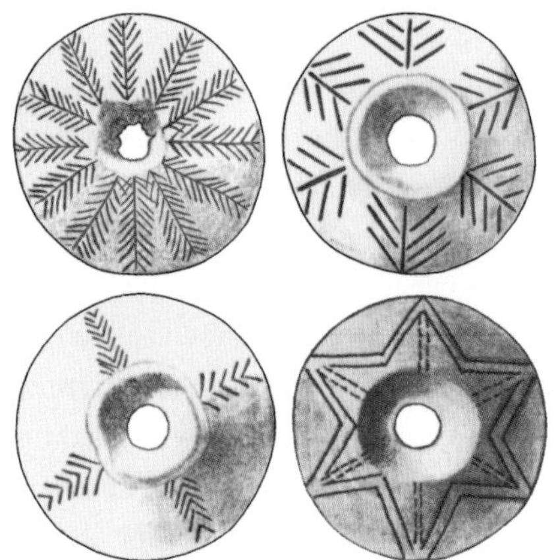

37.6 *The branch pattern is a recurring motif among the symbolic incisions on votive 'spindles' found at Troy; terracotta, c. 2100–1900* BCE.

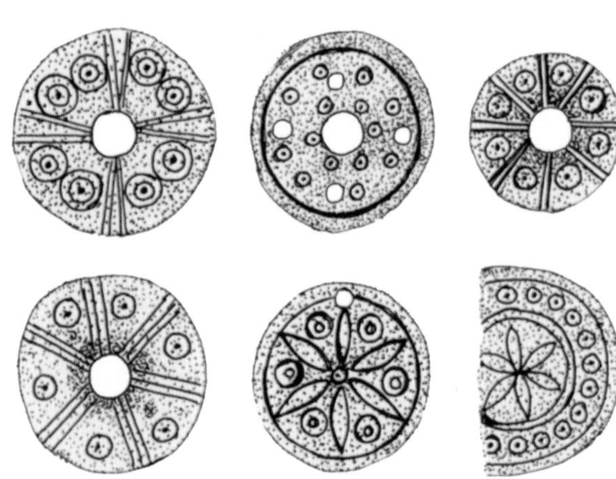

37.7 *Traditional spindles from Frisia, north-western Germany, have crosses or six-pointed stars, decorated with aril-like circular patterns.*

dandi, 'becoming'. Anglo-Saxon *Wyrd* or Norse *Urd* is an essential concept in Northern cosmology. It can be translated as 'fate, power, prophetic knowledge' but actually denotes the foundation of all existence, the force behind all life. It suggests a view of the cosmos in which everything is interconnected by a kind of three-dimensional web[25] – hence the mythological significance of weaving. The Norns are associated with carving fateful runes into (yew) wood, and with weaving the threads of destiny (see pp. 225–8). The Norns as well as the Fates stand not so much for the individual lot, but rather in a holistic sense for the harmonious ordering of the world.[26] The Kindly Ones are known in Germanic traditions as *die Holden*, 'friendly spirits, a silent subterranean people of whom dame Holde, so to speak, is the princess'; their main function was to guide the souls of the departed.[27] In Norse tradition, their queen is Huldre or Hel, the 'veiled one'.[28]

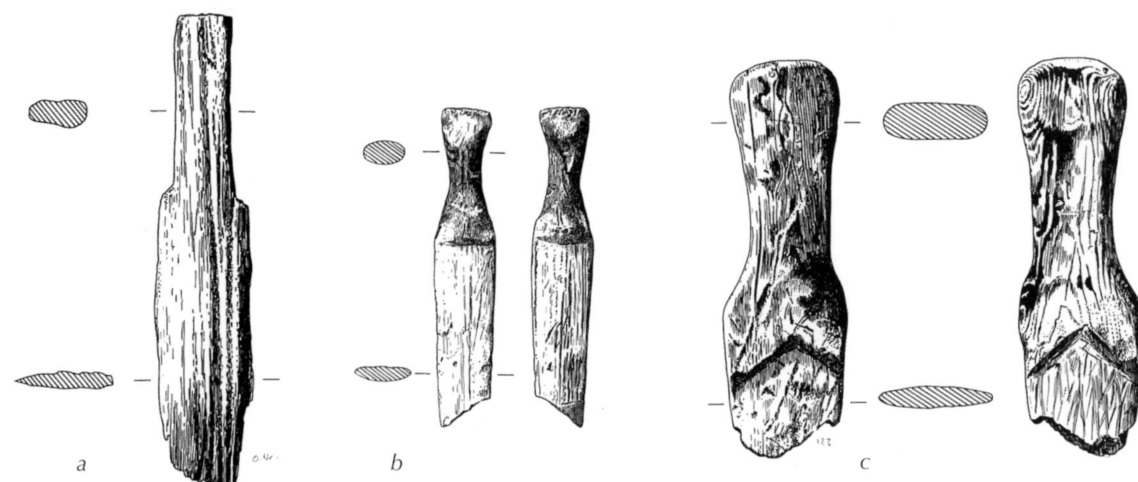

37.8 *A number of obscure wooden 'daggers' has been excavated in Neolithic sites in Switzerland. They have become known as* Webmesser, *'weaving knives', but this use is entirely hypothetical; their possible function has been a mystery since the first discoveries in the 1860s. The blades are lentoid in cross-section and usually a few millimetres thick; the size varies, but all samples are made from yew wood. The samples shown here are from Niederwil: a) 14.5cm long, b) 9.9cm long, c) 12.5cm long. (After Müller-Beck 1991)*

FROM ASIA MINOR TO CRETE

Crete's central position meant that it maintained strong cultural and religious links with Greece, Asia Minor, the Levant, and Egypt from an early time on. The Anatolian mother-goddess appeared in Crete under various names and in differing local cults.[29] Her mother-and-daughter aspect became predominant in the tradition of Demeter and her daughter Persephone (old Cretan *Phersoponé*), which spread from Crete via the Peloponnese to mainland Greece, especially to Eleusis (see Chapter 35). Almost the same relationship exists between the Cretan mountain-mother Diktynna and young Britomartis ('sweet maiden'), whose worship is connected with Mt Dikte (2,148m) in the eastern part of the island.[30] Coins from the last two centuries BCE show Diktynna with quiver and bow, a torch, and a dog by her side.[31] But the oldest tradition is perhaps that of Leto, which came to Crete from the land of Caria (modern south-western Turkey). Leto (also Lato) – her name has been connected with the Carian word for 'woman', *lada*[32] – has a daughter *and* a son: Artemis and Apollo. Many of her depictions present her as *kourotrophos*, a woman nursing a child on her lap, which is one of the oldest motifs in religious art. She was also the original divinity at Ephesus (see Chapter 40). From Crete her cult spread to Greece, namely to Delos and Delphi where Apollo would soon outgrow and outshine his mother, while in Crete and Ephesus, he remained a child.

The other sacred mountain of Crete was Mt Ida (2,456m) in the centre of the island, under the domain of the mountain-mother, Rhea. Its name means 'forest, wood',[33] but it is more than coincidence that it shares the same name with the Mt Ida (pronounced eedah) in the Trojan country where the rites of the Anatolian goddess were performed. Apart from a goddess cult, both mountains share a link with the Dactyls (Daktyloi), a group of advanced blacksmiths who are credited with the invention of ground-breaking iron-working techniques[34] and during the Hellenistic age became the 'mythological personification of a brotherhood of master metallurgists' (Eliade).[35] Both Hebat/Kybele in Anatolia and Rhea in Crete were not only deities of childbirth but also patrons of smiths.[36] For the sacred cave in the Cretan mountain, rites of death and rebirth are attested,[37] and Eliade states that the later cult of Zeus in the Idaean Cave still 'had the structure of an initiation into the Mysteries'.[38] A similar spiritual brotherhood lived at the Cretan Mt Dikte: the mountain-herding Kouretes who worshipped a goddess, guarded sacred mountain shrines, and very probably engaged in ascetic practices. It seems that the two main social functions of such brotherhoods were '(a) initiation of the youth and (b) cultivation of the sacred tree'.[39] Some also handled the initiation rites of specialised professions such as blacksmiths, musicians, or healers, like the followers of Asklepios.

Both mountains named Ida and Mt Dikte can be regarded as 'world mountains' in the tradition of the Muses, and both Cretan locations have yielded archaeological and textual evidence relating to the symbology of the Tree of Life.[40] Homer, for example, speaks of a conifer on the top of Mt Ida that reaches the ether (*aithér*),[41] the higher dimensions or heavens.

37.9 *(top left) A young multi-stemmed tree between two horns of consecration is watered by two genii (spirits) with libation jugs; gem from Vaphio.*
37.10 *(bottom left) Libation jug in front of an enclosure with two sacred trees; sealstone from Sphoungaras. A horn of consecration can be seen to the left (a right one is wanting).*
37.11–37.12 *In these unusual sealstones (on the right), the horns of consecration themselves have turned into boughs.*

The ancient Cretan depiction of the Tree of Life resembles *Taxus* more than any other species (see box pp. 186–7). Indeed, Theophrastus attests that yew (*milos*) was present around Mt Ida still in the fourth century BCE,[42] and he also mentions a fruit-bearing tree 'in which the dedicatory offerings are hung'[43] right at the mouth of the sacred Idaean Cave.

Under Hellenistic influence, both Cretan mountains, or rather the ritual caves located in their slopes, became associated with the birth and upbringing of Zeus. His name, however, replaced that of the native Cretan 'divine son'. But whatever the name – Zagreus, Dionysos, Zeus *Idaios* – the reborn god was also a tree-god, hence epithets like *Epirnytios* which means either 'set over the growing plants' or 'on the tree'. Zeus' cradle, in myth, hung in the sacred tree by the entrance to the Diktaean Cave. Or was he born from the cavity/womb of a hollow tree?

The mythical details of Zeus' upbringing (see box), in accord with archaeological evidence,[44] suggest that the actual ritual at the Diktaean Cave included the libation of the three substances that had been sacred since time immemorial: water, milk and honey (see p. 219). Water is encoded in this myth by Io being the Moon as rain-bringer,[45] milk is the gift of the goat, and the 'honey-man' (Melisseus)[46] is a remainder of other local traditions in which Zeus was fed by bees[47] – the Kouretes taught the art of bee-keeping.[48] Adrasteia's name means 'the inescapable one', and the later Greeks identified her with Nemesis,[49] the goddess of Fate (see below). The armed dance of the Kouretes possibly constituted an initiation ceremony of young men's brotherhoods.

Similarly, the traditions of Leto and her children, Artemis and Apollo,[50] point to the relevance of such groups of 'holy' women or men (Korybantes, Kouretes, Kabires)[51] who lived in or near the caves of these sacred mountains and teamed up with the priests or priestesses of the goddess to care for the coming of age ceremonies of the youth. The rebirth of the 'son' as well as the disappearance of the 'daughter' into the underworld (see the myth of Kore/Persephone in Chapter 35) are the mythological equivalents for rites in which the Cretan boys laid aside their boyhood garments before assuming adult or warrior costumes 'which each had received as a gift after his period of seclusion'.[52] Girls, too, disappeared; they were 'carried off and mourned'[53] by their relatives and their village; they were taught the secrets of womanhood in those remote spiritual communities – and none of them returned, for those who came back were women. While in the agricultural festivals Kore/Persephone might very well have represented the grain of corn descending into the dark earth, the disappearing maiden theme was of equal, social importance for how grown-up societies cared for the social and spiritual integration of younger generations.

The upbringing of Zeus

Rhea and Cronos (who in other Greek sources are the planetary powers of Saturn[54]) are the parents of a number of deities[55] but Cronos has devoured them all in order to avoid a prophecy coming true that one of his sons would dethrone him. When Zeus is born, Rhea gives the infant to Mother Earth who hides him at the cave of Mt Dikte, where he is nursed by Adrasteia and her sister Io, both daughters of Melisseus ('honey-man'), and by the goat-nymph Amaltheia. The Kouretes, who live on this mountain, 'have Cybele's cymbals fill the air',[56] or clash their spears on their shields instead, or both, to drown the baby's cries, lest Cronos might hear it from far off. Later, Zeus grows to manhood among the shepherds of Ida, occupying the Idaean Cave, from where he also plots his revenge and the release of his siblings from the belly of Cronos. After Amaltheia's death, Zeus sets her image in the stars, as Capricorn, and also takes one of her horns and gives it to Adrasteia and her sister Io; it becomes the cornucopia, the horn of plenty, which is always filled with whatever food or drink is needed – an ancient symbol of the abundance of nature.[57]

The Tree of Life in Crete

Two finds from Crete provide a rare glimpse into the Cretan mountain cult and the Minoan view of the afterlife. The first is a small bronze plaque from inside the Diktaean Cave (Mt Dikte). It shows three trees, each rising from a Minoan altar (typically depicted with a pair of horns signifying the sacred nature of what is between them), one of which (the top central one) is clearly bigger than the other two and elaborated with additional decoration. The trees are of the same kind and each have two neat rows of upward-rising branchlets or needles. A fourth tree rises from a rectangle that might represent a (sacred) enclosure of earth or a planting pot, but the branchlets of this plant point downwards. The other two dominant elements are a human figure, perhaps approaching the 'potted' tree or engaged in a ritual dance within the sanctuary; and a ring-dove or wood pigeon *(Columba palumbus)*, which is perched on the left-hand altar tree. The fish at first seems to be out of place in a mountain cult, but nevertheless was an important and recurring symbol in the religion of many goddesses (fish dwell in the life-giving waters as an embryo grows in the birth water of the womb, see figure 34.10). Less dominant but not less significant are the moon crescent and the solar disk at the top of the design. The presence of both celestial bodies, their equal balance or sometimes union suggests that this is a sacred space, a 'centre' (see pp. 155–6, 226) where the divine realms touch the mortal world. They also point to 'inner alchemy' and Yogic practices (see p. 196). The World Tree is the axis around which the world rotates, hence the crosses flanking the central altar tree: crosses and swastikas in many ancient cultures symbolise the rotation of the solar system (compare pp. 182, 195 Plato).

37.13 *The image of the votive tablet from the Diktaean Cave, Crete,* c. *1500 BCE; original size.*

EUROPA

In her ancient Cretan cult, located in the vicinity of Mt Dikte,[62] Europa is a parallel figure to Demeter, another emanation of the same archetype. She appeared as a latecomer among the Cretan goddesses, in all probability an import of the Phoenician Astarte.[63] Coins from both regions from the last five centuries BCE show the goddess associated with a bull and a tree.[64] In Crete she became a chtonian goddess of vegetation, sometimes merging with Demeter with whom she also shares the myrtle (a tree associated with midwifery).[65] Her union with Zeus (she bore him three sons, among them Minos, the legendary first king of Cretan Minoan culture[66]) suggests a sacred marriage tradition.[67]

37.15 *Europa in her tree; Aegean coins from* c. *430 BCE onwards.*

The other artefact is the so-called 'ring of Nestor', a gold ring found in a large beehive tomb above the Pylian Plain.[58] It weighs 31.5g and has been dated to the middle of the second millennium BCE.[59] Its elaborate relief design is centred on a tree with peculiarly twisted limbs and wide-stretching horizontal boughs (both phenomena being common in *Taxus*). The guardian at the base of the tree is the familiar underworld dog (see Hekate, p. 169). The bottom left quarter of the motif shows four human figures of whom the two nearer the tree resemble a couple being led to a ceremonial activity (which is depicted in the bottom right section). Their guiding priestess (behind them, to their left) is just warding off a third person (far left) who is not to be admitted. The actual ceremonial scene on the bottom right shows three more priestesses, all in typical Minoan short skirts and with bird heads or masks.[60] On the altar sits (the statue of) a griffin, whose piercing eagle-eyes might represent the inquisitive judgement that grants admission (or not) to the higher planes. Behind it (far right), stands the goddess represented by a high priestess in a long skirt. The ascent (top right quarter) leads to an encounter with the lion of the goddess, here residing on a horizontal branch and tended by (at least) two human figures. The beast is entirely peaceful, and reminiscent of a description of the Sumerian paradise: 'The lion does not kill, The wolf snatches not the lamb.'[61] The lion both keeps watch over the scene below and is the guardian of the gate to the eternal realms. Directly behind his head appears the only visible biological activity of the tree: a few branches bearing heart-shaped leaves or heart-shaped arils, which in this context are most likely to represent the fruits of immortality, the ultimate token to enter the Land of the Blessed (depicted in the top left quarter of the relief). Here, our couple emerges again (close to the trunk) and comes to witness the goddess and her consort themselves seated in peaceful yet animated conversation. The divine couple (or the human couple in an advanced state of being?) has no human mind or ego, hence the heads rather resemble the wings of butterflies. Two butterflies also hover above the 'goddess'. Next to the butterflies appear two chrysalises, which complete the symbology: the human couple, after passing the griffin, the lion and the fruits of immortality, are transformed and reborn like a butterfly from the chrysalis (see p. 166).

37.14 *The Tree of Life; Minoan gold ring,* c. *1500* BCE, *three times original size.*

Because in the ancient imagery the tree of Europa's meeting with the sky-god appears to have been cut (pollarded) and vigorously regrown, it has been interpreted as a willow. It is plain, however, that the early scholars, when examining the images of Europa and Zeus, were limited by their botany:[68] to them, only willows were capable of such regrowth – they did not know a thing about the growth patterns nor the regenerative abilities of *Taxus*, nor, indeed, of its wide distribution in the Mediterranean. Hence, another thing to come into being beneath the yew might have been the name of the European continent.[69]

37.16 *Fallen but thriving yew at Kentchurch, Herefordshire.*

37.17 *Yew tree on a cool and moist northern cliff at c. 1,000m altitude, Chilokastro, near Korinth, Peloponnese, Greece.*
37.18 *Ancient yew (shell stage) in the same location.*

37.19 *Aphrodite with dove and her divine son, Eryx; tetradrachme from Sicily.*

APHRODITE

Aphrodite, 'foam-born', has become known as the goddess of love whom the Romans called Venus. However, Aphrodite was also a goddess of vegetation, depicted in front of a tree and with a dove in her hand.[70] As Aphrodite *Urania* ('Queen of the Mountain'[71]) she is the goddess whose sacred union with the mortal shepherd Anchises on Trojan Mt Ida is the Greek version of the myth of Inanna and Dumuzi in Sumer, Ishtar and Tammuz in Babylonia, and Astarte and Baal in Syria.[72] Cook calls her a 'mountain-mother' and relates that 'many scholars have been content to regard this Aphrodite as a Hellenised form of the Phoenician Astarte'.[73] Perhaps surprisingly, the goddess of love was also a goddess of death and 'earned many titles which seem inconsistent with her beauty and complaisance', says Graves. 'At Athens, she was called the Eldest of the Fates and sister of the Erinyes: and elsewhere Melaenis ("black one") … [and] Scotia ("dark one")'.[74] And at Halikarnassos, on high ground beside a spring, she shared a temple with Hermes, the guide of souls.[75] Aphrodite is associated with 'apples', but in these high places there are no orchards in the domestic sense. Her underworld connection suggests that her apples could have been those of immortality.[76]

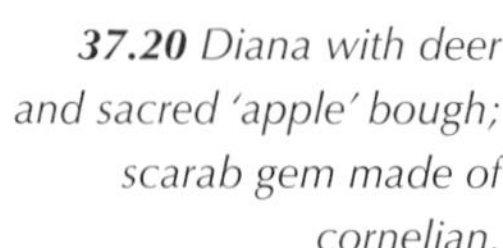

37.20 *Diana with deer and sacred 'apple' bough; scarab gem made of cornelian.*

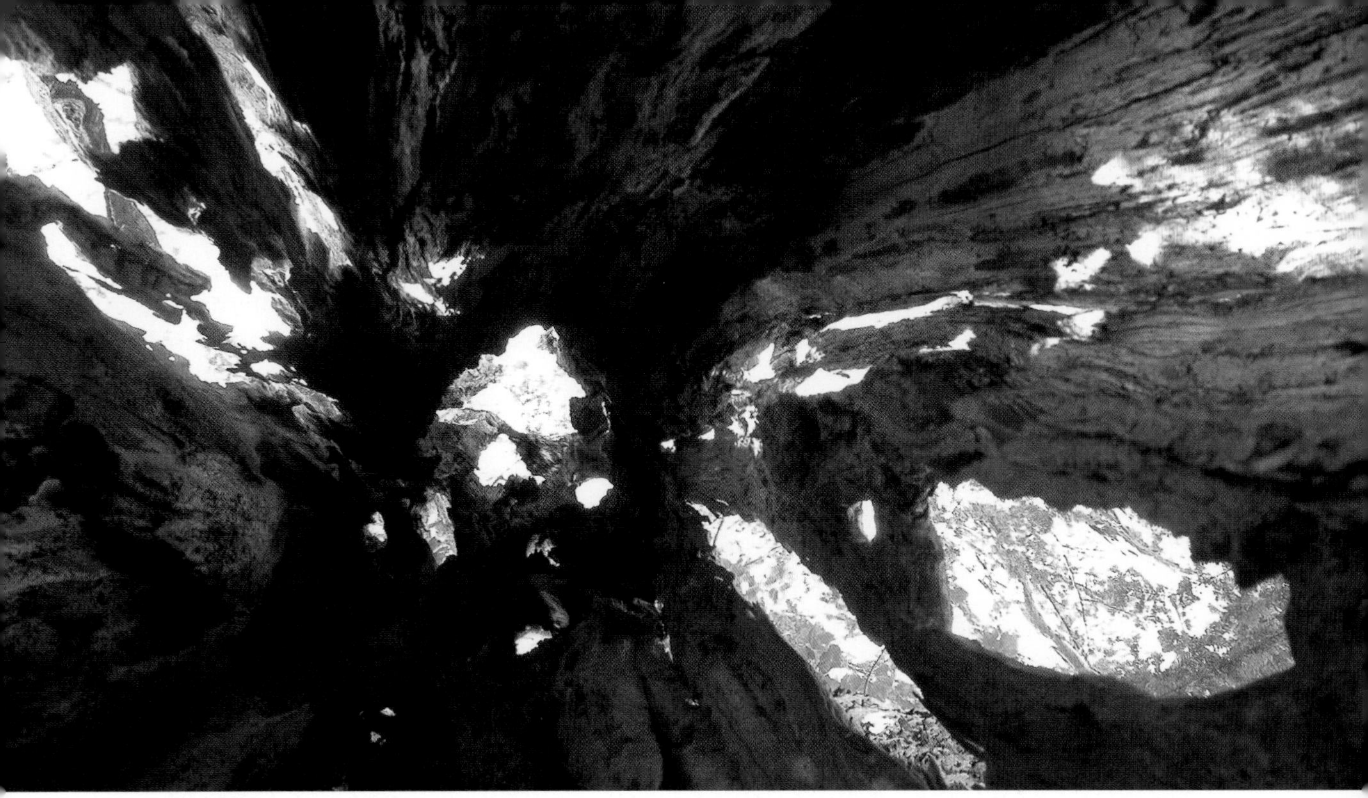

37.21 Inside an ancient hollow yew in Asia Minor.

DIANA AND NEMESIS

The mountain-mother and her sacred tree also moved to Western Europe; in Etruria (modern northern Italy), she became known as Diana. Like Artemis she roams the mountains and hills and is associated with the moon crescent, with bow and arrow, the stag, the dog and with 'apples'. Her counterpart in Germanic myth, Idun, provides the gods with apples of immortality (see p. 228). At Diana's cult grove at Nemi, she was called Diana *Nemorensis* which points to a fusion with Nemesis, a goddess of the greenwood (*némo*, 'I pasture', *némos*, 'glade'), whose name in turn corresponds with the Celtic goddess Nemetona (from *nemeton*, 'sacred grove, sanctuary'). But a Greek goddess Nemesis was the embodiment of divine vengeance (*némesis*, 'righteous wrath'), which would link Diana to the Erinyes, the avengers. The Romans adopted Nemesis under the name of Fortuna as a goddess of fate, and the emblem of Nemesis, a wheel, became the 'wheel of fortune' that remained popular throughout the Middle Ages[77] (and still in television shows of today). Nemesis or Fors Fortuna, however, was also thought to grant birth (*némo* also means 'I assign', and *ferre*, 'to bear') and was worshipped by women as Virgo or Virginalis. Sometimes depicted with a cornucopia (horn of plenty) and with multiple breasts, she was likened to Artemis/Kybele at Ephesus whom the Romans usually called Diana of Ephesus. In the tradition of Diana Nemorensis at Nemi, however, the sacred yew branch of the goddess is referred to as the 'golden bough' (see p. 210).[78]

***37.22** Winged Nemesis with sacred bough and serpent; Greek red jasper gem.*

ASHERAH AND ASTARTE

In the Levant (the region bordering the east coast of the Mediterranean, that is, modern Syria, Lebanon, and Israel), the Babylonian-Assyrian goddess, Ishtar, also known as *Sharrat Shame*, 'Queen of Heaven', diversified into Ashera and Astarte.[79] Her Hebrew name, *Ashtaroth*, means 'womb' or 'that which issues from the womb', while the Canaanite name *Ashtoreth* means 'She of the womb'.[80] She is first

Asherah and her tree[81]

In early religious art, the 'naked goddess' is typical of the Near East. The amulets of ancient Canaan and Israel, however, differ from their precursors, the Syrian cylinder seals, in three ways. On cylinder seals the goddess generally holds her arms across her abdomen or she holds her breasts, but on the scarabs the arms are usually hanging down or holding a branch. Secondly, while on the seals she is flanked by worshippers or shown opposite a partner, especially the weather-god, she appears alone on the scarabs. And, finally, the symbols for good luck, common on the seals, are absent on the scarabs. Instead, on thirty-six of the forty-four specimens (82 per cent) found in the Middle Bronze Age (layer IIB, which corresponds to the first centuries of the second millennium BCE) she is flanked by two branches or small trees – a frequency that also accounts for her having been dubbed 'branch goddess' by contemporary scholars. Her central position, particularly with the frontal view, emphasises that she is making an appearance, that is that she emerges from the tree (figures a–f). The direct frontal view (figures c, d, g) is extremely rare in the ancient Near East. Her overly large ears (figures c, g) are an expression of her willingness to listen to her worshippers. In one specimen (figure e), two small additional branches sprout

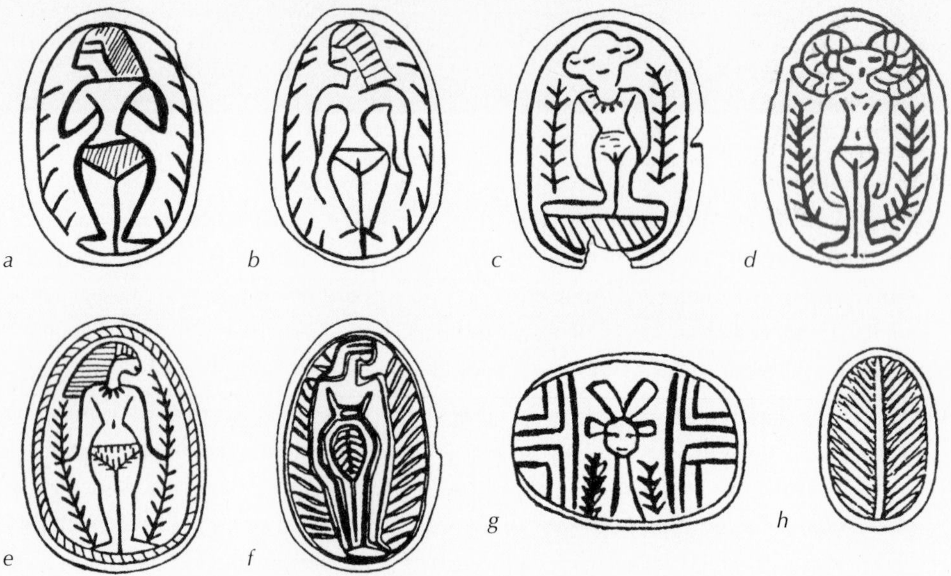

mentioned in the cuneiform tablets of Ugarit (northern Syria) dating from the fourteenth century BCE. Most of the chief deities of ancient Syria had two names, one was a proper name and the other an epithet. This was a requirement of classical Semitic ritual text composition and poetry, and allowed the reduplication of significant lines without a repetition of the same words.[82] Thus, Lady Asherah of the Sea was also referred to as *Elath*, 'goddess', and her consort, Father Shunem, was often styled *El*, 'God'. (Epithets were also widely used in the Greek world.) The children of this senior couple are Astarte, whose constant epithet is 'the Maiden/Virgin Anath',[83] and Hadd, usually referred to as *Baal*, 'Lord', and it was this second-generation couple that comprised the major fertility deities across the Bronze and Iron Age Levant. Both mother (Asherah) and daughter (Astarte) retained the epithet Queen of Heaven, as well as their association with mountains and hilltops.

Asherah was a chief deity in the Canaanite pantheon, and remained so for a long time to come despite the arrival of the patriarchal tribes of Israel. Ancient Hebrew religion was far less monotheistic than the Mosaic Law and the accumulated speeches and condemnations of the prophets make it seem.[84]

from her genitalia (compare figures 34.10–11); in another (figure f), the sacred branch seems to represent the life inside her womb while another creates a radiating aura around her figure. Unlike the seals, which contain an element of the eroticism of palace life, this solitary divinity is clearly a vegetation-goddess, and the function of the talismans is, in an agricultural sense, fertility and prosperity.

A few specimens place the branch in the centre, either solely (fig. h) or in an abstract scene (fig. i). In figure j the abstract head of the goddess appears top centre, flanked by two Egyptian-style falcons (compare hawk p. 226) on the outside and two *uraeus* serpents (Egyptian cobras) on the inside. Interestingly, inside these snake shapes (and in those in figure k) the branch pattern appears yet again. Below the reptiles occurs the sign for 'gold', and beneath this the sacred tree flanked by two priestesses. Another extraordinary design (fig. k) shows the tree itself protected by no less than four *uraeus* serpents. The principal beasts of the goddess, too, occur on scarabs, namely the goat (figures l, m) and the lion (fig. n). In figures m and n, their tails have transformed into a protective *uraeus*. One specimen (fig. d) shows the goddess wearing ram's horns, and flanked by branches.

None of these branches displays the conical shape of date palm leaves. Figure h could be the life-size rendering of a *Taxus* twig.

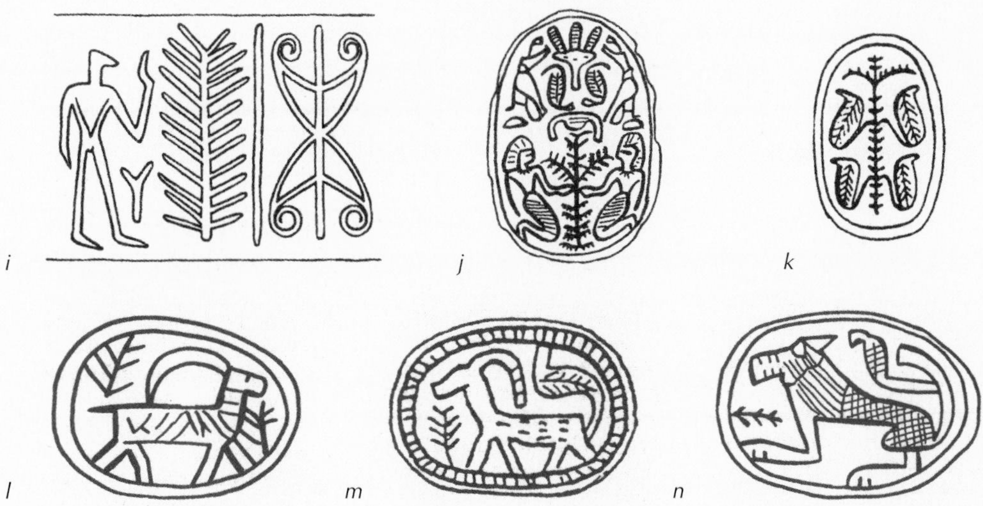

37.23 *The goddess Asherah and her sacred tree, motifs from talismanic scarabs, early second millennium* BCE.

Moses' chief deity, Yahweh, was originally a sky-god,[85] and his celestial qualities made him the perfect consort for the mountain-mother, Asherah.[86] For centuries, Yahweh and 'his' Asherah became the most popular divine couple in Israel and Judaea.[87] The archaeological evidence supports the notion that the goddess must have been extremely popular in all segments of Hebrew society, including the royal families. King Solomon 'burnt incense on the high places',[88] and even brought Asherah worship to Jerusalem. And during the reign of his son Rehoboam (*c.* 928–911 BCE), her cult pillar – also called *asherah* – was moved into the Temple itself.[89] However, the primary places for her worship, as the Bible and other sources state, were located in the wilderness, 'on the heights and the hills in the mountain country'.[90]

So what is the botanical implication of the Asherah cult? Why was the goddess of fertility and motherhood represented by a pillar of dead wood instead of a flourishing plant abundantly laden with fruit? The literary sources do not reveal much more about the asheras beyond their nature as poles or pillars[91] and their usual location in 'high places'. Luckily, there is a rich source of archaeological evidence regarding Asherah and Astarte: terracotta

***37.24** True date palm representations for comparison: the goddess Ishtar standing on a lion, imprint of a garnet seal cylinder, 720–700 BCE Mesopotamia; lion and bull flanking palm, scarab from Tell el-Far'ah, Israel.*

figurines, as well as scarabs and other small talismanic plaques[92] of the goddess have been found in every major excavation site in Palestine, their prevalence extending from Middle Bronze (2000–1500 BCE) to Early Iron II (900–600 BCE). In some sites these so-called Astarte plaques are the most common religious object,[93] and a good number of them show the divinity associated with the sacred tree or branch (see box pp. 190–1). Partly owing to the small size of the plaques, the tree or branch often appears as nothing more than a herringbone pattern, and since the early twentieth century archaeologists 'traditionally' interpret this as a *palmette*, a spray of a palm tree.[94] It is true, the date palm *(Phoenix dactylifera)* was a tree of extraordinary economical significance and also a sacred tree for some Near Eastern peoples.[95] But it does not quite make sense as the principal tree of Asherah: if the goddess had been the personification of the date palm, her cult centres would have been located at the agricultural hotspots in the lowlands, in the fertile irrigated valleys[96] where palm plantations were established since very early times. There were no date palms in the wild 'high places', nor is the soft and fibrous palm wood suitable for carving ceremonial pillars. Furthermore, the branches depicted usually do not look like palm leaves, which taper towards the apex and also have single leaflets thinning to a point. Therefore, when date palms are shown in ancient Oriental art they usually have distinct triangular leaflets. The leaves on the Astarte gems look more like coniferous needles with parallel sides, and the twigs themselves are usually neatly parted (as typical in *Taxus*), and either reminiscent of linear *Taxus* shoots (see figure 31.2) or are even forked (which is impossible for palms).

***37.25** In the greater part of its distribution area,* Taxus baccata *is a tree of the 'high places' …*

37.26 *The owl of mountain-mother Athena keeps watch.*

Of course it is rather daring even to consider the presence of yew in the history of Israel; for the last 6,000 years the regional climate has not changed and the summers are simply too hot and dry for wild *Taxus*. If migrating groups (see Chapter 43) had brought living yews (cuttings or young trees) to Israel, they would have withered and died quickly. Only in cooler and moister mountain areas, and if frequently watered (as sacred trees usually were), could they have lasted longer. It is not so unlikely that this scenario may have happened; the cult of the Mother of the Gods originally spread with its sacred tree as much as with its other associated attributes and symbols (see above, and previous chapter). And, in the Levant, it spread southwards, moving into ever hotter climates. The human-borne migration of sacred trees would explain perfectly why a dead tree was taken as the emblem of the Canaanite goddess of life. Taking into account the many parallels between the divinities described in this chapter, as well as their common origin in the yew–tree–goddess pattern (see p. 158), the question, perhaps, is not so much if Asherah/Astarte was connected to *Taxus*, but how and when their association faded out under climatic pressure. The cult of the *ashera* then, would be a unique case of, literally, holding on to a pole of dead wood,[97] while other cultures under hot and dry southern skies let go much more quickly and replaced the sacred tree with 'substitutes' (Assyria, for example, during the last millennium BCE did depict the Tree of Life as a date palm, while Egypt even had five sacred tree species[98]). After all, yews grew not so far from Israel; there were stands in Cilicia and the Taurus Mountains, and possibly even in the Lebanon mountains.[99] We also know that the clay moulds for the Astarte plaques were often made by non-Israelite potters and imported from the north;[100] one production facility that intensely exported faience scarabs around 1750 BCE was actually located in the region of southern Cilicia (northern Syria/south-east Anatolia).[101]

37.27 *Life-size yew twig on a dagger blade linked to the Bronze-Age warrior class Ben-Anat, which was located in northern Syria and Canaan.*

CHAPTER 38

GODS AND HEROES

THE FATHER OF LIGHT

Not long after writing was invented, about 5,000 years ago, the heaven-god made a first literary appearance. In the Sumerian clay tablets his name is An, and his sacred union with Ki, '(mother) Earth', brings forth *an-ki*, the universe.[1] Mythologically, An and Ki are the parents of the higher beings (*dingir*, 'gods') who order and rule the world. The seven leading deities live and also hold council to 'decree the fates' at the sacred centre of the universe where the World Tree connects all the worlds (see Chapters 32 and 41).[2] The concept of the union of sky and earth was widespread among the ancient peoples. In the Greek world, the great number of different regional sky-gods was eventually assimilated into the figure of Zeus. His name, 'the bright one', identifies him as the god of the bright or daylight sky. In many inscriptions, his name does not appear in the substantive form *Zeús* but in the genitive (as well as adjective) form *dios*, 'of Zeus' or 'belonging to Zeus', or simply 'divine'. Before he became increasingly personified as a male anthropomorphic deity (by and after Homer) he, or rather it, was *the* Zeus, the 'radiant sky credited with an impersonal life of its own', as Professor Cook phrased it.[3] This open name and wide concept facilitated a smooth assimilation of Zeus with many sky-, weather- and light-gods in many cultures within and neighbouring Greece (e.g. even with Yahweh among Syrian and Greek Jews[4]). But even the 'father of the gods' began as the son of a mother (see previous chapter).

In the precinct of the healing god Asklepios at Epidauros in the north-eastern Peloponnese, an oblong piece of limestone bearing an ancient carving of a yew branch was found, identifiable by its familiar neat parting of two rows of needles, enclosed in a circle and accompanied by a dedication to Zeus *Phílios*, 'the friend'.[5] A further addition says 'in accordance with a dream'. Usually, however, gods in the Old World were not 'loved' or 'friendly'; they were fearsome, unpredictable creatures, which did not attract a relationship closer than necessary. Zeus *Phílios* was one of the few exceptions. The concept of God as 'friend' or 'beloved' clearly came from the teachings of the mystics of the time; Plato helped to spread it.[6] 'Friends in general', says Cook in his 3,581-page study of the ancient sky-god, 'swore by Zeus *Phílios*, who came to be looked upon as the overseer and guardian of friendship, or ultimately as a god of love who would have all men dwell together in amity.'[7]

38.1 *Sunlight in the yew at Slaugham, Sussex.*

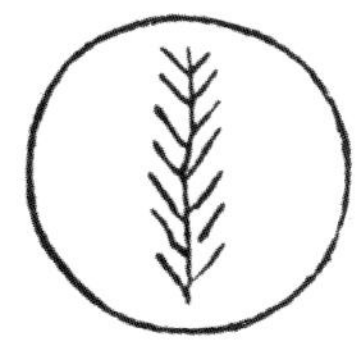

38.2 *The sacred branch. Main motif of an ancient limestone carving dedicated to Zeus* Phílios, *at the sanctuary of Asklepios in Epidauros, Greece.*

THE COLUMN OF LIGHT

The light that Zeus/Dios represents is ultimately not the daylight that regularly alternates with night, but a divine light from beyond the world of opposites, from the primeval source from 'where the sun [first] rose'. Plato shares a vision of this centre or 'navel' of the world, the World Tree as the world pillar or axis, except that in Plato's account it is not a tree but an astral column of light: on the twelfth day after his death, Plato's character in the *Republic* (who later will be revived to report to humankind about the other side of death), along with other departed souls, 'reached a place from where they could see a straight shaft of light stretching from on high through the heavens and the earth; the light was like a pillar, and it was just like a rainbow in colour, except that it was brighter and clearer. It took another day's

Orchil's loom. The Horn of Heimdal

travelling to reach the light, and when they got there they were at the mid-point of the light and they could see, stretching away out of the heavens, the extremities of the bonds of the heavens ... while stretching down from the extremities was the spindle of Necessity, which causes the circular motions'[8] of the planets in their orbits (see p. 182). Side by side with this scene of truly cosmic proportions, Plato describes two streams of movement, one going upward from a hole in the earth, and the other going downward from a hole in the heavens. Vast numbers of souls were travelling in these streams, going to or arriving from heaven or the underworld, the intermediary to lifetimes on earth.[9] The rainbow-coloured shaft of light as the supernatural bridge between the worlds has a parallel in the rainbow bridge in Nordic myth, faithfully guarded by the god Heimdallr. And Heimdallr (as will be shown in Chapter 41) *is* the World Tree.

The architecture of this world pillar with its upward and downward flow, described by Plato and mentioned in the *Odyssey*, is stunningly reminiscent of how the ancient Yogic teachings from India describe the structure of the body's energy flows. This 'miniature' Tree of Life inside the human body is also associated with a serpent: *kundalini*, the Sanskrit term for the highest human energy, stems from *kundal*, 'coiled', and has often been described as a sleeping serpent, coiled up three and a half times. It is depicted in old Indian documents, while a serpent in exactly the same posture occurs on the omphalos stone at Delphi, Greece.[10] As the awakened *kundalini* rises, it unfolds full human potential by liberating it from the dominion of mind and body. This following ecstatic flight of the spirit has often been symbolised by an eagle (Europe, Siberia), a great mythical bird (Sumer, India), or a winged disk (Hittite, Mesopotamian, Egyptian). From the Yogic point of view, the Tree of Life, as well as the serpent at its base and the eagle at its crown, is the translation of concepts relating to energy flow into mythical language.

***38.3** The human energetic body according to Indian Vedic sources.*

Yoga

Like most Asian traditions for the advancement of health and personal development (e.g. Traditional Chinese Medicine, Taoist alchemy), Yoga acknowledges not only the physical body and its function but also the interplay of energy or life force. According to the ancient teachings of *kundalini yoga*,[11] there are two main kinds of energy flowing through the human body: *prana*, the nourishing life force which is absorbed through the breath, food, and the senses; and *apana*, which leaves the body mostly with the physical waste products. *Prana* rises from the base of the spine in a 'channel' called *pingala*, and *apana* flows downwards through a channel called *ida*. Normally, both energies do not meet, flowing separately up or down their respective channels that spiral around the spine.

Yogic exercises and breathing techniques facilitate a meeting of these two basic energies, which brings about an intensive white 'light' in the navel area. This in turn generates a heat that 'wakes' the *kundalini* energy at the bottom of the spine. It will rise inside its own channel – the straight, central *sushumna*. After a series of transformations within the six *chakras* ('wheels', 'circles'), it reaches the 'thousandfold lotus' in the top of the skull, which leads to 'enlightenment', resulting in a blissful trance. The final step is to 'awake' within the trance and integrate this state of 'being in God', as many mystics and yogis across the ages have called it, into everyday life. *Kundalini* is regarded as the highest potential of the human being.

Zeus in Olympic myth is almighty because of his 'thunderbolt' *(keraunos)* but the origin of this symbol goes far beyond the plane of sensual natural phenomena. It reveals the power of focused consciousness, just like its Himalayan counterpart, the Tibetan *dorje* and Indian *vajra*. In the centre of these symbolic cult objects is a sphere that represents *bindu*, the seed or germ of the universe. A spiral issuing from its centre symbolises the potential capacity of the *bindu* to grow and to become the whole world of consciousness.[12] From this centre, two stylised lotus blossoms expand, signifying the polarity in conscious existence. In the Yogic system, the sixth chakra, located between the eyebrows and connected with the pineal gland inside the skull, is associated with the powers of conscious presence, focus, and deeper insight. The dye for the red spot, the *bindhi*, that in Hindi tradition marks this point on the forehead, used to be made from ground yew bark, at least in Himalaya border regions such as Nepal.[13] In ancient Greece,[14] Aeschylus called Zeus' *keraunos* 'unsleeping'.[15] Reborn into a new life, the initiate entered a new state of awareness that could only be gained by mastering the *kundalini*. In the myths of western Asia this is expressed as Zeus' battle against the giant serpent monster, Typhon – a fight he almost loses.[16] Typhon is the offspring of Mother Earth and the lord of the underworld, and lives in the Corycian Cave[17] in Cicilia (a limestone formation). The motif of the taming of the serpent power goes as far back as the proto-Hittite myth of the serpent Illyunka who temporarily overcomes the storm-god, Telipinu.[18] (The dragon-slaying motif – a plain hero-versus-monster story – became popular in the European Middle Ages.[19]) The motif of the child of light being born in a cave recurs in Byzantine belief, where Jesus Christ is born not in a stable but in a limestone cave.[20]

38.4 *Typical shape of the Tibetan* dorje *(left); stone carving of the* keraunos *from western Asia,* c. *900* BCE.

The initiate, like the infant Zeus in his cradle on Mt Ida, hangs in the World Tree at the calm centre of the earth, in a mythical time before the eternal light has split into opposites (sun and moon). He is nursed by the (double)*helix*, 'spiral', of ida and pingala, and the dove (the female spirit of the tree) brings him ambrosia from heaven.[21] In the scripts of non-initiates, however, the *helix* became misinterpreted as *helike*, the nymph of the willow tree (Pliny, again!).[22] *Helix*, the world pillar, is also the place, of course, where Zeus unites with Europa (see previous chapter).

THE KEEPER OF THE GATE

The god Apollo became the very embodiment of the best in Hellenistic culture – cosmic order, divine harmony, beauty, music, art, poetry, justice and moderation – and yet he was not even Greek in origin.[23] Essentially, the Hellenistic Apollo was a migratory god of cosmopolitan character,[24] being 'invited' to or invoked at a great number of places; he had strong connections, both in myth and cult, with the Hyperboreans north of Greece as well as with Asia Minor. As for the ethnobotany of this deity, the laurel (bay) of his sanctuaries at Tempe and at Delphi is clearly a later Greek addition. No *direct* yew association can be found. Apollo is the son of Leto, however, the ancient Anatolian mother-goddess associated with *eya*, the yew, and she gives birth leaning on her tree:[25] hence, Apollo, after all, is born at the tree, perhaps in the tree, and – with the goddess being the spirit of the tree – ultimately, from the tree. So is his sister Artemis who remains connected with yew in her cult (see p. 167). Like his sister, he is an archer, which also links him with the yew tree (although Olympian religion does not dwell much on his archery). Other points of contact with the yew–tree–goddess pattern are that his priesthoods were often female, that they were at times named *Melissae*, 'honey bees',[26] and that his figure is intimately linked with the serpent power and prophecy (see pp. 159, 165).

Apollo's cult centre at Delos received annual offerings – certain unspecified sacred objects firmly

38.5 *Apollo with bow, hound and tree, Artemis with bow and deer; coin from Eleuthernai, Crete, fourth century* BCE.

38.6 *Figure of an archer; votive offering from Bronze Age Sardinia.*

wrapped in wheat straw – from the Hyperboreans in the north.[27] The identity of the Hyperboreans is a mystery. The common Greek interpretation of the name *Hyperborean* was 'beyond the north wind', although the root of the word might have come from a pre-Greek Balkan word for 'mountain', and Boreas might not originally have been the north wind but the 'wind from the mountain'.[28] There is also a high peak in Macedonia, Mt Bora (modern Mt Nidje). However, in myth the north wind often signifies the direction of the Pole Star, and hence of the world axis.[29] Thus there are allusions to the Hyperboreans as a heavenly race dwelling in Elysium (which is reached by vertical ascent along the world axis).[30] But from the fifth century BCE the understanding definitely shifted towards the geographical, from *above* Mt Bora to *beyond* Mt Bora, and *Hyperborean* came to denote either the regions directly bordering Greece in the north (Macedonia and Thrace), or the Balkans and Central Europe.[31] The Hyperborean offerings came to Delos via the amber trade routes, and a great number of placenames[32] suggest Apollo's presence along the Adriatic amber route, which was already extensively used in the Bronze Age.[33] This particular amber route followed the Danube river across the Balkans, and the Elbe river into Germania where we find another deity associated with yew and of a strangely similar name: Ull or Ullr.[34] He is one of the oldest gods in northern tradition, and later presented as one of the twelve *asir*, the principal gods who hold council with the Norns every morning at the foot of the World Tree and 'decree the fates'.[35] Ullr's particular feature is that he is the prime archer who, conveniently, lives in 'yew valley', Ydalir. In the winter, he moves on (yew-wood?) skis.[36] Ullr is generally thought of as an alter ego of Odin (p. 224) and is therefore, like Apollo, linked to prophecy. An oath sworn by Ullr represented the strongest pledge of loyalty.[37] The name of a more obscure Nordic deity, Ivaldi, has been suggested to have been an epithet of Ullr, and a form of *Iwa-waldan*, meaning 'the husband of the yew (goddess)' or 'the deity ruling/working in yew'.[38]

38.7–38.8 *These yews (up to* c. *3m girth) escape sheep and goat browsing on dramatic cliffs at* c. *1,300m altitude at a location called Itamo, 'Yew'; Lidoriki near Delphi, southern Greece.*

Most of Apollo's greatest holy places, however, are located in Asia Minor: in Lycia (between the Mediterranean Sea and the westernmost Taurus Mountains), in neighbouring Caria, and northbound towards Ephesus. He bore the epithets *Lykeios*, 'the Lycian', and also *Letoides*, after his mother Leto, because – still in Herodotus' time – the Lycians named themselves after their mothers, not their fathers.[39] Apollo's matrilinear epithet has been put forward as a strong argument for his Asiatic origin.[40] This original,[41] Oriental Apollo, was called Apulunas,[42] and he was the protector of gates. This was one of the sacred functions of yew in Hittite culture: 'at whose gate the yew-tree is visible [his house is free from imposts]'.[43] In classical Greece, too, Apollo *Agyieus* stood as a wooden pillar in front of houses to ward off evil from the doors.[44] (For the guardian function compare Heimdallr, Chapter 41.)

Although the legend of Delphi relates that Apollo on his arrival killed the giant serpent Python, serpent symbology remained central at the sanctuary, and the seeress kept the title of pythoness. Where Zeus almost fails – to master the serpent Typhon (see above) – Apollo succeeds with the first arrow. Bow and arrow are also common images in the quest for higher consciousness in Indian symbology: according to the *Maitrayana Upanishad*, he who practises yoga pierces through the darkness (of the illusionary world) and aims at Brahman, the immortal root of reality, with *manas*, 'thought, mind', being the sharpened arrow laid upon the bow of the human body.[45] Thus, the yew bow can be understood not only as practical but also as highly symbolic: the Tree of Life, the human body, and the yew bow represent each other.

'GOD'S TREE'

By the time of Roman rule the Greek god Dionysos had long degenerated into the rather shallow god of wine (Roman Bacchus), but he goes back even further than the actual introduction of the cultivated vine plant.[46] In myth, Dionysos is torn into seven pieces but reborn afterwards (this is reflected by one interpretation of his name as 'twice-born'), a motif reminiscent of the dismembering and reassembling of the 'spirit body' of the novice shaman in a number of Siberian traditions.[47] The crowns of the maidens in the Dionysic procession consisted of ivy, vine leaves, and '*smilax* (*milax* of the Attic speech)'[48] – the appearance of yew in the cult of a deity of initiatory rebirth, however, comes as no surprise.

Dionysos' upbringing seems to be a Thracian parallel to that of the Cretan divine child (Zeus): Dionysos is the son of Zeus in serpent form; he was born in a cave, and at his festival his priestess led his followers to the mountain for nocturnal dancing.[49] His cult came to Greece in the fifth century BCE, representing a reversion to pre-Hellenistic nature religion. Dionysos was nursed by a number of dryads (tree nymphs), especially Nysa, whose name means 'tree'. Regarding the search for his origin this rules out the other two plants,[50] vine and ivy, which were later associated with his cult. A dryad originally is the spirit of a sacred tree, and *Dios-nysos* consists of *Dios*, 'divine' or genitive 'God's', and *nysos*, the masculine form of 'tree'.[51] A classical source states that Dionysos was so named because he flowed from *Zéus* onto the trees (*nysai*).[52] This may indicate 'in-spiration' or pollination. Interestingly, Nysa recurs as Mt Nysa, the place where the three Fates give the 'ephemeral fruits' to Typhon (see above).[53]

HE WHO SAW THE DEEP

Preserved on clay tablets inscribed mainly in Mesopotamia during the two millennia BCE[54] is the story of Gilgamesh, the world's oldest epic. Gilgamesh[55] is the first figure after the Sumerian goddess Inanna (see p. 235) to descend to and return from the realm of the dead,[56] hence the Babylonian title of his story, *Sha Naqba Imuru*, 'He who saw the Deep'. Gilgamesh became the prototype of the mythical hero with supernatural strength, which appears in many myths and legends across the world. For the people of Babylonia, the legendary Gilgamesh was identical to a king who featured in the Sumerian king lists as the fifth ruler of the First Dynasty of Uruk[57] (which would place him in about the twenty-eighth century BCE).[58] The epic is not a myth, but contains many mythical motifs (the most famous being the story of the Deluge, see below). Furthermore, various details of the epic reflect ancient Mesopotamian customs, most noteworthy Gilgamesh's mourning rites for his close companion, Enkidu:

At the first glimmer of brightening dawn,
Gilgamesh opened [his gate.]
He brought out a great table of *elammaku*-wood,
he filled with honey a dish of cornelian.
He filled with ghee [purified butter] a dish of lapis lazuli,
he decorated ... and displayed it to the Sun God.[59]

At the current state of evidence, the interpretation of *elammaku* as 'yew'[60] (see p. 143) cannot be ultimately verified, but is highly probable: yew suggests itself because of its worldwide role in mourning and funeral customs (see Chapter 31), and because the other main woods for Mesopotamian palace furniture – cedar, cypress, juniper and box – have already been identified under different names.[61]

There is another episode in the story that is worthy of closer scrutiny. One of the deeds of Gilgamesh and Enkidu is their venture to the Mountain of Cedar, a place described as the seat of the gods, and the godesses' throne.[62] In the older Sumerian sources it is called the 'Living One's mountain'.[63] The sacred place has a mysterious guardian, the powerful giant Humbaba; his mother, according to the tablets, 'was a cave in the mountains'[64] (earth-womb symbolism, compare Zeus and Dionysos). Gilgamesh and Enkidu slay the guardian and search for the sanctuary to destroy it. They discover 'the secret grove of the gods, Gilgamesh felling the trees, Enkidu choosing the timber'.[65] They choose an especially magnificent 'cedar' to make a great door for the temple of the god Enlil, and build a raft to ship it down to his city, Nippur.

The cedar forest is actually a mixed woodland, also containing myrtle,[66] *ballukku*-trees and other species (whose names in the clay tablets have been damaged),[67] with thick undergrowth beneath a dark or shrouding canopy[68] – all in all not unsuitable for yew. King Solomon, we are told, ordered yew wood from the Lebanon, and the palace-builders of Nineveh used yew wood labelled as 'cedar' (see Chapter 30), a trade practice also known to Theophrastus. The possibility of *Taxus* in Lebanon, however, is disputed by botanists for good reasons:

the hot climate in the Levant has not significantly changed in the last 6,000 years. But micro-climates have.[69] And, since Lebanon in this context has long been suggested as a later Babylonian alteration, the original legend (Sumerian) could just as well have referred to the mountains to the north – the Syrian-Anatolian borderlands.[70] The fact that the tablets specify the Euphrates river as the transport route of the timber raft is even more telling for our enquiry: the upper Euphrates runs through the Amanus and Taurus Mountains, the very post-glacial 'yew heartland' of Asia Minor (see Chapter 36).[71] This location would also support the notion that the giant's name, Humbaba, could be a derivation of Kubaba, an ancient southern Anatolian name for the mother-goddess who in Phrygian times became known as Kybele.[72]

This analysis indicates that the Gilgamesh legend might contain a record of a historical raid by a Sumerian king of a Syrian-Anatolian mountain sanctuary where a goddess was worshipped under the local name Kubaba, and where the sacred grove inside the surrounding cedar forests consisted of or contained a cluster of yew trees. With or without yews, this incidence marks the historic shift of the religious focus from nature to the city and from living wood to dead wood. And if it is true that the giant Humbaba was originally identical to the goddess Kubaba, the Gilgamesh material would also incorporate a notion of the (pre)historical shift from female to male divinities,[73] a development that without doubt also had a dramatic effect on the ancient religious status of yew.

Tablet XI of the Gilgamesh epic introduces Utnapishtim, a legendary Sumerian who, like Noah in the Hebrew version much later,[74] survived the Deluge by heeding divine instruction to build an ark (see water symbolism, pp. 163, 168). The Flood legend was widespread in the Old World. In the light of the traditions of the Tree of Life, the motif of the dove returning with the sacred branch in its beak[75] is of much interest. The fact that even the biblical version still depicts the ark landing on Mt Ararat in eastern Anatolia strengthens the assumption that the story of the ark can be related to the ancient yew–tree–goddess pattern. The biblical 'olive' branch, however, is not likely to have grown in such a high altitude. Interestingly, the wife of Deukalion, the Greek flood-hero, and hence the ancestress of the post-Deluge Greeks, is called Pyrrha, which Graves[76] translates as 'fiery red' (compare the Knights of the Red Branch, p. 209).

CROSSING BOUNDARIES

The figure of the famous Greek super-hero Herakles (Roman Hercules) acquired many stories during his long life of popularity, among them motifs from the Gilgamesh epic. Herakles was most probably of Cretan Minoan origin, and his fame spread to Mycenaean Greece. Originally the spouse of the mother goddess, Hera, Herakles may have served as a mythical role-model for rituals of sacred kingship (see next chapter). His well-known Twelve Labours have been interpreted as variations of a common theme in legend: the set of 'marriage tasks' to define a worthy and fit candidate for the role of sacred king.

Herakles' Eleventh Labour consists of presenting to Athena some of the golden apples of the Hesperides. This magical tree grows in a paradisiacal garden in the west, where the sun sets, and is guarded by the giant, Atlas, his three daughters (the Hesperides) and the never-sleeping dragon, Ladon.

***38.9** Suitable hardwood for Herakles' club: dense* Taxus *grain grown on pure rock.*

Herakles kills the reptile guardian with an arrow (compare Apollo slaying Python), deceives Atlas and brings some apples to Athena, who smilingly returns them to their rightful owner, Hera, the real mistress of the garden. The tree of the Hesperides can be easily identified as a variant of the World Tree: the location at the world axis is encoded in the presence of Atlas who holds the pillars that keep the sky and earth apart; Apollodorus[77] even places the garden in the land of the Hyperboreans, although the common location is Mt Atlas where the chariot-horses of the sun complete their journey[78] (compare the 'gates of the sun', p. 171; see also the next chapter).

Two other elements in Herakles' story are also of interest: his famous club and his funeral pyre. Herakles usually fights with a wooden club cut from the wild *elaos* tree from Mt Helicon. During his adventures, he replaces it twice, always from the same species. When he leans the third club on a statue of Hermes it strikes root and sprouts.[79] When his death approaches, Herakles instructs his son to take him to the highest peak of Mt Oeta in Trachis, and burn him on a pyre of oak branches and trunks of the male wild *elaos*.[80] *Elaos (elaios)* is generally 'olive',[81] but neither the cultivated nor the wild olive has 'male' trees; olives are monoecious, and are not to be found as fine trunks in mountainous regions.[82] Could this be another hidden occurrence of *Taxus*? Could it be that *elate*, 'soft fir', existed already in ancient times as an epithet for (a female) yew, and *elaos, elaios* as an alternative name for a male yew tree? *Elaos*, after all, is the name of the tree that Herakles was said to have brought (see below) from the Hyperboreans, that is, from the polar axis 'beyond the north wind', where neither the domestic nor the wild olive grow. It was Apollo's tree. Cook[83] discounts olive because the wood that Herakles uses (his 'personal' sacred wood) for his sacrificial fire to honour Zeus at Olympia is wood from an *acherois* growing beside the Acheron river; usually translated as 'poplar', *acherois* might generally have denoted 'river tree', the river in question being not the one in Thesprotia but the underworld river.[84]

38.10 *The dark tides of the underworld river ...*

This river is the border to the realm of the dead, and in other texts the guardian tree by this river's side is called *leuke*, 'white',[85] as a synonym for fearfully pale and dead-looking – in the eyes of Homerian religion, that is. Or did it signify the white column of light that vertically connects all the worlds? However, as a verification of its nature is not possible, the exact identity of the tree of Herakles, son of Alkmene, remains unknown.

The Olympic Games

When the other legendary Herakles, the Dactyl from Crete (see p. 184), founded the Olympic Games in Elis (Peloponnese), he wanted to plant a sacred tree at the end of the race course. He is said to have ventured into Hyperborea, as far as the source of the Danube river 'where Artemis welcomed him among the fine trees'[86] of the sanctuary to Apollo.[87] 'The trees that he saw there filled him with wonder'[88] and Herakles persuaded Apollo's servants to give him the 'grey' or 'wild olive' to take back to Olympia. Whether he really went as far north-west as the source of the Danube (which is located in south-west Germany) or merely travelled up a tributary anywhere between the Balkans and the Alps, none of this is olive country. Also, while the establishment of the Olympic Games is pre-Mycenaean, a recent pollen analysis for the area does not show the presence of olive even in the Mycenaean levels: the introduction of olive-farming to the region of Elis happened only after the Mycenaean collapse.[89] The identity of Apollo's tree in this context was lost over time … and perhaps in translation: its fruits might once have been called *drupepes*, 'what matures on the tree', or *drupetes*, a term denoting fruits that fall off the tree of their own accord. Later, the meanings of both words became limited to olives and figs, because of the increased economic importance of these trees. And in Latin, *drupe, druppa* even came to mean solely the black (i.e. ripe) olive. But the Greek words came from *drus*, 'sacred tree',[90] and would have meant nothing more specific than a fruit-bearing sacred tree.

The religious environment of the early Olympic Games, too, is distinctly remote from the olive of Hellenistic times, and is instead steeped in pre-Achaen motifs from Crete:[91] at the site of the games, Herakles dedicated a hillock to Cronos (Saturn, El); on the northern side of it was a shrine to Eileithyia (Artemis as patron of birth), which housed a sacred serpent (with the most curious name Sosipolis, 'saviour of the city'[92]), and this snake was fed with honey-cakes by a white-veiled virgin-priestess; these ceremonial priestesses where called 'queens' (an allusion to queen bees).[93] Furthermore, the original Olympic Games consisted of a foot-race between young women for the honour of becoming priestess of Hera, and had been established originally by a priestess Hippodameia whose name is reminiscent of the Amazons (because of *hippo*, 'horse', see Chapter 40).[94] According to Graves,[95] the winner's prize of these early days was 'an apple-spray'[96] as a 'promise of immortality' (compare the Hittite kingship rituals, next chapter), and not before the Olympics were re-established in 776 BCE, after the civil war, did the prize become a wreath of olive – this time the olive proper, we should add, for as Pausanias says,[97] when the games were restored 'people had still forgotten the old days; little by little they remembered' – but obviously not the yew.

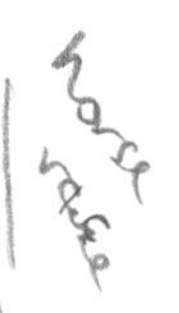

38.12 *Young yew tree in the mists of 'Hyperborea'.*

38.11 *Yew tree near the Upper Danube river.*

CHAPTER 39

THE MAKING OF KINGS

When in about the middle of the fourth millennium BCE the fundamental arts of high civilisation – such as writing, mathematics, monumental architecture, systematic scientific observation (of the celestial sphere) – were invented, the 'kingly art of government' was part of the package.[1] Initially, kingship was a social institution to ensure harmony between the human group and the land on which it lives. Until the Bronze and Iron Age cultures began to inflate and abuse the possibilities of kingship, it was about service rather than power. In various traditions, the king was appointed for a limited period of time only,[2] and an important, if not the central, part of his duties was the *hieros gamos*: the sacred marriage with the earth-goddess, the female personification of the land who granted and denied abundance and prosperity.[3] The sacred king would serve as a high priest representing the sky-god, while the goddess was represented by a high priestess; their ceremonial union could be literally sexual, or purely symbolic, depending on the different traditions. This was not only part of the fertility rites ensuring survival, but also an attempt to ask 'permission' for the changes to the landscape and effects on its creatures that human occupation (via hunting, farming practices, forest management, etc.) inevitably asserts. The earth belonged to herself and not to man – a concept which today we may call profoundly ecological.

HATTIAN-HITTITE ANATOLIA: THE GIFTS OF THE THRONE-GODDESS

In Anatolia, the Indo-European Hittites retained much of the older, indigenous religion of the Hattians. Only the goddess of the land, or 'throne-goddess', had the power to 'adopt' a candidate for kingship and to bestow the royal insignia.[4] In their ritualised pact the king agreed to administer and protect the inhabited land (which belonged to the storm- and the sun-god) and to respect her territory, the mountains[5] (see mountain thrones, Chapter 37). Their ritual union was sanctified by the power of the sacred tree: 'As the yew-tree is ever verdant and does not shed its leaves, even so may king and queen be thriving.'[6] This line from a ritual text also shows perfectly why the earthly representative of the Tree of Life had to be an evergreen.

39.1 *Golden armlet from a Celtic male burial, Rodenbach, Germany. The head at the apex is crowned with five yew arils, while thirty-six smaller chalice-like arils are placed between the four ram heads (a further two are missing in the bottom left sequence). Late fifth century BCE, diameter 6.7cm.*

39.2 *Scenes from the ancient Hittite kingship ritual: (centre) the king (on the left) receives the royal insignia from the goddess/high priestess who stands on a lion; (left) the front section of the entourage of the king, known as the twelve gods; (right) the high priestess's following. Motifs from the rock carvings at Yazilikaya, Turkey.*

In Old Hittite myth, Telipinu, the son of the great storm-god and fosterer of agriculture, restores the fertility and well-being of the land and its people: 'Before Telipinu there stands a yew-tree [*eya*]. From the yew is suspended a hunting bag (made from the skin) of a sheep. In (the bag) lies sheep fat. In it lie (symbols of) animal fecundity and wine. In it lie (symbols of) cattle and sheep. In it lie longevity and progeny. In it lies the gentle message of the lamb … In it lie plenty, abundance, and satiety.'[7] The reception of these gifts of prosperity by the king on behalf of his people was periodically re-enacted in ritual.[8] But the ritual also seems to have worked with living sheep: other tablets say 'he tied a wild sheep under an *eya*-tree', and 'set [it] free under the *eya*-tree'.[9] In Hittite cuneiform script the glyph for the name of Telipinu is a tree.[10] As for the divine 'owner' of the fleece and bestower of its goods, in one text the fleece occurs in the sanctuary of the war god Zababa (Greek Ares), but usually it belongs to the goddess Inara. In some texts the sign for fleece even stands for her name; she also bore the epithet 'Fleece-Inara'.[11]

Another part of the annual renewal of nature and human society was the so-called palace-building rituals – activities for the spiritual cleansing and renovation of the palace – which probably took place as part of the celebrations for the new year or the beginning of spring. On this occasion, the king encountered the female divinity as the goddess of fate as she reconfirms his reign and prophesies his lifespan:

39.3 *Portrait of Queen Teye; gold-plated yew wood, Egypt, New Kingdom, Eighteenth Dynasty,* c. *1355* BCE.

> But when the king enters the palace the throne-goddess calls the eagle, 'Come, I am sending you to the sea! But when you go, look which divinities are present in the forest.' And the eagle answers, 'I have looked. It is Istustaya and Papaya, the infernal, the ancient goddesses, who are kneeling there.' The throne-goddess replies, 'And what are they doing?' And the eagle answers, 'One holds a spindle, both hold a filled mirror. And they prophesy the life years of the king.'[12]

Istustaya and Papaya (in Hattic, Estustaya and Wapaya) are two Hittite goddesses of fate;[13] as in Greek and Norse myth they spin the life threads of mortals. The Hittite word for their class of deity, *guls-*, denotes 'to mark, to carve letters, to decree'[14] (compare Norse runes, see p. 223). Mirrors are common attributes of goddesses (e.g. Kubaba, Aphrodite) and symbolise the ability to see hidden things. Before the use of metal, mirrors consisted of a shallow bowl filled with water. Interestingly, the Hittite word for 'mirror', *huisa*, is derived from *huis*, 'to live'.[15] (Compare the mirror in Japanese tradition, below.) With the help of the eagle who represents highest vision or an insight into the spirit world, the infernal divinities appear in the forest, among the trees – or are they the spirits of the trees? The location of this sacred forest, however, is clearer: it stood just south of the shores of the Black Sea, in the 'Amazon' kingdom of Zalpa/Themiscriya (see next chapter).

These are the two things the Hittite king received at or from the sacred tree: (1) the sheep-bag that guarantees abundance and fertility during his reign, and (2) the gift of 'long years' (or even eternal life) for himself, ceremonially bestowed upon him with the 'authority' of the longest-lived tree of them all.

GREECE: 'THE GOLDEN FLEECE'

Greek myth abounds with indications that local sacred marriage rituals were once widespread on the Greek mainland, in the Aegean, and on the islands.[16] Ritual contests of the applicants for the offices of king and high priestess were part of many traditions, although at some locations they later lost their connection with kingship and developed into immensely popular and grand athletic games such as those of Olympia or Delphi. In myth and ritual, however, the tests of strength also incorporated the retrieval of magical objects, such as the sheepskin bag symbolising well-being and fertility. As the Hittite texts say, 'From the *eya*-tree is hung a sheepskin':[17] hanging in a male yew at the time of the renewal of nature in early spring, the fleece became gold-dusted by the pollen-releasing branches. Across the ancient Near East, the ram, for all its male vigour, was associated with kingship because both king and ram were representatives of the fertilising sky-god.[18] Furthermore, the ram symbolically supplies the very yarn for the goddess of fate, to spin the life threads of individuals and weave them together on her loom of destiny. A ram dusted with a golden pollen cloud by the Tree of Life itself would have been the most sacred animal in ritual.

This part of the Greek heritage from Asia Minor became known as the 'golden fleece'. Its most famous account today[19] is the legend of Jason and Medea.[20] In a nutshell: Jason, an exiled prince, has to go to the far-off land of Colchis (at the far end of the Black Sea) and retrieve the golden fleece, which hangs from a sacred tree in the sanctuary of Ares. He meets the hostility of the local king who does not want to part with it (understandably). The help of the king's daughter, Medea, however, enables Jason to get hold of the sacred item, and together they board a ship and flee from Medea's homeland, fiercely pursued by the fleet of the Colchian king. Thus, the hero gains the fleece to secure the vitality of his homeland, wins his kingship and finds his queen as well.

It is Medea, however, who holds the sovereignty and chooses Jason to become king. Medea is no mere princess but the daughter of the king and his first 'wife', the Caucasian nymph Asterodeia, which suggests that Medea is the child of a sacred marriage – a good position to begin a career as high priestess. As such she serves Kirke – the divine guardian of the world pillar, and the first mistress of magic in

39.4 *Medea in her serpent chariot; Greek vase painting from Policoro, southern Italy,* c. *410* BCE.

Hellenistic view (see Chapter 41). Like Demeter and Persephone, Medea is often depicted riding a snake-drawn chariot,[21] and her legend associates her with the cauldron of rebirth. She is a powerful witch who has mastered incantations and herbal knowledge, both of which she uses to soothe the guardian of the sacred tree, the immortal dragon of a thousand coils who was born from the blood of Typhon (the arch-serpent from the Corycian Cave – another hint at the Anatolian origin of the golden fleece legend, see previous chapter).[22] Medea also heals the wounded in Jason's band. In Greek tradition, she is said to have never died but to have become an immortal in the Elysian Fields as the goddess who presides over the cauldron of regeneration.[23] Perhaps she was this goddess all along ...

Regarding the journey of the Argonauts (Jason's team of heroes on board the *Argo*), geography alone (see map p. 214) provides enough hints to base this legend firmly in the old religious strata of the yew–tree–goddess pattern; there is no hint whatsoever of the (Indo-European?) oak-lightning-thundergod theme:[24] (1) The Argonauts board at the island of Samothrace, an initiation centre for the north, and become initiated into the mysteries of Persephone/Hekate.[25] The 'Samothracian Mysteries' were closely linked with those of Eleusis and Mt Ida,[26] and the Samothracian priestesses were referred to as the sisters of the Dactyls.[27] (2) The *Argo* stops over at three other islands associated with the Amazons (see next chapter): Lemnos, Sauromatia, and the islet of Ares,[28] and (3) takes harbour at Mariandynia,[29] the legendary homeland of the Amazons. Last but not least, (4) Medea and Jason marry in the sacred cave where Dionysos ('God's tree') was nursed.[30]

Hence it can be said that Jason and the Argonauts hardly missed a location connected with goddess worship and with Amazons; the voyage, one might say, actually seems like a pilgrimage on the goddess trail. It is unlikely, however, that this route reflected an actual ancient ritual journey; the legend as we know it is a fabrication of later times, after c. 750 BCE to be more precise, when Greek colonies around the Black Sea coasts had been established, and Colchis was known. However, being a world prime stand for monumental yew trees (some of them now regarded as up to 3,000 years old), Colchis was an excellent choice of the early mythographers. Originally, the easternmost location of the tree and fleece was probably simply a symbol for the point of sunrise as the magical gate between the worlds, just as the 'Apples' of the Hesperides were located in the far west at the point of sunset (east and west being the points where the sun enters

39.5 Taxus *pollen cloud – the golden fleece?*

for and emerges from its nightly underworld journey). The fleece symbolises the life force rising with the sun, from the eternal underworld to the realm of time and creation (pollen, spring), into this world; the 'apples of immortality' appear in autumn and in the West and are a token for the soul's journey out of this world. Both myths, that of Jason and the 'golden fleece' and that of Herakles and the 'golden apples' of the Hesperides, complement each other and are centred upon a sacred object that in ancient Greek would have been called *chrysomelon*, from *chryso*, 'golden', and *melon*, which means 'sheep' as well as 'apples'.[31] Can this be coincidence, especially when both myths originated in Asia Minor with its ancient veneration of *eya*, the 'apple'-bearing female and gold-dusted male yew tree?

It seems that in some local traditions the 'fleece' was not even a sheepskin but the pure pollinating branch. Apollonius Rhodius (born c. 295 BCE) describes the fleece as a 'cloud' glowing in the sunrise; when Jason picks it from the tree it throws a 'blush like flame' over his forehead and cheeks; and as he walks with it the ground begins to sparkle brilliantly.[32] In an Old Hittite legend, the mother-goddess Hannahanna sends out the bee to bring her the fleece, and the insect successfully returns to the sacred grove and puts the fleece into a bowl in front of the sitting goddess.[33] Bees carry pollen (not sheepskins) ... In the second century, Lucius Apuleius, a Platonic philosopher interested in contemporary religious initiation rites and himself an initiate of the cult of Isis, wrote down a tale in which the goddess Venus sets the beautiful princess Psyche the task of bringing her a tuft of gold from the trees by a certain spring where a group of sheep is grazing. She has to 'strike the foliage of the neighbouring wood' of the big plane tree in the field and that will turn the wool 'which is everywhere clinging and cleaving to the undergrowth' golden. Psyche is careful in everything and brings the golden treasure 'back to Venus'.[34] In the same century Pausanias related that at Corinth the priest was sacrificing once a year on the 'altar of winds', singing the 'incantations of Medea'.[35] Wind carries pollen ...

CELTIC IRELAND: 'THE LASTING ONE'

The concept of the divine female as the personification of the earth's abundance was central to Celtic religion,[36] not only in Ireland. Evidence suggesting the sacred marriage between her and the king abounds.[37] The closeness of the divine feminine and the land is beautifully illustrated by the Paps of Anu, a sacred double mound in Co. Kerry (located not far from the largest surviving yew wood in Ireland: Reenadinna, Killarney). Anu was an ancestral mother-goddess of Ireland (*an*, 'to nourish'), though particularly associated with Munster. Anu, also Dana, Danu, Ana, Aine,[38] was also a solar deity (*Aine*, 'delight, joy, brightness, radiance') as well as the Queen of the Underworld (*bean sí*, today known as banshee).[39] Dana is the ancestor of the first generation of gods of Ireland, the Tuatha dé Danaan.[40] She has three mythic sons and one brother, all with yew names. Of the four mounds on top of Cnoc Aine in Munster, one is dedicated to Dana herself, two to her sons, Eogabal, 'fork in a yew tree', and Uainide, 'yew foliage', and the fourth to her brother, Fer I, 'man of yew'.[41] In the tale *The Yew Tree of the Disputing Sons,* Fer I is able to create a 'yew tree of incredible beauty' out of nothing and in no time.[42] Dana's father, the great Dagda, 'good god', also bears a yew epithet: Eochu, 'mighty', derived from *ivo-katu-s*, 'who fights with yew'.[43] Hence it is fair to assume that Dana herself is a personification of yew

spirit; one of her epithets is *Búannan*, 'the lasting one'.[44] Another personification of the tree, this time from the Ulster Cycle, seems to be Scáthach, the extraordinary female battle-teacher and seer living in a mythical northern place called Alba. When the great warrior-champion, Cú Chulainn, arrives for his martial education he finds her 'in a yew tree', surrounded by her listening sons.[45]

Cnoc Aine is the sacred hill that contained the genealogy of the kings of all Ireland. The Celtic invaders respected the ancient 'yew deities' from the very beginning. The dynasty that ruled Munster from the seventh to the tenth century CE, for example, was called the Eoganacht, and gained the political right to rule from their association with the sacred tree. *The Exile of Conall Corc*, written down in the eighth century CE, describes the foundation stone of royal dynasties: 'I saw a wonder today on these ridges in the north, I beheld a yew bush on a stone and I perceived a small oratory in front of it and a flagstone before it. Angels were in attendance going up and down from the flagstone. "Verily," said the Druid of Aed, "that will be the residence of the King of Munster for ever, and he who shall first kindle a fire under that yew, from him shall descend the kingship of Munster."'[46]

In a tenth-/eleventh-century Irish poem, 'Baile in Scail' ('The Rapture of the Hero'), a maiden called Eriu, 'Ireland', gives every prospective king of Ireland a golden cup with 'red beer' *(derg-laith)* that bestows 'red rulership' (*derg-flaith*).[47] In Irish legends dating from about the fourth to the tenth century, the warrior elite protecting the royal house of Tara, the residence of the high king of Celtic Ireland, is called the *Craeb Ruad*, the Knights of the Red Branch. Later in the Middle Ages, the composers of Arthurian legend would use the fellowship of the Red Branch as the prototype for the Knights of the Round Table.[48] Their quest is for the Holy Grail, a development from the inauguration cups of Celtic and Germanic tradition, and still a symbol for what the spiritual aspects of sacred kingship had always meant: a harmonious mystic union of man and nature, of matter and spirit.

ROME: THE 'GOLDEN BOUGH'

The idea of kingship and sacred marriage originally comprised 'an extinction of ego in the image of a god (mythic identification)', but eventually degenerated into 'an exaltation of ego in the posture of a god (mythic inflation)'[49] – the emperors of Rome clearly belonged to the later psychological stage.

39.6–39.10 *Botanically, there is only one 'red branch' native to the British Isles; its colour especially comes out when wet with rain. (There is no colour manipulation on these images.)*

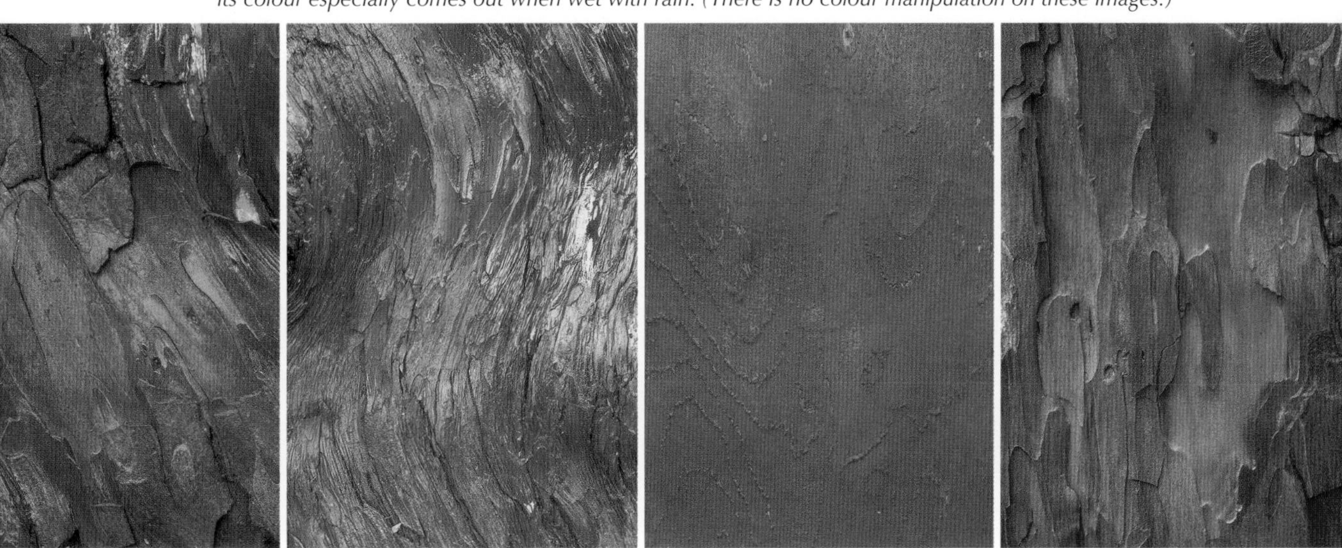

Already from about the eighth century BCE, Greek colonies in Italy and Sicily had been bringing Greek myth and legend to the indigenous populations. The story of Aeneas, the son of the goddess Aphrodite and a legendary hero of Troy, was one of them.[50] In Italy he was regarded as the legendary leader of the Trojan survivors, who came to Italy after Troy and was taken by the Greeks. As an enemy of the Greeks, Aeneas was particularly fit to serve as one of the celebrated founders of Rome's greatness, and Roman patriotic writers began to construct a mythical tradition that would dignify their state. The family of Julius Caesar claimed descent from Aeneas, and so did Augustus, the patron of Virgil's excellent version of the story, the *Aeneid*, in which the hero symbolised not only the course and aim of Roman history but also the career and policy of Augustus himself.[51] What a coincidence!

Aeneas, like other semi-divine heroes before him, has to enter Hades, and asks the help of the Cumean Sybil, who is a priestess of Apollo attending a remote sacred cave known as a 'gateway of the underworld'.[52] She tells him to go deep into the sacred forest of Diana (Artemis) that surrounds her cave and to get the 'golden bough – gold the leaves and the tough stem – held sacred to Proserpine [Persephone] … Fair Proserpine ordains that it should be brought to her as tribute'[53] – without it there would be no return to daylight. Aeneas would not have found the branch if it were not for two doves in whom he recognised 'his mother's birds'; when they settled on the wished-for tree, its branches 'gleamed a bright haze, a different colour – gold … And in a gentle breeze the gold-foil foliage rustled'.[54] Nowhere does Virgil indicate the identity of the tree that brings forth the golden bough.[55] But since Aeneas is credited with having brought the religion and the shrine of the Anatolian mother of the gods with him from Troy (whether subsequently after the fall of Troy, in the twelfth century; or later, in the time of the Greek colonies, in the eighth),[56] it can be assumed that the Italic priests of Apollo and the grove of Diana (Artemis, Kybele, Leto) knew about the goddess's tree. Surprisingly, the yew is indeed named in this context, about a century after Virgil, by the Roman poet Statius who says that 'Amphiarus had descended into Hades so abruptly that there was no time to purify him by a touch of the yew branch of the Eumenis [a 'Kindly One', see Furies p. 182].'[57]

39.11 *Yews guard the path to the Gola Orbisi cave, Gennargentu, Sardinia.*

Closely linked to the legendary tree of Aeneas is the sacred grove of Diana at Nemi. It was here that the legendary first king of Rome, Numa, in about 700 BCE had received the laws of the city from the *dryad* (spirit of a sacred tree), Egeria. This was an epithet of Diana, short for *Dea Aegeria*,[58] which originally denoted the 'goddess of the gateway tree to the underworld' (it became misunderstood as 'poplar', however). King Numa's mystical union with Diana *Aegeria* points to a *hieros gamos* tradition which, however, seems to have been long forgotten in the days of the empire. Diana's grove at Nemi is often associated with oaks, but on Roman coins and gems Diana is frequently shown with 'apples' (see figure 37.20).[59]

JAPAN: 'NUMBER ONE RANK'

The history of the monarchy in Japan is inevitably linked with that of the Japanese yew. According to legend, the top of Kuraiyama (Mt Kurai) in Gifu prefecture in the Chubu region was the abode of the sun-goddess Amaterasu, the progenitor of the imperial bloodline. A crucial ceremonial object in the enthronement ceremonies of the early kings was the

***39.12** Contemporary statue of a Japanese emperor, yew wood, c. 25cm high, Hida district, Japan.*

A statue of this sort and size takes about two months to make and costs about £3,000/US$4,500. Carved by Shigeru Murayama.

shaku, an oblong staff made of yew wood grown on Kuraiyama. The *shaku* or emperor's sceptre has traditionally been made of yew wood ever since. The tree is therefore called *Ichii*, 'number one rank'.[60] Only experienced master carvers are allowed to carve in centuries-old *Ichii* wood. The artist manufacturing a *shaku* partakes of the title *Sho-Ichii*, which translates roughly as 'rank of Imperial Court public servant': this is a name given to very important and prestigious positions only. In Gifu today, the Hida district is still famous for its *Ichii* crafts (see also figures 29.13 and 45.12), the *shaku* shown in figure 39.13 was carved from the same piece of wood as the *shaku* for the current emperor's inauguration ceremony in 1989. The *shaku* is also used by Shinto priests during ceremonies.[61] Interestingly, the word *shaku* also has the meaning of a 'Japanese foot' as a measure unit with the length of $^{10}/_{33}$ of a metre (= 30.303cm).[62]

An old text[63] about the age of the gods says that the deities arrived at Kuraiyama by heaven's floating boat, and made the mountain their abode from where civilisation spread across the world. The town of Miya at the foot of Kuraiyama has the *Ichii* as its village tree. Not far from Kuraiyama – less than 30km (19 miles) almost due east – and also in the Hida district in Gifu prefecture is another unique shrine: Ichiimori Hachimanjinja. It is considered to be the birthplace of the Japanese yew. The stone at the entrance informs us that 'here was the first *Ichii* in the whole of Japan!' Ichiimori means 'yew forest' – today there are about 160 yews between 300 and 600 years old. The entire complex, the shrine and surrounding yew forest (with mixed *sugi*, 'Japanese cedar'), is legally protected as a National Cultural Asset. Hachiman is the Shinto 'god' of war, that is, the divine protector of Japan and its people; his symbolic animal and messenger is the dove. It should be noted in this context that Jimmu Tenno, Japan's legendary first emperor and direct descendant of the sun-goddess Amaterasu, is traditionally depicted holding a crude and knotty wooden longbow (surprisingly, as Japanese archery is very refined[64]) on top of which sits his 'divine crow'.[65] In the language of the Ainu, the indigenous population of Japan, the word for yew, *Onco*, means 'Tree of the bow' and also 'Tree of God'.

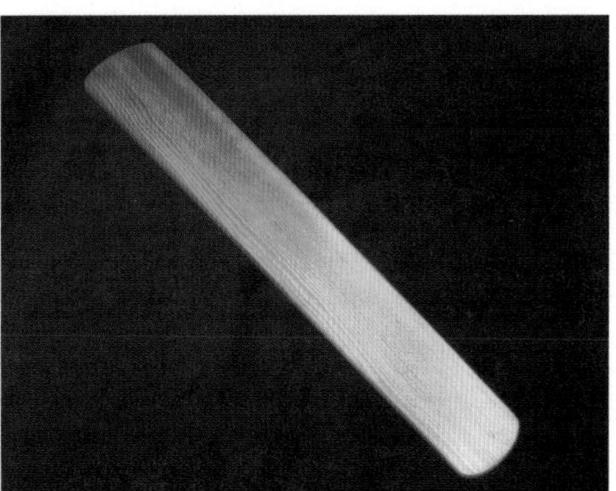

***39.13** A* shaku, *sceptre of the Japanese emperors.*

39.14 *The entrance to Ichiimori Hachimanjinja, the 'birthplace of yew', Japan.*
39.15 *Torii gate and sacred water bowl at Ichiimori Hachimanjinja.*

In Japanese myth, the World Tree is adorned with a mirror (possibly representing the sun and moon, as in Siberian Altai cosmology), so-called 'soft offerings' that consist of straw ropes, *nawa*, and paper strips, *gohei,* as are still used in Shinto today (see p. 133 and figure 29.11), and certain jewels representing the stars and which are also referred to as the 'pearls of the Pleiades', worn as a necklace by the goddess Oto-Tanabata, 'She who dwells in the skies' (compare Freyja's necklace *brísingar men*, p. 228). Together they comprise the regal insignia of ancient Japan – and still do today, although the 'soft offerings' have long been replaced by a sword.[66]

The symbols and mythological motifs surrounding the Japanese yew are surprisingly akin to those from Europe: at the dawn of history the divine powers descend onto a sacred mountain (e.g. Mt Olympos in Greece, Mt Ida in Crete, Mt Ida in Anatolia) from where they spread civilisation and heavenly order over the earth. The yew is there from the very beginning (the heavenly descent), and a sacred yew object or symbol is given by the deities to the first human kings as a sign of sovereignty over the land. In Ireland, Greece, Anatolia and Japan alike, the boundary between deities and yew trees frequently blurs, and it is impossible to know whether the deities came to earth through the trees, or whether they were their inherent spirits, only given human form in the mythical stories. Either way, in the political sphere the meaning of the myths continued in the inauguration rituals in which the representatives (priestesses and priests) of the deities (or yews) bestow the power to rule. The early ruling dynasties justified their claim to the throne by claiming descent from divine beings.

NATIVE AMERICA: A HUMBLE TALE

A particularly special position of yew in any of the traditions of the many indigenous cultures of North America is not discernible. There is a remarkable myth from the north-west coast, however, which blends nicely with the European as well as Japanese traditions mentioned above:

> O friend … I came sent by our friends to let you know the end of their speeches last evening when chief Yew tree called in those who have him for their chief, all the trees and all the bushes. And this was the speech of Yew tree, 'You have done well to have called for mercy all the trees and all the bushes, and also done well that you have purified yourself twice and rubbed your body with hemlock branches in the pond.'[67]

CHAPTER 40

THE DANCE OF THE AMAZONS

The patriarchal culture of Hellenistic Greece had no problems with recognising female divinity, as long as it was contained in legends and myths, or shaped into graceful marble statues. Even the free spirit of true and independent 'virgins' such as Artemis and Athena, with their fiery and incorruptible attitude and mastery of serpent power (*kundalini*), were eventually tamed into characters almost as domestic and predictable as the powerless and right-less housewives and female slaves of Athens. In the Greek *polis*, the ecstatic dances in the cult of Dionysos and other outbursts of the wild and irrational side of women (and men!) were likely to get bad press. The rare event of a small band of female warriors from Sparta, who might have casually passed through Athens on horseback, would probably have caused a palaver for weeks on end. The flame of unease would have been kept burning with recurring stories and military reports of female fighters among the Scythians,[1] a people that dominated the vast lands north of the Black Sea, confirming that a wild world lurked, a world where anything was possible, even a degree of social status for women.

40.1 *Goddess, framed by serpents and lotus flowers; painted terracotta plaque, late eighth century* BCE, *Athens.* This plaque, *c.* 24 by 12.5cm in size and with two holes for suspension at the top, was found at the Agora of Athens and probably brought there as a votive offering from the adjacent shrine of the Furies (Cook 1940, p. 189).

The Amazons are described by the Greek writers[2] as fierce and powerful women warriors, excellent horse riders, fighting with *bow and arrow* as well as with the *double axe*. They worshipped Artemis, and considered themselves (say the Greeks) daughters of Ares, the god of war. Amazons are generally found to be fighting on the 'other' side, that is, against the famed heroes of the confederation of Greek city-states, and they are usually located east of Greece: to the north (Scythia) and to the south (Asia Minor, see map next page) of the Black Sea. Greek literature added many fanciful and even shocking traits to their character, which shall be ignored here; with a lack of space for a detailed analysis of Amazon literature it shall suffice to say that the male Greek fantasy of the 'Amazon' partly overlaps with what we may call female warriors of neighbouring cultures, as well as (possibly armed) priestesses of the older, archaic religions.[3] It is the second group that is of interest to our study, as the geography of their occurrence considerably overlaps with the areas of yew distribution (see map).

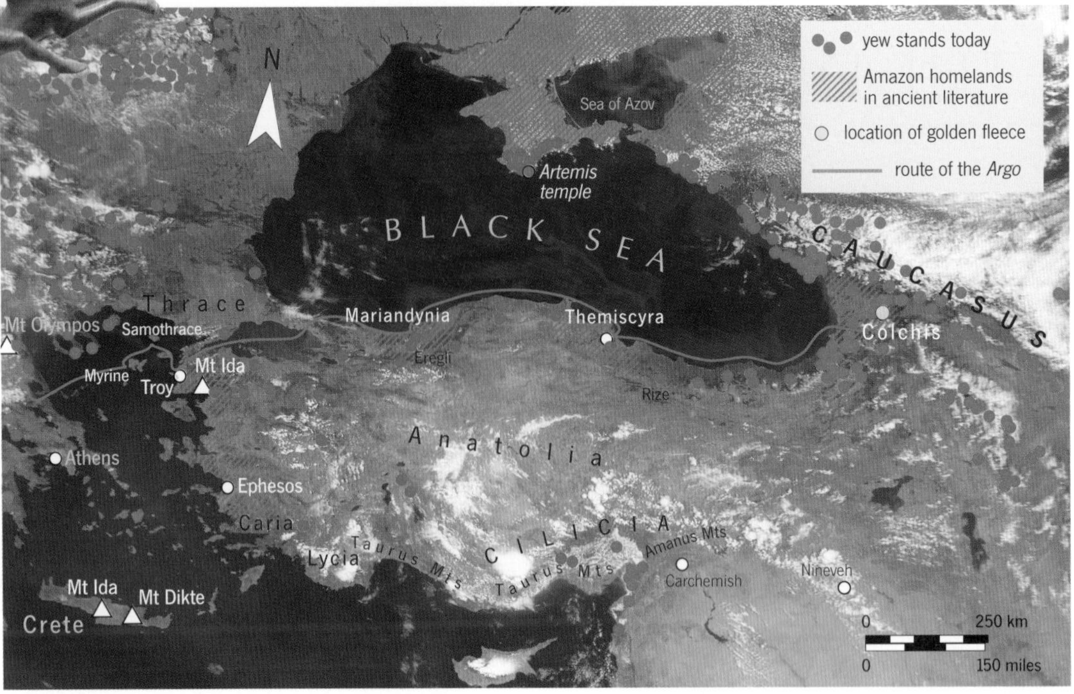

***40.2** Distribution of places related to Amazon legend and of present yew stands in the Black Sea region. (yew distribution after Browicz and Zielinski 1982) (For inset see 40.11)*

Going from east to west, there is Colchis (modern Georgia and south-western Russia) at the far end of the Black Sea. Here, the oldest grave of a female warrior has been found by archaeologists.[4] In Greek legend, Colchis is the location of the Golden Fleece. Today, it is one of the most important regions of ancient yew forests worldwide.

***40.3** Close to legendary Themiscriya, the monumental yew at Rize, Lazistan, Turkey.*

Secondly, there is Themiscriya (Zalpa), located in the hills of the eastern part of the Anatolian Black Sea shore.[5] This is described as the tribal homeland of the Amazons.[6] Its presentation in Greek legend would position its heyday sometime before the Trojan War in c. 1200 BCE.[7] According to Hittite tablets, it was in the first centuries of the second millennium BCE that the coronation rituals of the Hittite kings (see previous chapter) were performed at Zalpa; in all probability it was under the ancient trees of Themiscriya that 'the infernal, the ancient goddesses' spun the life-threads of the kings. Today, this part of Turkey is still known for scattered stands of ancient monumental yews which, theoretically, could just be the first- or second-generation offspring of trees that stood at the time of Troy.

Thirdly, there is the ancient land of Mariandynia, located in the hills of the western part of the Anatolian Black Sea shore. Greek literature does not mention it often, but modern Turkish research links the region's earliest traceable inhabitants (c. 2400 BCE)

40.4 *Today, ancient monumental yews (the dark, elongated oval shapes) rise over a 2m-thick Rhododendron cover in the hills of Mariandynia, modern Alapli, Bithynia, Turkey.*

40.5 *The second-greatest of the monumental yews at Alapli (girth 796cm, height 20m, alt. c. 1,200m).*

to 'Amazon' (i.e. matriarchal) culture. Graves relates the name, *Mariandynia*, to Sumerian *Marienna*, 'high fruitful mother'.[8] The region of Mariandynia encompasses the modern nature reserve of Alapli that holds the largest ancient yew population in Turkey, if not Europe. The modern name of the nearby district capital, Eregli, is the Turkish form of Greek Heraklia. It was here that, according to legend, Herakles in his Twelfth Labour dragged up the beast Cerberus (Hekate's dog) from the underworld, which once again links yew with the gate to the subterranean realms. The cave of Herakles, Cehennemagzi, has been sacred since ancient times. In the first century CE, one of the Apostles, Andrew, converted the cave and turned it into one of the first Christian churches.[9] The unique status of yew in ancient ceremony, however, still leaves its mark, and not only in rural folklore (see p. 179): in November 2004, the foresters at Alapli explained to our expedition that according to old management practice, the mature oaks and beeches had been successively removed from the woodland, but the yews had not been touched 'because you don't cut a yew if you don't have to'. (Rhododendron, however, has become a pest on some slopes, growing some 2m high between the ancient monumental yews and suffocating the natural regeneration of all trees; figure 40.4.)

And, most famously, there is Ephesus, the city founded by Amazons as a sanctuary of Leto. Originally, the goddess was worshipped in a sacred tree, being represented by a cylindrical wooden image that was narrower at the bottom (reminiscent of similar Minoan-Mycenaean columns;[10] compare the *ashera*, p. 191), and set up at a sacred tree of 'noble girth'.[11] Later, the upper part was given a human shape.[12] (It should be remembered in this context that still in Pausanias' time, the surviving religious images of early Greece were mostly made of cypress, juniper or yew wood.[13]) The earliest shrine at Ephesus

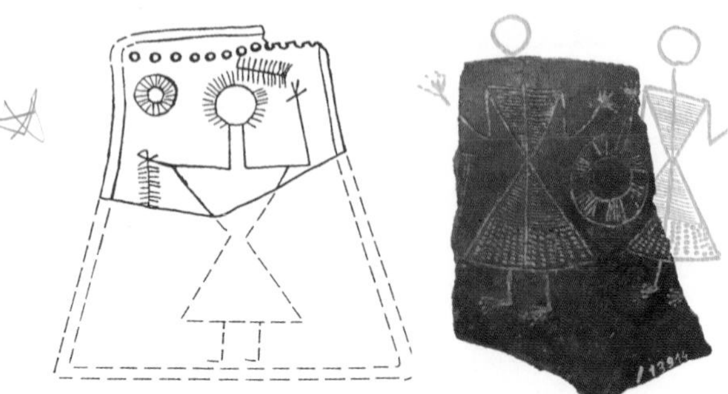

40.6 *Female dancers (carrying the sacred branch); motifs from pottery shards, Sardinia, late fifth millennium* BCE.

40.7 *Circle of women or priestesses, with doves; terracotta, Palaikastro, Crete.*

40.8 *Monumental yew at Rize, Lazistan, Turkey (same tree as figure 40.3).*

40.9 A labrys *(double axe) with sacred branch; Caria, first century* BCE.

40.10 Labrys, *tree and serpent symbolism on Minoan pottery.*

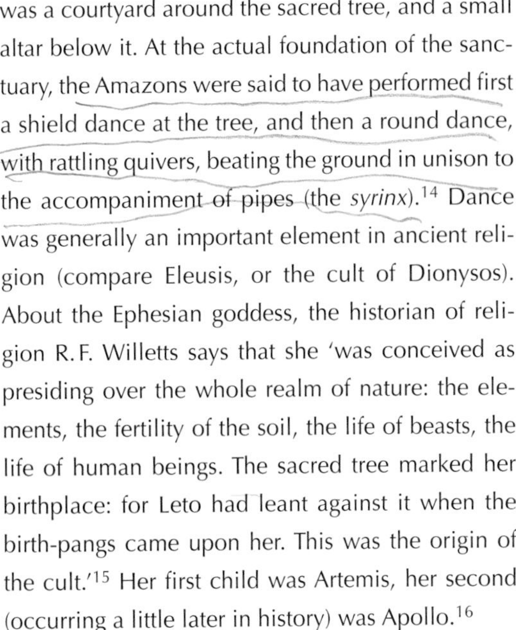

was a courtyard around the sacred tree, and a small altar below it. At the actual foundation of the sanctuary, the Amazons were said to have performed first a shield dance at the tree, and then a round dance, with rattling quivers, beating the ground in unison to the accompaniment of pipes (the *syrinx*).[14] Dance was generally an important element in ancient religion (compare Eleusis, or the cult of Dionysos). About the Ephesian goddess, the historian of religion R.F. Willetts says that she 'was conceived as presiding over the whole realm of nature: the elements, the fertility of the soil, the life of beasts, the life of human beings. The sacred tree marked her birthplace: for Leto had leant against it when the birth-pangs came upon her. This was the origin of the cult.'[15] Her first child was Artemis, her second (occurring a little later in history) was Apollo.[16]

The botanical identity of the Ephesian tree can safely be assumed to have been *Taxus*, the evergreen *eya*-tree of Leto, Artemis and Apollo (see Chapters 34, 36 and 38).[17] Evidence for the actual tree shrine has been found on the *northern* slope of the hill (this site is not identical to the location of the later temple). The northern exposition is typical for yew in southern climes, a fact already remarked upon by Virgil[18] (compare p. 166). The earliest remains of a shrine have been dated to the seventh century BCE,[19] but the site might date at least as far back as the thirteenth century.[20] In the sixth century BCE a nearby spot was chosen to build the first temple,[21] and eventually, after its rebuilding in 334 BCE, the temple of Artemis (now under her Greek name) became one of the largest Greek temples of all, unrivalled in magnificence.

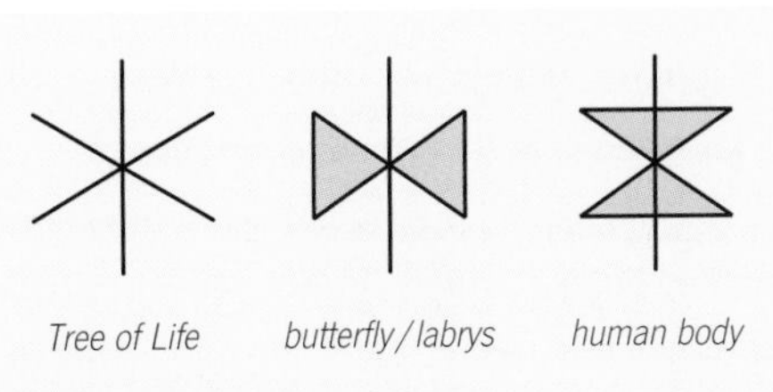

Diagram 14 Metamorphosis of the six-pointed star (Tree of Life)

The *labrys* or double axe had predominantly been a ceremonial item since the Stone Age[22] (particularly in Asia Minor and Crete), and in Hellenistic times was often associated with the Amazons. A diagram of the basic structure of the symbols reveals the relationship between the Tree of Life (a six-pointed star,[23] left), the human psyche as a butterfly, moving vertically (with additional vertical lines, centre), and the human physis, moving on the horizontal plane (with additional horizontal lines, right). The double axe has the same structure as the butterfly symbol, but it has turned into metal, and it can kill. In ancient cult practice, the double axe is intimately associated with sacred trees.[24] Its northern equivalent is the hammer of Thor.

40.11 The Amazons were famous for their archery on horseback; Graeco-Etruscan bronze work which decorated a cinerary urn, Capua, Italy, c. 510–490 BCE.

CHAPTER 41

THE WORLD TREE

HEIMDALLR

The concept of the World Tree was superseded over the ages by other religious and philosophical concepts. In the various cultures, it declined and disappeared at different times and at different speeds, apparently lasting longest in northern Europe. While Celtic myth and artefacts only contain hints and fragments of the World Tree tradition, a more coherent image has survived in Norse myth – the World Tree, Yggdrasil. Poems and songs preserving at least some knowledge about it were committed to writing in the Middle Ages.[1] But setting aside the *kennings* of Skaldic poetry, which do not refer to anything directly by name (see p. 149), the translators of the Icelandic sources have interpreted the references to the World Tree, *askr*, literally, as a common ash *(Fraxinus excelsior)*, and the 'myth' of the 'world ash' has persisted ever since. However, the case for yew as World Tree is quite clear. In the old Icelandic texts the World Tree is described as *barraskr*, 'needle ash' and *vetgrønstr vidr*, 'wintergreenest tree'. The main references, however, are to be found in the

41.1 *Another epithet of the Norse World Tree is* Lærad, *the 'glossy one' (*Grímnismál *21f, in Doht 1974, p.154).*

Voluspa, an intense and haunting poem that unfolds an apocalyptic drama told from the viewpoint of a prophetic sybil or pythoness whose vision encompasses the beginning and end of the world, and the dawning of a new cycle of creation.[2] Today, this poem is honoured as the national poem of Iceland.

Voluspa opens by addressing the human race as the 'hallowed seed, greater and humbler, sons of Heimdallr' (the translations in this chapter follow Ursula Dronke 1969–97).[3] *Dallr* is a rare, probably archaic, word for 'tree'. The term also occurs in northern Iceland as *dallr, dallur* to denote a wooden bowl for liquid food, sometimes with lid and handle. Elsewhere in Iceland such a bowl was called *askr/askur*. None of these terms, however, is limited to a single tree species. *Dalr*[4] was also a poetic term for 'bow' (beside *almr*, 'elm-wood bow', and *yr*, 'yew-wood bow'). By the Viking age, Heimdallr had developed into an anthropomorphic deity, a stout warrior guarding the rainbow bridge to the abode of the gods, but he really *is* the world tree because *heim* means 'world' – and as the world axis, he himself is the light-bridge to the divine world (see 'The column of light', Chapter 38).[5] The opening of *Voluspa* addresses the human listeners as children of the World Tree. The seeress then begins her journey to the early days of the world:

Nío man ek heima, níο ívidiur,
miotvid mæran, fyr mold nedan.
'Nine worlds I remember, nine wood-ogresses,
glorious tree of good measure, under the ground.'

These lines from stanza 2 give essential information about the tree. 'Wood-ogresses', however, might be a rather misleading term and may be better explained as tree nymphs or spirits (feminine, plural). For her interpretation, Dronke draws on two other Icelandic poems, *Hyndluliód* (35) and *Heimdallargaldr*, 'Incantation of Heimdallr', one of which says

41.2 The awe-inspiring trunk of the ancient yew at Fonte Avellana, Italy.

that 'one of the race of divine powers' was born from nine *ívidia*, 'giant girls' – giants, like the ancient Greek titans, being the 'old', later somewhat defamed nature deities; and in the other text Heimdallr himself says he is the son of nine mothers, all of whom are sisters.[6] Usually, *vid* means 'tree', but because the 'nine worlds' in this verse specifically refer to nine subterrestrial worlds Dronke suggests that these tree mothers may instead be roots.[7] The even more significant point for our study, however, is held in one single letter: *vid* is 'tree', but *ivid* would be 'yew tree' or 'root'.[8] This suggests that the ancestor of humankind, *Heim-dallr*, 'world tree', is emerging from nine yew roots who were seen as divine sisters. Dronke points out that this creation myth does not appear here as a narrative legend on which the poem elaborates, but instead is cited only briefly and 'articulates a religious mystery'.[9] This might help to explain why this cosmology has remained so obscure for centuries.

The 'tree of good measure' is of interest, too. *Miot* denotes precisely measured estimates and correct calculation for practical purposes, which makes the *miotvid* a tree of 'measured sufficiency, supplying the existence of the universe ... Yet by its nature all miot is finite, and fate itself may be seen as a measured thing.'[10] (And indeed, it is named as such, *miotudr*, in *Voluspa*, verse 45.) Referring back to the subterranean realm ('under the ground'), *miotvid* may also confirm the numerical exactitude of the nine root divinities. The inherent notion of a harmonic mathematical order of the universe and its movements directly leads into the field of the music of the spheres (see next chapter).[11] This line points out that the maintenance of the harmony of creation (also?) happens 'under the ground'.

'HONEY-DEW'

The notion of the Tree of Life producing divine ambrosia or nectar[12] is generally seen as a shared heritage of the peoples speaking Germanic, Persian and Indian languages (Indo-Europeans). This liquid is called *soma* in India, *haoma* in ancient Persia, and (among other names and kennings) *œdrerir*, the 'exciter of spirit', in Germanic tradition.[13] In Old Indian literature, *soma* is described as an extract from the fruits of the World Tree[14] – which suggests a rather juicy and sweet fruit (once again excluding the ash tree). Symbolically speaking, the World Tree is not only a radiant column of light but also an overflowing receptacle (for the divine ambrosia or soma). In Yogic tradition (see p. 196), the *chakras*, particularly the crown *chakra*, are receptacles of the continuously flowing stream of divine nectar, and often symbolically depicted with moon crescents. A number of Near Eastern traditions present the Tree of Life with a cup or bowl in its top.[15] This relates to the Jewish shofar (see p. 232) and to the symbolism of the cornucopia (see box p. 185), the cauldron of rebirth and its later variant, the Holy Grail. On the physical level, the liquid element was incorporated into ritual in the form of libations for sacred trees (that is: to water them); while the sacred trees in dry southern climes actually needed the water, drink offerings in the farmyards of Scandinavia (which lasted into the nineteenth century CE[16]) were poured into a bowl beneath the tree. In Norse myth, the Norns water Yggdrasil daily with liquid from the sacred well at its foot. Only at the two times when it relates to the context of liquids does *Voluspa* call Yggdrasil *askr* (commonly interpreted as 'ash') and it might very well be that the original meaning was the above-mentioned 'wooden bowl'. 'An *ask*', says the seeress, 'I know there stands … showered with shining loam. From there come the dews that drop in the valleys. It stands forever green over Urdr's well.' (*Vol.* 19)[17] (For receptacles, also compare the yew-wood buckets as grave goods, p. 152.)

In the physical world of sacrifice, the spiritual concept of the 'honey-dew' usually took the form of fermented honey mead,[18] the first alcoholic drink to be produced by humans (see 'The bee', p. 164). In none of the old sources is yew directly associated with ambrosia/*soma* – this assumption rests on the World Tree as a mutual link. Whether *soma* is the extract from the fruits of the World Tree, as the old Indian texts say, or the product of the 'honey-fall' dropping from Yggdrasil, as the Scandinavians tell us, both can be related to *Taxus* whose sweet arils usually stay on the tree well into winter while single overripe ones drop to the ground like little honey-balls and split open[19] (or did the 'honey-dew' for the production of consecrated mead come from bees living inside trees? – compare Virgil's warning, p. 166). Arils can be fermented on their own[20] or mixed in during the honey mead production process, but since they are the only part of yew which does not contain poisons (alkaloids), *soma* from yew arils would have had to contain other toxic ingredients. In which traditions and to what extent other substances were actually combined with the alcohol to create the psycho-active effects of *soma* is not known, but natural intoxicates such as hemp *(Cannabis)*, opiates, fly agaric *(Amanita muscaria)* and the members of the nightshades family (Solanaceae) have been known to humankind from early times,[21] and, of course, so have the toxic needles of yew. However, there is only one historical piece of evidence from Europe that mentions yew in a psycho-active context: a treatise by the German chemist Friedrich Hoffmann (1660–1742) lists *Taxus* poison, together with opiates and nightshades, as an ingredient of the 'witch balms'[22] (drug creams to rub on the skin for mind-altering experiences, fairly widely known in the Middle Ages; compare 'A hallucinogen?', p. 49).

***41.3** Aril in the dew drops.*

Whatever the details, mead in ancient Nordic tradition was central in a major sacrificial rite, the drink offering of the *disablót*. This particular festival was held in honour of the Dísir, female vegetation-spirits or land-spirits. The *disablót* was a collective tribal activity involving the consumption of huge quantities of mead. However, it was not a mere opportunity to carouse, it involved the presence of people of high rank (such as priests and priestesses, aristocrats, and at times envoys from the king), the consecration of an altar, and the incantation of sacred songs while the drinking horns were raised. The aim of the *ritu libationis*, the holy drink offering, was not to get out of one's mind, but to attune to a higher reality than the personal: the *Havamál* says that 'the best about intoxication is that everybody *gains* his consciousness'[23] (my italics). The *disablót* lasted for a period of nine days. Its position in the calendar is not certain, but thought to have been between 21 January and 20 February.[24] Originally, the Dísir were motherly earth spirits, fertilised by the sky-god, and the *disablót* seems to have been connected with the sacred marriage rite, with Freyja holding the sovereignty of the land, and either Thor or Ullr being the divine spouse[25] (and ceremonial title of the king). At Uppsala, the old Swedish capital, a major *disablót* under the attendance of the high king of Sweden was held every nine years.

41.4 *Ancient yew at Chillingham Castle, Northumberland. Was this the coronation site of King Ida?*

At the heart of the sanctuary stood a huge evergreen tree, widely accepted today to have been a yew.[26] Two intriguing placenames near Uppsala are *Uller-åker* ('Ullr's field', for the bow god's association with yew, see p. 198) and *Ulltuna* ('Ullr's yew grove'; pronounced *toonah*, not like the fish).[27]

THE NINE

Among the Germanic and Celtic peoples, the cults of female deities and guardian spirits were important in everyday life, and were by no means confined to women and children.[28] It seems that the above-mentioned *ívidiur,* the nine yew-root spirits that nourish the World Tree, belong to the class of Dísir or land-spirits. In Ireland, Dana and her divine yew-clan may represent an equivalent, and just as the early settlers in Iceland made a voluntary 'treaty' with the local Dísir[29] (to avoid misunderstanding: there are no yew trees in Iceland, this is referring merely to the custom of respecting guardian spirits), the first Celtic invaders of Ireland made their treaty of respectful co-existence with the Danaans (see p. 208). A single female Dísir is a Día, and this raises the possibility of a connection to the names of the goddesses Diana, Dana and Dione. Furthermore, if a single 'motherly' yew-spirit was called *ídia,[30]* and the abode of the twelve divine powers (Asgard) is in Idavelli, 'yew valley'[31] (and similarly in old Irish, *ida, idha* means 'yew'), is there an ancient connection to the name of the Trojan and Cretan mountains *Ida* which were connected with mountain-mothers and the reverence of a sacred evergreen tree – when even the classical sources state that Mt Ida was named after its trees?[32] Do Yggdrasil's nine *ívidiur* share the same roots (literally) as the nine Muses from the Taurus and Amanus Mountains?

Is the name of Ida (reigned from 547, died 559), the first king of Bernicia (modern north-western England), related to this? When his Anglians arrived in Northumberland, he took up royal residence at Yeavering Bell, an ancient tribal capital and hillfort in Northumberland. Yeavering Bell also contained a sacred precinct, and its name means 'place of the (sacred) yew tree' (see p. 140). The earth-goddess Nerthus was worshipped in the Anglian homeland, and also many men from various Germanic tribes (e.g. as Roman soldiers) would have come across the goddess Kybele who was worshipped throughout the Roman Empire[33] as *Mater Idaia*, 'the mother from (Mt) Ida'.

The Celto-Germanic world was never as far removed from western Asia as one might expect: being the westernmost 'appendix' of the vast Asian continent, the European peninsula received continual influx from Asia throughout its prehistoric period.[34] And from the Bronze Age on, the amber trade routes were the southern extension of an intricate trade network spanning the entire pre-Roman Celtic and Germanic north, connecting it with the cultures of Greece, western Asia and the Near East.

The number nine appears to be the number of 'worlds' or dimensions connected by the World Tree in a great many traditions across Eurasia.[35] Whether they refer to nine altogether or to nine heavens and nine underworlds (and nine realms in this world), the recurring theme is the ninefoldedness. There are nine yew root mothers of the World Tree in *Voluspa*; nine Muses or mountain-mothers connected with trees in Anatolian-Minoan-Greek culture. In myth, Demeter searches for Kore for nine days; Persephone releases the souls of the dead for rebirth after nine years;[36] the Greek ark floats for nine days until a new world dawns;[37] Artemis chooses 9-year-old nymphs for her attendants;[38] it was said that it would take a falling anvil nine days to reach the bottom of Tartarus (the lowest region of the underworld);[39] at his death, Thor takes nine dying steps through the underworlds.[40] The sacred number is reflected in cult practice, of course: at Athens, the annual Festival of the Wild Women (the *Lenaea*) was presided over by nine priestesses, and a bull, representing Dionysos, was cut into nine pieces; the annual purification ceremonies at Lemnia, also known as the festival of the Kabires, lasted for the space of nine days, during which offerings were made to the dead;[41] the Cretan Kouretes are associated with this number;[42] the Greater Mysteries at Eleusis

41.5 *An Iron Age wheel with the unusual number of nine spokes was found at Ryton, Tynedale, north-east England – like nine roots, the spokes support the central 'trunk', that is, the axis.*

lasted nine days, so did the Germanic annual *dísablot*, while the major *dísablot* was held every nine years; the Anglo-Saxon farmer humbly bowed nine times before uttering his harvest prayer.[43] In Jewish Kabbalah, nine resembles the womb in which new life grows, also in a spiritual sense of renewal and regeneration.[44] The system of Chinese numerology is entirely built on the nine, which also is the luckiest number; Chinese numerology associates nine with humanitarianism, and self-sacrifice[45] (compare Odin). The Siberian Buryat shaman's 'career' involves nine successive initiations;[46] Odysseus' boat journey to the underworld takes nine days, and it is for this same period that Odin hangs in the World Tree, sacrificing 'himself to himself' in the quest for higher knowledge ... eventually he receives the rune lore in 'nine lays of power'.[47]

VISION QUEST

The figure of Odin is the prototype or first initiate of the quest for wisdom that can only be gained by sacrifice and deprivation.[48] Information on three of his attempts to gain deeper knowledge has survived in the Nordic material. Least known is Odin's theft of the 'mead of the

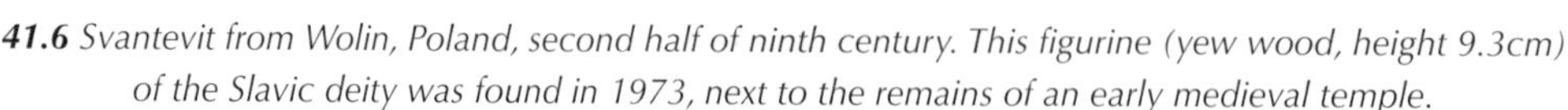

41.6 *Svantevit from Wolin, Poland, second half of ninth century. This figurine (yew wood, height 9.3cm) of the Slavic deity was found in 1973, next to the remains of an early medieval temple.*

poets' (identical with *œdrerir*), for which he stays three nights with the giantess Gunnlod at her abode deep inside a mountain: 'Gunnlod gave to me, on a golden chair, a sip of the magic drink.'[49] The golden chair emphasises a solemn, ceremonial atmosphere of an underworld ritual.[50] Odin leaves the (world) mountain in the shape of a serpent.[51] In another journey he visits the giant Mímir who guards the well at the foot of Yggdrasil. For a drink of the magic mead, Odin gives one of his eyes which, symbolically, leaves him with one eye for the physical world and one for the spiritual.

The best-known of Odin's endeavours in the search for higher truth, however, is his vision quest spent hanging in the World Tree for nine days and nights, wounded by a spear. His extreme physical exhaustion and deprivation eventually leads to an altered state of consciousness in which a magical horse takes him through the different worlds – the mystical horse can be understood as another form of the mystical tree itself,[52] when human perception shifts from the physical, static dimension to an energetic one. The Tree's name, *Yggdrasil,* is generally taken to mean 'the horse of Odin',[53] but also, 'yew pillar' has been suggested.[54]

Of interest in this context is an anthropomorphic statue made of yew wood that was found at Ralaghan, eastern Ireland. The artefact is 113.5cm high and has been radiocarbon-dated to 1096–906 BCE.[55] Like one of the figures of the Roos Carr ensemble (see p. 164),[56] it has a left eye socket that is less deeply cut than the right eye socket.[57] For this reason, Bryony Coles, specialist in wooden artefacts at the University of Bristol, has linked these figures to Odin, or rather the prototype of the one-eyed shaman. The fact that the Ralaghan figure is both chronologically and geographically outside the radius of the Germanic tribes suggests that the figure of the one-eyed shaman was a Eurasian archetype from very early times onwards. His connection with the yew tree might be just as old. In ancient times, this predecessor of Odin might have been called Ullr.[58]

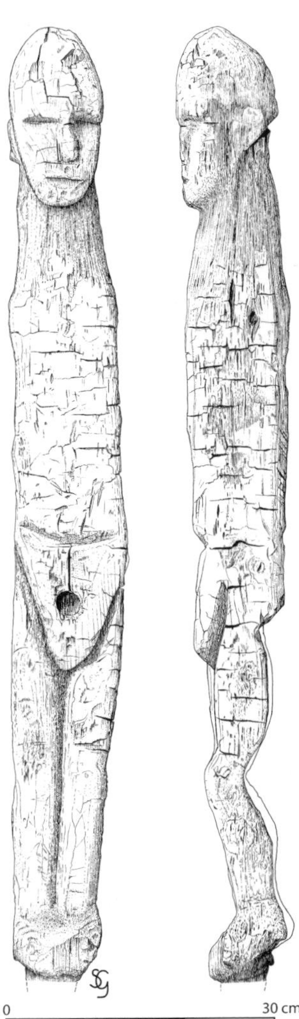

***41.7** The Ralaghan figure, eastern Ireland, eleventh to tenth century BCE.*

THE RUNES

Odin brought back the runes – a powerful means of sharing secret knowledge with his people. Systems of writing are a significant step in the civilisation process, and in the Old World, alphabets were deemed magical as they have the power to immortalise the temporary, spoken word and overcome time. Systems like the runes, however, were more than 'alphabets'; each sign was a symbol conveying *meaning* (unlike our letters). They were used for divination (soothsaying) but can also be used in a systematic way to study the laws of nature. The oldest surviving set of runes, the Older Futhark, was in use between *c.* 200 BCE and *c.* 500 CE.[59] All of the shapes of its single runes can be found in the matrix of the six-pointed star (enclosed in a hexagon), an ancient symbol for the World Tree. Out of the twenty-four symbols, two are named after trees: *berkana*, 'birch', signifies motherhood and protection, and *eiwaz (ihwaz)*, 'yew', symbolising death and transformation.[60] The Younger Futhark has a different rune for yew, *yr*, which has the shape of the three roots of the World Tree. In the sacred calendar of Northern tradition, *yr* is

41.8 *After days of fasting and deprivation, the patterns of needles might begin to look very different.*

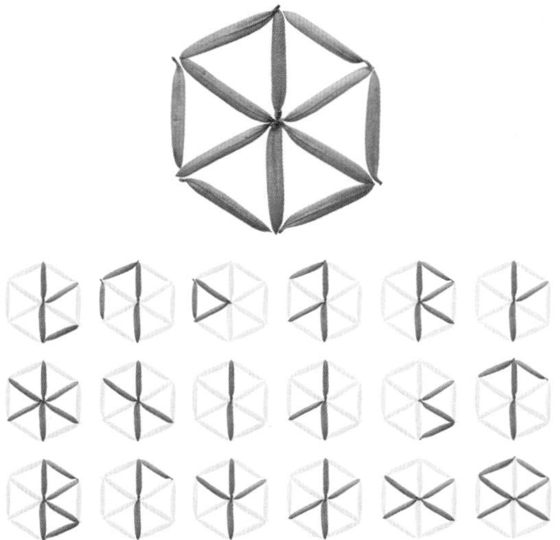

***41.9** The six-pointed star (compare figure 32.3) as a possible matrix for the runes.*

associated with the winter solstice, which marks the transition into the new year and hence was celebrated with rites of renewal and rebirth:[61] the Saxons celebrated the three nights of the winter solstice as the *modraneht*, 'mother nights', with a ritual descent into a cave; among the Scandinavian tribes the winter solstice marked the beginning of yuletide (from *hjól*, 'wheel', meaning the wheel of the year). Yuletide was a joyous feast that lasted for thirteen days;[62] thirteen is also the number of the yew rune *eiwaz* (for the symbolism of numbers, see

***41.10** The Norse yew runes* eiwaz *and* yr*; and the Irish ogham rune* idho, *'yew'.*

the next chapter). In some of the Anglo-Saxon and Nordic rune calendars, the World Tree appears as growing upwards (at the position of the winter solstice) as well as downwards (at the position of the summer solstice); the same theme appears in old Cretan seal stones (figures 41.11a).

A number of old runic talismans have been found in the coastal regions of the North Sea, and they are mostly made of yew wood. A yew wand from c. 600 CE, found at Westeremden, Netherlands, holds a spell to calm the storm and the waves of the sea. The inscription on a wand from the ninth century, found at Bristum, Frisia, has been translated as 'Always carry this yew. It contains strength.'[63] A short wooden sword found at Arum, Netherlands, dates from the first half of the seventh century.[64] And a 'weaving sword' (made of yew wood) from the Westeremden find may be compared to the 'weaving knives' from the Neolithic Swiss lake villages (figure 37.7). Similarly in Celtic Ireland, yew was the principal wood for carving ogham runes.[65]

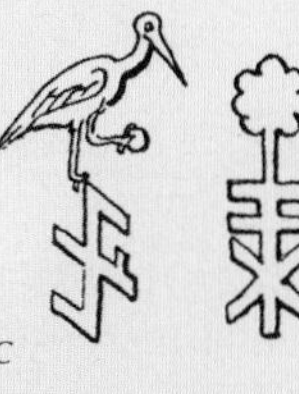

***41.11** The yew rune in different traditions.*

The shape of the yew rune denotes a tree that grows downwards and upwards at the same time, which – curiously – is also reflected in these sealstones from ancient Crete (a). The Germanic tribes along the North Sea coast also associated the life force *(od)* – that is so vividly expressed in the yew tree – with the stork as a bringer of life, hence the bird's Old High German name *odobero*, 'od-bearer' (later *odebaar, adebar*). In various (Saxon, Dutch, Hessian, Swabian) dialects, the bird's name remained akin to that of the yew: *euver, iwwer(i)ch, oiber, auber*. Until the early eighteenth century, the stork appears in pictorial traditions in conjunction with the yew rune and/or the Tree of Life rosette, for example, in apex stones from the March of Brandenburg near Berlin (b) and Twente, Netherlands (c), or local heraldry from Groningen, Netherlands (d). Sample (e) from Utrecht, Netherlands, combines the Tree of Life rosette and the yew rune.

THE WEB OF LIGHT

The World Tree as world axis is naturally located at the centre of the world. This is a mystical 'centre' (as described in Chapter 32), the access to which lies within the human soul; hence there is no competition among sanctuaries in 'real space' to be the one and only true centre (there are as many true centres as there are people in the world). The very heart, the holy of holies, is located within a protective sphere, which in many traditions is symbolically depicted as a circle (around a smaller circle or dot). In myth, the 'centre' can be found encoded in various ways; if it is not a tree, mountain or pillar it may be called an island at the 'navel of the earth' – with the term 'navel' signifying, as in the embryo, that the entire earth plane is nourished by a greater reality surrounding it. In the *Odyssey*, this has survived in the islands of two goddesses, Calypso and Kirke. Both these goddesses are guardians of the sacred circle. Both sing while they are weaving on a loom, like the Fates and Norns. Calypso weaves with a golden shuttle (possibly the sun as a kind of cosmic shuttle, see below). She is the daughter of the giant, Atlas, who holds the sky and earth apart, that is, he is the world pillar. As for Kirke, she herself is the daughter of Helios, the sun, and on her island, the sun is always in the zenith above the island, hence no one

41.12 *Designs of ancient Greek 'praying-wheels' made of polished bronze and set spinning with a jerk (Cook 1914, p. 256).*

casts a shadow,[66] and Odysseus, visiting on his mystical (inner) journey, becomes disorientated and does not know where east and where west lies.[67] The sun in constant zenith is another allusion to the cosmic axis (compare the column of light, p. 195). Here, at the still 'centre'[68] and under Kirke's wise guidance,[69] Odysseus can visit the underworld – and return safely.[70]

Kirke's very name might have derived from *kírkos*, 'circle' (Latin *circus, circulus*), which is also the meaning of the Sanskrit term *chakra*.[71] This would confirm her function as the goddess of the sacred circle. Because of her solar character,[72] however, it is widely agreed that her name may have derived from Greek *kírkos*, 'hawk', a bird which in many religions is a symbol of the sun,[73] and in Greek literature frequently associated with Apollo.[74]

41.13 *Weaving the light …*

41.14 *Another weaver in the yew tree …*

(Hawks attack from the direction of the sun: they come with and suddenly emerge from the blinding light.)

Gold is a well-known symbol of light, of the sun, but not only the physical sun: it represents the *divine* light, and hence eternity.[75] The golden apples of the Hesperides, the golden fleece, the gold pendants of butterfly and cicada 'souls' (see p. 164), the glyph for gold on the Astarte scarab (see figure 37.23) must be understood in this light. The fleece gives the raw wool for the weaving of the cosmic tapestry, and because the real threads are made of light – eternal light – the symbolic fleece is golden. In Scandinavia, Heimdallr was also called *Hallinskidi*, 'ram', and seems to have been symbolised at times by a golden ram, while his horse's name is 'gold-top'.[76] The deep mystical meaning of gold is perhaps best summed up in a line from the Indian *Isa Upanishad:*[77] 'The face of truth is covered with a golden disc.'

The cosmic loom

In Japanese myth, the most sacred place in the heavens is *imi-hatadono*, the 'sacred weaving hall'. Here, in the centre of the universe (also the location of the Japanese World Tree[78]), the world's destiny is determined. The weaveress is the sun-goddess herself, Amaterasu, or her helper, Wakahirume, 'Young Day-Woman', another guise of the solar deity. In a third version,[79] she has 100 (female) helpers weaving the 'divine fabric' from the light of the stars. But never is the receiver of the divine garment specified; hence it has been suggested that it is made for the sun-goddess herself: her cloak of light. Encompassing the entire universe, it is not a random adornment of mother nature, but an 'expression of her innermost being', and at the same time, her veil.[80]

In Norse myth, the Norns have set up a loom of equal proportions,[81] which incorporates the solar shuttle moving the weft from east to west, and the

Pole Star to which the warp is aligned as an unbreakable cord:

It was night in the dwelling. The Norns came,
those who shaped the life of the prince …
They twisted firmly the threads of fate …
They set in place the strands of gold,
held fast in the midst of the hall of the moon.
East and west they hid the ends;
the prince's lands would lie between;
while Neri's kinswoman knotted a cord
fast to the north, and forbade it to break.

FREYJA

The Germanic goddess Freyja shares most aspects of a divinity of nature and the wilderness with Artemis, and like her she presides over fertility and birth. Also like Artemis, she was immensely popular among her people. Furthermore, Freyja has a magic girdle or necklace made of gold and named *brísingar men*, 'the fiery, flaming ones'[82] (the Japanese queen of heaven has a similar necklace called the 'pearls of the Pleiades'). With Kirke, Freyja shares the association with the hawk – she has a feather coat and is able to fly as a hawk, she is the 'mistress of the hawk's plumage'[83] – as well as that of spinning and weaving.[84] Freyja is also the arch-witch, the mistress of *seidr*, the shamanic art of northern tradition[85] – her power lies in the realm of transformation (compare Medea, p. 206). In Iceland and Scandinavia, brewing the (sacred) mead and beer was performed by women, and Freyja was the divine patroness of this activity.[86] Despite combining solar and terrestrial links (as Aine/Dana, p. 208), she is a chtonic deity, too: she is the goddess of love as well as of death (like Aphrodite); 'until I come to be with Freyja' means until one is dead.[87] Originally it was she who received the souls of the dead (in an otherworldly marriage, like Persephone); only much later, in Viking times, did the warrior aristocracy propagate the idea of half the share of the dead for Odin's warrior hall, Valhalla.[88] The only godesses who appear frequently in Norse myth and are actually known to have received worship in the Viking Age are Freyja and the more domestic Frigg (goddess of marriage, Odin's wife and hence Queen of Heaven). The boundaries between them are blurred,[89] however, and it is likely that the two have developed out of one original Germanic goddess. The material regarding Freyja gives clues to the former spiritual and political importance of a great goddess of the north, and her power to determine the rise and fall of kings.[90]

Was this goddess of 'unfailing regenerative power' (Dronke[91]) in any way associated with the yew tree? In the surviving sources, nothing can be found under her name. But then, giving many different names to one goddess was a northern custom[92] just as it was in the ancient East, if not more so. Let us turn to Idun (Iduna), who has been suggested to be another guise of Freyja.[93] She is the goddess said to supply the apples of youth to the Asir, that is, she knows the secret of regeneration. The apples, of course, are 'golden'. When she is abducted by the giants (the chtonic powers, compare Hades and Kore), Loki flies her back (borrowing Freyja's feather coat, significantly) after she changes into 'nut shape'.[94] Is this figure who holds the secret of immortality, supplies magical 'apples', and regenerates herself from a 'nut' perhaps a tree? Is her name, Iduna, related to Ida, Dana, Diana? A glimpse into Norse genealogy reveals that Idun is the daughter of no other than Ivaldi[95] whose name means 'the deity ruling in yew' (or 'the husband of the yew (goddess)') and who is generally associated with Ullr.

Another incident suggesting an intimate connection between Freyja and the yew takes us back to *Voluspa* and refers to the supersession of an older generation of deities (the Vanir) by a younger one (the Asir), the latter having often been linked to an invading horse-riding warrior aristocracy. When Odin flung his spear into the host (the Vanir) the first war had begun (*Vol.* 24). The two classes of gods made their truce and created a single divine 'guild',[96] and the patriarchal sky-god Odin absorbed Ullr's shaman function and his association with the yew; but not without challenging the goddess's claim on the World Tree. The first divine victim of

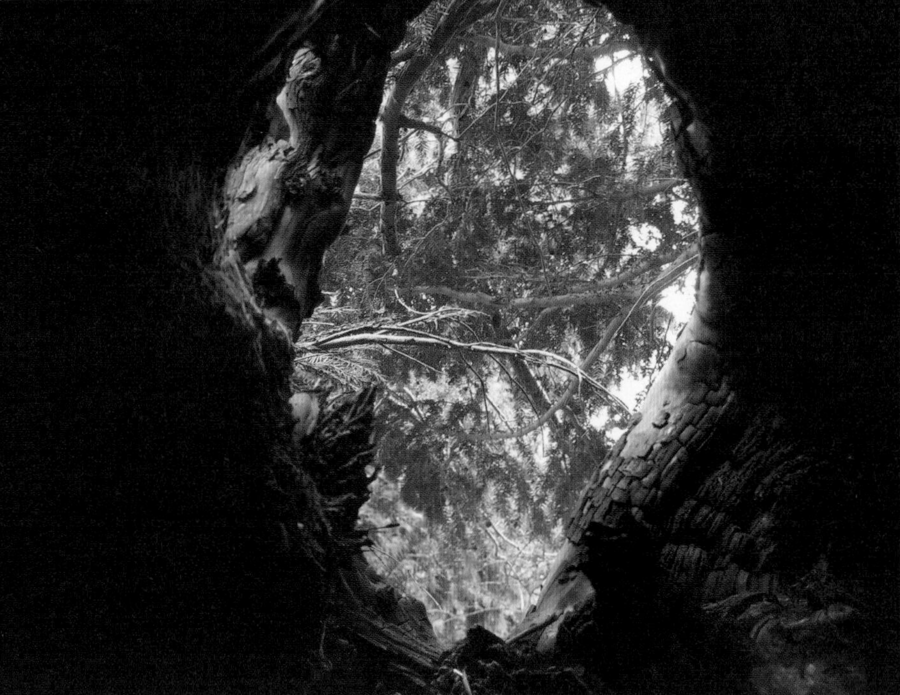

41.15–41.16 *The ancient yew at Linton in 2006, still green seven years after the disastrous arson attack.*

this war – the only one we ever hear of, in fact – struck by the spears of Odin's followers (spears made of ash wood, by the way[97]) is none of the known Vanir deities:

> She remembers the war, the first in the world, when Gullveigo they studded with spears and burned her in Hárr's hall – three times burned her, three times reborn, often, not stinting – yet she still lives.
>
> Bright Heidr they called her at all the houses she came to, a good seer of fair fortunes – she conjured spirits who told her. Seidr she had skill in, seidr she practised, possessed. (*Vol.* 21: 22)

Gullveigo literally means 'gold brew'[98] and may refer to the World Tree as supplier of the sacred mead. Dronke suggests that Gullveigo denotes a golden idol of Freyja,[99] regardless of the fact that metal statues are not known from Germanic peoples, and in any case they would not be 'reborn' or yield mead. A living and reviving image of a goddess suggests a tree, and one that has (like Freyja), tremendous abilities of regeneration (see figures 41.15–16).

The location of the sacrilege is Hárr's hall, a sacred grove 'tall' and 'wise with age';[100] until the crime, it was a place of mysteries and wise councils.

In the subsequent stanza, the goddess assumes human form again and appears as a powerful sorceress: her name, Heidr, derives from *heidr*, 'bright', a term used almost exclusively in relation to the sun and stars (*heid* is 'radiant sky'). *Heidr* also means 'honour'. And Heidrun is the name of the sacred goat that feeds in the crown of Yggdrasil and yields the 'shining mead' of the gods.[101] The plot thickens: there is a network of connections between the tree, the mead, and the web of light penetrating the universe.

Vitod ér enn, eda hvat? – Do you still seek to know? And what?[102]

41.17 *A one-eyed face appeared naturally on a dead piece of yew wood at Borrowdale (in 2002 and has decayed since).*

CHAPTER 42

HARMONIES

THE WORLD'S OLDEST WOODEN INSTRUMENTS

A sensational find was made in May 2004 in Co. Wicklow, Ireland. Archaeologists discovered six wooden pipes during excavations for a housing development site near the coastal town of Greystones, south of Dublin.[1] The pipes, which are made of yew wood, were discovered in the bottom of a wood-lined trough where they had been covered and possibly concealed when the site was abandoned. A wooden peg from the trough has been radiocarbon-dated to between 2120 BCE and 2085 BCE (Early Bronze Age), which currently makes them the oldest wooden instruments in the world. They appear to be 1,000 years older than anything on record.[2] Of course, a number of prehistoric musical instruments made from bone, like simple flutes and whistles, date back more than 100,000 years. But the oldest wooden instrument has previously been a 2,000-year-old sophisticated wooden pipe organ (no

42.1 *(above) Yew arils.*

42.2–42.3 *The yew pipes in the excavation site at Greystones, Ireland, and fully cleaned up.*

wood specified), dating from Roman times, that was discovered in Hungary. Ireland, however, does not lack musical instruments of prehistoric date: most notable are the cast bronze horns (dords) of the later Bronze Age and Iron Age. The only other wooden instruments, also all made of yew wood, are a set of four curved pipes from Killyfadda, Co. Tyrone (400 BCE), the Bekan Horn from Co. Mayo and a short conical wooden horn from the River Erne in Co. Fermanagh (both 700 CE).

The hollow yew pipes from Greystones measure between 30 and 50cm and are tapered at one end. Experts have been able to play a series of notes, including E flat, A flat and F natural, on the yew wood pipes. E flat is a common pitch for many ancient Irish bronze horns (see above).[3] The yew pipes have no finger holes such as can be found on a tin whistle, recorder or flute. Instead, these instruments utilise the different lengths of the air column in the pipes to generate a note. They are not joined together but have other features that suggest they are a set, and that they were attached to something that no longer survives. Some sort of flexible bag, such as a bagpipe, is unlikely because of the absence of a blowpipe or chanter, and because of the way the pipes were disposed. This leaves two possibilities: either a pan-pipe or some kind of wooden tube organ.

THE WORLD IS SOUND

The web of light of the universe is woven with song. Kirke, the guardian of the sacred 'circle' around the World Tree, weaves on her loom singing in a sweet voice as she fares to and fro before 'the great web imperishable'[4] and all the hall makes echo. Calypso, too, sings while working on her loom with her golden shuttle; her cave is surrounded by a wood, and four streams issue from it and run in different directions (compare the four directions of the compass – and the symbol of the cross; also the four paradisiacal streams in Genesis).[5] In Plato's vision of Lady Necessity and the Fates, whose spinning causes the rotation of the planetary spheres (see p. 182), a siren sits on each spindle and sounds a single note, and all notes together make a single harmonious sound – the music of the spheres.[6]

It is obvious that this original 'song' that echoes through space and reverberates in everything that exists resembles a kind of hum or drone. This was another reason why bees and cicadas were sacred in many ancient traditions (see p. 164); bees particularly were thought of as bringing to humankind the gift of song, the 'Muses' Honey' as Pindar, among others, calls it.[7] And when Sophocles says that '[t]he swarm of the dead hums and rises upwards',[8] he might have been referring to a tuning in to this cosmic music. In Hindu cosmology, the entire universe is sound: *nada brahma*. *Nada* means 'sound' and is related to *nadi*, 'stream, river', also '(re)sounding'.[9] This constant stream of sound emanates from Brahma who is the pure root of the World Tree; he is the creator who is present in all of creation, and is also described as the immortal self within the human being (*Katha Upanishad*, VI, 1).[10] This divine 'stream of sound' is manifest in the sacred syllable *aum* or *om*; the *Mundaka Upanishad*[11] relates that in order to become one with the inner self, and hence the whole universe, the human can use the syllable (and intonation of) *aum* as a bow and the mind or consciousness as an arrow, with Brahman as the target. Apart from drums, pipes, flutes and string instruments, the sacred music of ancient Anatolia and Crete also featured bull-roarers, which create a constant hum or drone sound (Sanskrit *nadá* also means 'bull') – the music of the spheres, after all, is not only gentle like the humming of bees, but also powerful

***42.4** Woman (priestess?) blowing a conch (triton shell) before an altar with three sacred branches between the horns of consecration and outside of them. The woman's skirt also suggests a* Taxus *twig. Gem from the Idaean Cave.*

and mighty. Other ritual instruments include the conch shell in Crete, bronze horns from Ireland (so-called dords)[12] and Germania (lures), all of which can create a permanent drone like the Australian Aborigene didgeridoo and the shofar in Jewish tradition. A shofar was made, according to its ritual purpose, either from a ram's horn or that of a wild goat.[13] At least in the latter case its symbology corresponds with that of the cornucopia; the abundance of nature's fruits begins with the 'abundance' of cosmic music, with the immanent sound of very existence.

In Norse myth, Heimdallr, the personified World Tree and ancestor of man, has a horn called Gjallarhorn, from *giallr*, 'resounding' (with a sense of a mighty sound). In the Icelandic poem, *Voluspa*, Heimdallr also is the watchman of the divine order and he will blow this horn when the end of the world approaches in the form of the final attack by the 'giants'. He has miraculous hearing that enables him to catch all sounds. Dronke says, 'his "ear" is to the ground, where his tree-roots in the nine realms of the underworld attend the vibrations of the giants' coming'. Until then, however, his horn serves Mímir, the guardian of the primeval well of wisdom at Yggdrasil's base, as a drinking horn for the very mead of wisdom from this well. And it also supplies Yggdrasil itself with the divine liquid; in this context the tree is called *heidvanr*, 'accustomed to bright mead'.[14]

In the context of ancient sacred music, the fact that the Bronze Age yew pipes found in Ireland are drone instruments is not surprising. In ceremony, drone instruments can create an awe-inspiring atmosphere for ritual activities (the church organ, too, fulfils this function), and also work well with melody instruments (including the human voice) playing over the top. Music and dance have always played a significant role in world religion in general, and so they did in the various faiths and traditions mentioned in this study. For example, the followers of Dionysos went to sacred hilltops and danced by torchlight to the music of the flute and the tympanon (kettledrum); an all-night dance was part of the Great Mysteries at Eleusis (see p. 173), probably with a similar instrumentation; the Kouretes on the Cretan Mt Dikte had 'Cybele's cymbals fill the air', a tradition that goes back to the mountain-mother cults of Asia Minor – in Phrygian times, the goddess Kybele was mostly associated with her musicians. Singing Artemis was worshipped throughout Arcadia from very early times on, and sometimes depicted playing a stringed instrument herself.[15] When the founders of the ancient Greek city of Messene (see p. 173) had found the scroll with the Mysteries buried underneath the yew tree, they invoked the ancestors and 'worked with no other music but Boeotian and Argive flutes' (Pausanias).[16]

In the first century CE, Apollonius of Tyana noted that on one of Apollo's temples at Pytho hung golden *íynges* (prayer-wheels, see figure 41.12) 'which echoed the persuasive notes of siren voices'.[17]

42.5 *Music in ancient Anatolia.*
Kybele's cult is well known for its emphasis on music; two bronze figurines of musicians from west or central Anatolia, eighth to sixth century BCE. Background: figurine of a female frame drum player (coloured black-and-white photo). This type can be found across Mesopotamia, Greece, Thrace and also in Anatolia where Kybele herself was often depicted as a drummer. Uruk, Seleucid period, 323–140 BCE.

NUMEROLOGY AND SACRED GEOMETRY

At the root of music lies the harmony of numbers. In various Old World traditions a sequence of sacred numbers was conceived as a key to the secrets of creation; the same mathematical proportions that can be found in musical scales and chords also appear in the shapes and proportions of nature (e.g. the spirals of shells, the dimensions of butterfly wings, the proportions of the human skeleton) and in the proportions of ancient temples (e.g. in Egypt or Greece).[18] Architecture as well as the annual rhythms of the agricultural and sacred year were aligned with the rhythms of nature and the celestial movements. Pythagoras (*c.* 580–*c.* 500 BCE) was the first to systematically render a single science of harmony that incorporates art, music, psychology, philosophy, ritual, mathematics and even athletics.[19] An exploration of Pythagorean thought, or indeed the vast field of numerology, is beyond the scope of this book. However, a brief reflection on the symbology of certain numbers that are immanently embroidered in the traditions of the World Tree is necessary (the number nine has already been discussed in the previous chapter).

Diagram 15 The relationship of six and seven in geometry
A circle of six always creates a centre as a seventh element. For the 'rosette' motif (left) compare figures 32.1, 32.3e–h and 44.12–13.

Diagram 16 The relationship of six and four in geometry
The six-pointed star illustrates the World Tree with the four directions emanating from the world pillar (left). The structure of the Celtic cross (right) might be understood as illustrating the same thing but looked at from the top: the view of the world pillar then changes from a vertical axis to a central point. The same may be said about the ancient Greek praying-wheels (figure 41.12).

The six and the seven

In Indo-European cosmology, where the number three is of special importance, the World Tree has three roots and three main branches. This corresponds with pre-Indo-European symbols from different parts of the Near East, Asia and Europe that show the World Tree in this way, the matrix of this shape being a six-pointed star (see figures 32.1, 37.7 and 41.9). Not by chance does this design also signify the six directions of three-dimensional space (the four directions of the compass plus above and below). The secret of the six-pointed star, however, is that it has a centre, which represents the seven: a gate to the inner world that goes beyond three-dimensional space. In Judaeo-Christian tradition, this relationship of six and seven has been preserved in the six days in which God created the world and the seventh when he rested.[20] The sixfold rosette often appears in cathedrals and churches dedicated to the Virgin Mary; the Cistercians of the Abbaye de Sylvanès in southern France call it a symbol for Mary's compassion.[21] The Vedic hymns speak of *rta*, the cosmic rhythm of the cycling starry sky, which also incorporates the rhythm of day and night. The lion is the symbol of solar light which is eternal, and the serpent represents the rhythmic round of lunar tides.[22] The snake has seven folds for the seven folds of temporality; these were identified in the Hellenistic world with the seven celestial spheres after which our days of the week are named[23] – an inheritance from the ancient Mesopotamian astronomers.

The twelve and the thirteen

The twelve as a cosmological number goes back to the Mesopotamian astronomer-astrologers who conceived the twelve signs of the Zodiac as cardinal cosmic forces influencing life on earth.[24] In the Old World, the sacred status of this number was honoured in various ways: ancient Sumer consisted of a confederation of twelve city-states; the Etruscans

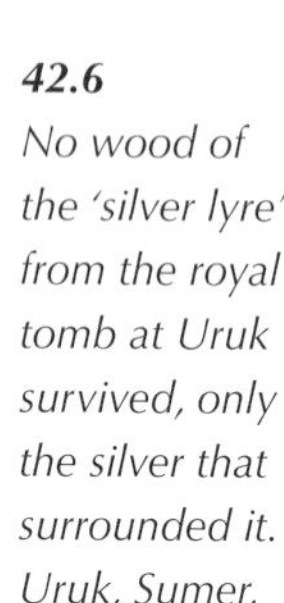

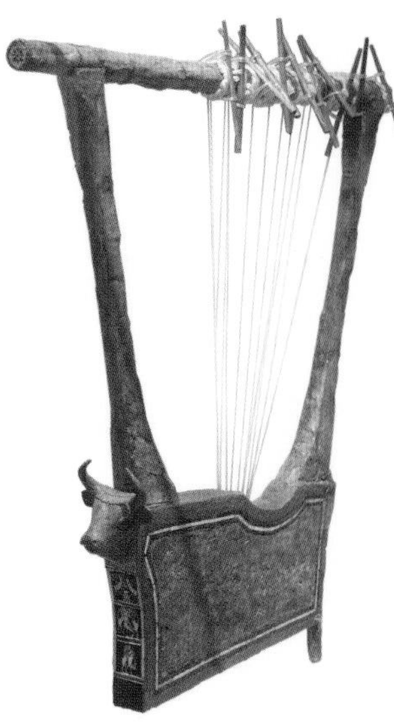

42.6 *No wood of the 'silver lyre' from the royal tomb at Uruk survived, only the silver that surrounded it. Uruk, Sumer, c. 2450* BCE.

in Italy founded twelve cities around a sacred lake and grove;[25] the Greek Olymp has twelve gods; the Germanic Asir are twelve divine powers to guide the earth – to name but a few. Furthermore, the day has twelve hours, the year twelve months, the circle 360 = 12 x 30 degrees (or 5 x 72 – with 72 being 6 x 12, as multiples of 6 and 12 were sacred in Sumer). With the circles of space and of time being divided into twelve sections, the number symbolising the central axis of the cosmic wheel – the point of stillness (the gate from time to eternity, the gate from physical space into psychic space) – is thirteen. *Eiwaz*, 'yew', is the thirteenth rune in the Older Futhark. There is nothing superstitious about this number: when Christ taught his – twelve – disciples (in the sacred grove of Gethsemane, 'as was his custom'[26]) the master was the thirteenth person among the twelve.

Long ago, the spectrum of musical pitches within the octave in Western music was divided into twelve halftone steps. The thirteenth note is the octave of the first – symbolically speaking a return to the roots, but on a higher level the musical ascent is not a circle but an unfolding spiral.

In the model described by Plato (see pp. 195–6), the world consists of nine concentric spheres. These are the seven celestial spheres known in the Old World (Sun, Moon, Mercury, Venus, Mars, Jupiter and Saturn), the eighth and outermost sphere is the universe (with its twelve regions of the Zodiac), and the ninth, innermost one is the earth.[27] Nine is the number of the earth (and hence of the World Tree notions of earth's inhabitants). In Hellenistic tradition, possibly earlier (Pelasgian Greece, Anatolia?), the nine Muses or mountain-mothers were associated with these spheres.

Music, like geometry, requires precise 'measurement': that of frequency (sound pitch) and that of timing (metre and rhythm). There is a parallel between the Norse 'tree of good measure', the Japanese yew sceptre *(shaku)* also serving as a length measure (see p. 211), and the Greek goddess of fate, Nemesis, who carries the sacred branch in one hand and a measuring rod in the other:[28] the personal fate she assigns is part of the 'great symphony' of the music of the spheres.

And the nectar of the gods, the mead of the poets, might not be a mythical liquor after all, but the 'divine music' that underlies all physical matter and vibration. (Modern views on this subject can be found in the field of quantum physics.)

42.7 *Muse playing the kithara; Hellenistic marble bas-relief from Manyas, second century* BCE.

CHAPTER 43

MIGRATIONS

The idea of moving a consecrated tree appeared long before 1880, although the transplantation of the yew within the churchyard of Buckland-in-Dover (see p. 123) was a uniquely huge undertaking. However, the first occurrence in literature of the transplantation of a tree is some 4,000 years older.[1] It is in a Sumerian creation myth, *The Huluppu-tree*, that Inanna, the Queen of Heaven, who had just descended and assumed the life of a young (mortal) woman, was walking by the lower Euphrates when she came upon a struggling young tree. 'In the first days, in the very first days', just after the creation of the world and the separation of the heavens, the earth, and the underworld, this tree had been planted (presumably by the gods) by the banks of the Euphrates, nurtured by its waters. But the south wind had pulled at it and the waters finally had carried the tree away. Inanna …

43.1 *Sacred tree in a cart procession; Babylonian seal.*

Plucked the tree from the river and spoke:
'I shall bring this tree to Uruk.
I shall plant this tree in my holy garden.'
Inanna cared for the tree with her hand.
She settled the earth around the tree with her foot.[2]

Whether the tree in question was a common riverside species (e.g. willow or poplar) – but why give such a special treatment to a struggling specimen of a common species? – or a rarely seen species (like a *Taxus* seedling miraculously washed down from the Amanus range), has to remain anybody's guess; the term *Huluppu* has to date given no clues as to its meaning. In Inanna's temple garden, however, the seedling becomes an image of the Tree of Life, with a serpent 'who could not be charmed' in its roots and the mythical Anzu bird (usually depicted as an eagle with a lion's head) in its crown. Furthermore, 'the dark maid Lilith built her home in the trunk' – Lilith is a figure who much later became demonised in Hebrew tradition[3] but her name goes back to Sumerian *lil*, 'spirit', and it is not difficult to see the parallel to the Cretan goddess who dwelt in the tree as the spirit of the dove. Later in the narration, the hero Gilgamesh helps Inanna to reclaim the sacred tree and carves for her a throne (symbolising queenship, sovereignty) and a bed (symbolising womanhood) from its trunk (or does this mean that she begins to occupy the tree like Lilith before her?). The goddess gives him two ceremonial objects that are also made from the tree, and are possibly insignia of kingship.[4]

Trees in transport appear in images and texts throughout the ancient world.[5] A Babylonian seal, for example, shows a procession of a sacred tree with the plant being carefully fixed onto a wheel-cart, followed by a priest and a winged horse. The scene is reminiscent of Tacitus' report about an annual procession among Germanic tribes in the first century CE: a consecrated boat-shaped cart was kept in a sacred grove until the priest 'noticed the arrival of the goddess in the sanctuary',[6] who then rode on her cart pulled by cows in a joyful procession

43.2 *Minoan seal from Crete.*

43.3 Conifer on a ship – Bronze Age ritual scene or migration legend? Bohuslan, Sweden.

through the villages, until 'she was given back to the grove by the priest'. Nowhere does Tacitus state what actually was on the cart but everybody since his time has assumed it to have been a human-like wooden idol – although Tacitus elsewhere claims that the Germans did not represent their deities 'under any kind of resemblance to the human form. To do ... [so was] in their opinion, to derogate from the majesty of superior beings' (*Germania*, 9).

In this context of trees outside their soil, a motif found on Roman coins struck *c.* 41 BCE is noteworthy. Three female figures are standing in a ceremonial manner, supporting a stand or tray from which rise five young trees. The left figure holds a bow, the right one a lily. The front of these coins shows the word *lariscolus*, a diminutive of *larix*, 'larch', but no Italian larch cult is known. The image has been linked by mythologists to the grove of Diana (see pp. 189, 210), with the caveat, though, that the trees do not bear the least resemblance to oaks.[7] The scene might refer to a time and meaning that had already largely disappeared by the first century BCE.

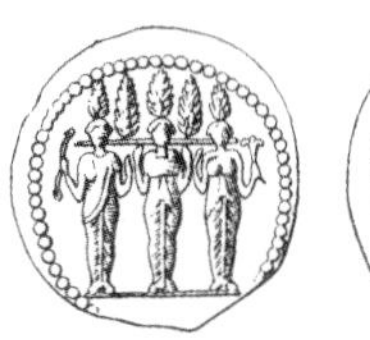

43.4 Roman denarii *showing three women and five trees in a ceremonial scene; first century* BCE.

Ceremonies inside a sacred grove or beneath a mature tree did not require any alteration to the tree or its location, but rituals in caves did. If we are to take the Minoan images (figures 37.9–13) from the caves of Mt Ida and Mt Dikte literally, seedlings or cuttings of the sacred tree (presumably growing as one or more individuals somewhere outside the cave) were put in pots upon the altars. A similar practice might have been in use for the Hall of Initiation at Eleusis, and certainly is reported on a Hittite cuneiform tablet from Anatolia: *ta* [GIS]*eye siunas parna petanzi*, 'they bring yew-trees to the god's house'.[8]

The moving of young trees was not only an element of annual ceremony, but also occurred when folk moved to new lands. Early human migrants (discussed in Chapters 35 and 36) are likely to have taken their sacred trees with them, just as they took, along with their livestock, seeds and seedlings of their food plants: a majority of the species (or their ancestors) that nourish humankind today originated in the rocky habitats surrounding the caves of early man.[9] The agricultural revolution took place in the river valleys – but its seeds came from the mountains. Hence we can safely assume that seeds (or cuttings) of a sacred tree would have been taken as well. The scene depicted on the above-mentioned Roman coins might give an idea of how the ceremonial 'departure' (or arrival) of sacred trees might have looked.

In mythological language, a new tree growing from a seed is a child of the parent trees[10] while a cutting is the tree itself (only a few tree species propagate successfully from cuttings – yew is one of them). Given the common Neolithic heritage of the mythologising and deification of plant species, it is not surprising that trees engaged in ritual would be referred to by their mythological names, that is, personal names (for example, the family of the Irish 'goddess' Dana, p. 208). Perhaps the mountain-mothers originally were the ancient trees on their respective mountains – which would also explain the repeated occurrence of pollen (see below) in sacred marriage and inauguration rites. A common motif in world mythology is that of humans having descended from

43.5 *Danaë receiving the 'golden shower'. How else to draw pollen? Carved amethyst, fifth or fourth century* BCE.

divine beings via a sacred tree or animal; in Norse myth, for example, it is Heimdallr, the World Tree, who was born from the nine root-mothers and who is the ancestor of all humankind. In later times, the accounts of such tree 'personalities' became mistaken for human-like 'gods' or legendary humans; an interesting example of this process is Danaë of the Danaeans.

The Danaeans were a seafaring folk that (possibly) occupied parts of Libya/Egypt and arrived in the eastern Peloponnese, Greece, in 1510 or 1509 BCE.[11] According to Pausanias,[12] Danaos and his descendants were associated with a sacred grove at Mt Pontinos, a limestone hill that closes the southwest corner of the plain of Argos (Peloponnese). At nearby Lerna, their rites were concerned with Dionysos' descent to the underworld as well as with the Mysteries of Lernaean Demeter, which were celebrated in an enclosure that marked the place where Hades and Persephone were said to have descended to Tartaros. Danaos' daughters, the Danaids, were priestesses and even credited with having introduced the Mysteries of Demeter to the early (Pelasgian) Greeks. In Greek myth, the mother of the great hero Perseus was Danaë, a fourth-generation daughter of Danaos. She was impregnated by Zeus who had come to her in the form of a 'golden shower'. This has often been interpreted as a shower of rain, but at that moment Danaë was situated in a bronze underground chamber where rain could not have fallen sideways through the tiny window. The strange location appears to be a reference[13] to the particular character of Danaean tombs, which were sometimes bronze-decorated, and suggests that the Danaean fertility rites were held underground or in caves. Similarly, Homer says that when Zeus embraced Hera[14] on the summit of the Trojan Mt Ida, glittering dew-drops fell from the golden cloud that encompassed them (compare the golden fleece, p. 206). In the case of Danaë, we meet Zeus, 'the bright one', again in his guise as a golden, fertilising cloud. Later in the myth, Danaë and her son are sent away from Argos in an ark or chest. In a similar container, Apollo (as the Delphic laurel tree) was said to have first come to Delos. The Danaeans, however, took to the sea again: a division of the Danaoi, the Daunioi, are said to have settled in Apulia in Italy. The link with the Irish Tuatha dé Danaan does not only exist in the name, but also in a shared association with burial mounds and trees.[15]

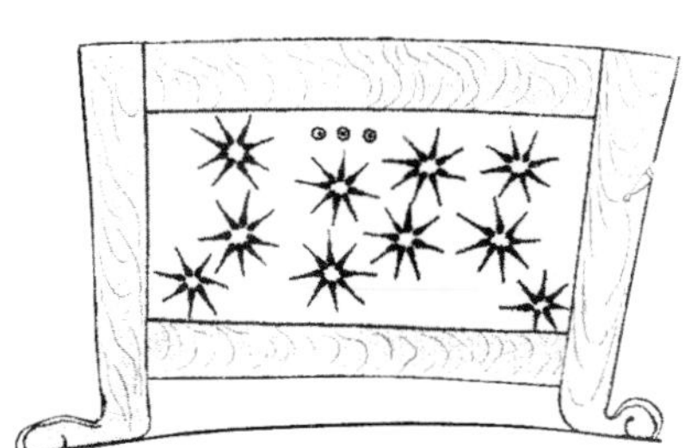

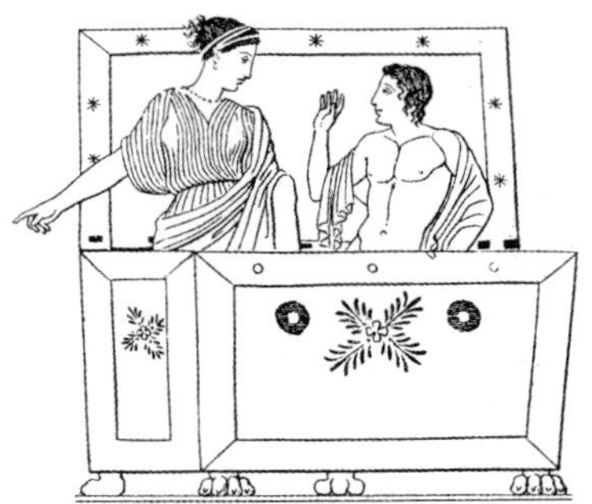

43.6–43.7 *The chest of Danaë and Perseus is usually decorated with needle-bearing twigs and small 'apples'. Greek vase paintings, fifth century* BCE.

It was during the 'reign' of the Tuatha de Danaan that the five sacred trees of Ireland were planted. Their very names – Eo Rossa, Eo Mugna, Bile Tortan, Bile Uisnig and Craeb Daithi – suggest that at least two of them, if not four, were yews, as *eo* is 'yew' and *bile* 'sacred tree'. In the records of the Middle Ages they are described as ash and oak trees, but legend has it that all five grew from the seeds of a legendary branch with three kinds of fruits: apples, acorns and nuts. In the Irish *Dindshenchas*, one of these five trees, Eo Mugna, is praised as 'blest with various virtues, with three choice fruits, The acorn, and the dark narrow nut, and the apple'.[16] An Irish legend, *The Settling of the Manor of Tara*, relates how a mysterious messenger from the otherworld appears at a great assembly of Irish nobles. His mission concerns the sun; he is responsible for its rising and setting, and himself surrounded by a radiance like 'a shining crystal vein'.[17] He carries the magical branch of which he gives Fintan a few fruits before he departs again. Fintan plants them and they grow into the five sacred trees of Ireland. One in each of the five provinces, they are widely regarded today as having represented the *axis mundi*, the world axis, for each province. Irish legend describes this ancestral bough as a 'golden many-coloured branch of Lebanon wood'.[18] This may well be an interference of well-educated, classically trained medieval writers; but on the other hand, if there ever was a human

43.8 *'Acorns' and 'apples'.*

migration (e.g. the Danaeans) that reached Ireland with young plants of a particular genetic lineage, it would be exactly this kind of notion one would hope to find in legend. (The medieval redactors should have said Taurus or Amanus instead of Lebanon, though.)

A surprising motif appears in the seventh century BCE on red relief-ware found in Rhodes and Caria[19] (modern south-western Turkey), the motherland of Leto (see p. 184). It shows the sacred tree being handed over by a centaur to a man with a double axe in his hand. There is nothing festive about the scene, and the centaur holding up his empty other hand emphasises the threat imposed by the armed man. The threat, indeed, is even more obvious in a fragment from the Carian coast that shows the axe-

43.9–43.10 *Centaurs handing over the sacred tree; pottery fragments from Caria, Asia Minor, seventh century* BCE.

bearer yielding a sword as well. It is secondary to our enquiry whether the centaurs were merely a mythological invention or a monastic brotherhood skilled in the art of healing[20] (compare the Dactyls and Kouretes, pp. 184–5) and possibly retaining an ancient tree cult. What is of interest is that suddenly there occurs the idea of the sacred branch moving hands by force. What was formerly part of peaceful ceremony – initiation or inauguration rituals – was now taken at sword-point. We do not know who these aggressors were, but they mark the end of an age.[21] In the same century, the shrine of the mother-goddess Leto (Kybele, Artemis) at Ephesus was moved from the sacred tree into a man-made temple, which women were not allowed to enter (!). Over the centuries, the tree increasingly fell into oblivion.

In the third century CE, coins from the same region show the sacred tree of the goddess being felled.[22] End of story.

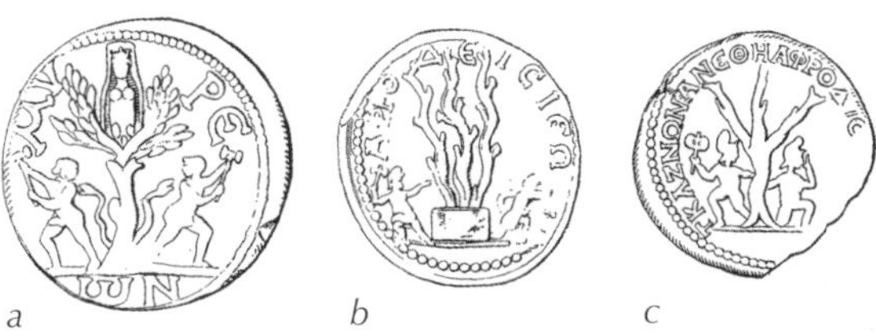

43.11 *a) Two snakes defend the sacred tree of a goddess, which is felled nevertheless by two naked men, one of whom is running away; third-century copper coins from Lycia (a) and Caria (b, c), Asia Minor.*

43.12 *(below) Pure yew wood in Wiltshire.*

***44.1** Idyllic scene at Preen Manor, Shropshire.*

CHAPTER 44

TEN HUNDRED ANGELS

The coming of Christianity changed certain aspects of the relationship between humans and nature. Most noteworthy in our context is that it stopped the veneration of trees and springs (much less discussed in recent times is that it also ended blood sacrifice – huge numbers of animals had been slaughtered on the altars of antiquity). The yew, however, in the long run has maintained some of its special status within the churchyards and graveyards of many countries. Let us look at the relationships of a few Christian groups and individuals with this tree.

Some of the early monks in Ireland and Wales lived in ancient hollow yew trees, which served them as home and shelter as well as an oratory. For example, in the Dee valley near Rhydyglafes, a hermit lived in a hollow yew near a Bronze Age mound, and St Kevin is said to have lived in a hollow tree for seven years from about 510. Some hermitages eventually grew into flourishing monasteries, and the significance of trees in their early history might have contributed to Irish Christian placenames such as Killure, Cell Iubhar, Cell-eo (all meaning 'church of

44.2 Brickwork steps lead to a (long disused) oratory in the hollow trunk of the ancient yew at Nantglyn, Denbighshire, Wales.

the yew') and Killeochaille ('church of the yew wood'). Clonmacnoise had a huge yew tree, possibly planted by St Ciaran himself (unless the site was chosen for a tree that was there already), which was struck by lightning in January 1149, a disaster said to have also killed no less than 113 sheep sheltering beneath it. In 1162, the Cistercian abbey of Iubhar Cinn Trágha ('the yew tree at the head of the strand', modern Newry, Co. Down), founded by St Malachy just about eighteen years before, burnt down, including an old yew said to have been planted by St Patrick himself.[1] St Columba (Columcille), the founder of the monastery on Iona ('yew island'), is reckoned to have been particularly fond of *Taxus*. His church was built next to a yew tree about which a poem attributed to Columba says:

This is the Yew of the Saints
Where they used to come with me together
Ten hundred angels were there
Above our heads, side close to side.

Dear to me is that Yew tree
Would that I were set in its place there!
On my left it was pleasant adornment
When I entered into the Black Church.[2]

Columba – the Latin name means 'dove' – is also recorded as having preached under a large yew tree on the island of Bernera in the Hebrides. The people of Mull and Morven crossed the water in their skin coracles, while the people of Lismore came on foot, and gathered beneath the yew to hear Columba preach. The tree stood right at the edge of a cliff overlooking the sea. Its ability to shelter a thousand people might be somewhat exaggerated but various sources confirm an unusually great size.[3] It is reported to have eventually been felled in about 1770 by the Laird (clan chief) Campbell of Loch Nell.[4]

In contrast to the sad loss of the ancient yew on Bernera, the ruined abbey of Dunkeld in Perthshire, Scotland, is still surrounded by fine old yews. Dunkeld emerged as a centre for Christianity in the seventh century, after Columba had come there, and in the ninth century was made the centre of the Celtic Church by Kenneth MacAlpin, the first king of Scots after the union of the Scots and the Picts.

44.3 The old yews on the north side of the ruined cathedral of Dunkeld, Perthshire, Scotland.

44.4 *One of the windows of Dunkeld Cathedral framed by yew boughs (viewed from the tree in the previous image).*

Dunkeld received cathedral status in 1325 and was extremely important in ecclesiastical terms until it was destroyed in the Reformation in the sixteenth century. A few old yews survived the troublesome times, and others have been planted at the site since.[5]

Still in the High Middle Ages, the yew gave shelter, for example, during the first years of Fountains Abbey in Yorkshire, founded in 1132 by Thurston, Archbishop of York. A group of monks from the Benedictine abbey in York, intending to adopt the more austere discipline of St Bernard[6] (i.e. the Rule followed by the Cistercians as well as the Knights Templar), chose to build their new abbey in a remote field overlooked by seven old yew trees. Until the first buildings were ready the monks took shelter, ate and prayed beneath these trees which, according to Dr John Burton in his *Monasticon Eboracense* (1758), 'stand so near each other as to form a cover almost equal to a thatched roof'.[7] The largest of the seven trees came down in 1658, showing a ring count of *c*. 1,200 years, which suggests an age of about 700 years for the trees at the time of the abbey's founding. Today, two of the ancient trees still survive, the biggest has a girth of *c*. 663cm (at 30cm above ground in 2002). Very recently, five new yews have been planted in an attempt to restore the original seven.

Like Fountains Abbey, the Cistercian colony at Ystrad Fflur (Anglicised as Strata Florida) in Ceridigion, Wales, was originally equipped with an abundance of yew trees. John Leland, in the late 1530s, mentions 'xxxix great hue trees' within the enclosure of the ruined abbey.[8] Of these thirty-nine, only two have survived today, their girth being similar to that of the large tree at Fountains Abbey. A third, ruined yew was last seen in an etching from 1874, the other thirty-six probably had been removed – natural loss of such a high percentage is unlikely – before 1741, as an etching of that date shows no trees identifiable for species type. It is possible that three of the Ystrad Fflur yews were left alone because of the venerable age of this group, and also because the national poet of Wales, Dafydd

44.5 *Fountains Abbey and its yew trees.*

44.6 *The ancient yew at Waverley Abbey, Surrey, the first Cistercian colony in England.*

44.7 The memorial stone of the celebrated Welsh master poet, Dafydd ap Gwilym, at the base of one of the two surviving ancient yews at Ystrad Fflur, Wales.

44.8 A Knights Templar cross on a gravestone at Ystrad Fflur.

ap Gwilym (*c.* 1320–*c.* 1380), was, and is, believed to have been buried beneath one of them.

Another ancient yew location that drew the attention of the Cistercians is the valley of Borrowdale in the Lake District (see figures 27.4–5), which was used by Cistercians from Furnass Abbey (Cumbria) from the twelfth century until the Dissolution under Henry VIII. In southern England, an ancient female yew that is famous for the extraordinary shape of its trunk (girth at ground level *c.* 640cm) stands in the grounds of Waverley Abbey, the very first Cistercian colony in Britain, founded in 1128. The founder of the monastery, Bishop William of Winchester, was related by marriage to Walter de Clare, Lord of Chepstow, who three years later, in 1131, founded the second Cistercian colony in England and the first in Wales: Tintern Abbey. It is located in one of the areas remarkably abundant with *Taxus*: the banks of the Wye river in Monmouthshire.

***44.9** The Cistercian coat of arms from the Abbaye de Sylvanès, Aveyron, southern France, shows the dove and the fleur-de-lis (possibly not meaning 'lily' originally but* fleur-de-luce, *'flower of light', a symbol that developed from the Tree of Life 'rosette', see figure 32.3).*

Less obvious than the Cistercian links with *Taxus* are those of the Knights Templar, the most famous of the medieval Christian military orders and an organisation closely associated with the Cistercians.[9] Of course, no significant references to botany have been found in the manuscripts of any of the two orders, and yet modern yew field research comes across the Knights Templar (full name The Poor Fellow-Soldiers of Christ and of the Temple of Solomon, from Latin *pauperes commilitones Christi Templique Solomonici*) time and time again, especially in the British Isles. Hugo Conwentz (1855–1922), the celebrated pioneer in yew research and conservation[10] who dug for fossil and semi-fossil remains of *Taxus* in many localities in the British Isles and Germany, stated that he never found yew planted in the ramparts of prehistoric forts, but often at the fortifications of the knights of the Middle Ages.[11] The modern traveller can still find yews at Templar sites, even without digging.

Possibly as early as 1128 – a year before the Templar Rule was even drawn up at the Council of Troyes (Champagne, France) in the Order's ninth year, and at a time when there were supposedly only nine Brothers[12] – Dabid mac Maíl Choluim (David I, reigned 1124–53) by royal decree granted the Templars the Chapelry and Manor of Balantravach in Midlothian, Scotland. The Templars established their Scottish headquarters here and the place became known simply as Temple, which is still its

44.10 The yew avenue at Temple, Midlothian, Scotland (September 2004).

name. At the back of the ruined church, towards the river, is the famous avenue of yews. However, the trees are considerably younger than the Templar foundation itself – at least those parts of them that can be seen above the ground.

44.11 *The firm trunk of the single yew at La Couvertoirade, southern France, appears as solid as the stone tower of the Templar castle.*
44.12 *An ancient symbol above the altar of the original Templar church at La Couvertoirade. (compare figure 32.3.)*

In southern France, there is a fine yew at La Couvertoirade, a Templar castle that was part of the Commandery of Sainte-Eulalie, one of the earliest Templar headquarters in Western Europe. Although the tree is hardly older than some 400 years, it is still a feature to marvel at: the Ayveron *département* is virtually bare of (known or recorded) yew trees, so who planted it, and why?

The presence of the Templars at Iubhar Cinn Trágha ('the yew tree at the head of the strand') in Ireland (see above) is suggested by a number of local placenames such as Templegowan, Templetown and Templepatrick. Another ancient yew stood at the regional Templar headquarters for the north of England at Hartburn, Northumberland. J.C. Loudon recorded the tree at a girth of 495cm in 1836. Some ten years later, however, the tree was destroyed by another tree falling on it.[13]

The official history of the Knights Templar ends with the dissolution of the Order of the Temple on 2 May 1312 by Pope Clement V. The sudden disappearance of an unknown number of the Brothers, and particularly of the supposed 'treasure' of the Order, has given room to endless speculation and modern 'myth'-making. While five years after the dissolution two new military orders were officially created in the Iberian Peninsula,[14] the events in Scotland – another country where surviving Templars are supposed to have found refuge – are not so easy to discern. The Papal bull regarding the dissolution of the Order of the Temple gave all Templar assets across Europe to the Knights Hospitaller; in Scotland, in a sense, the two orders 'combined' and continued as the Order of St John and the Temple.[15] One site frequently associated with the 'post-dissolution' history is Roslin, a small village in Midlothian, south-eastern

44.13 *Medieval gravestone of a pilgrim to Santiago de Compostela, Languedoc, southern France.*

44.14 The old yew at Rosslyn Castle, Midlothian, Scotland.

44.15 The young yew at nearby Rosslyn Chapel.

Scotland. Its fifteenth-century church was designed by William Sinclair of the St Clair family, a Scottish noble family descended from Norman knights; their birthplace is generally thought of as St Clare in Pont d'Eveque (fortuitously, another yew placename) in Normandy. The St Clair (later Sinclair) family, were the Earls of Roslin. Some sources link them with the Scottish Knights Templar but their only well-documented connections are with Scottish Freemasonry, which has a Templar degree. William Sinclair, First Earl of Caithness (1455–76), Third Earl of Orkney (until 1470), and Baron of Roslin (1410–84) was the builder of Rosslyn Chapel. A later William Sinclair of Roslin became the Grand Master of the Grand Lodge of Scotland, and also a 'Knight of the Golden Fleece'.

While it is almost impossible to reconstruct the events and connections of fifteenth-century Midlothian, it is certainly clear that the medieval Templars underwent a remarkable revival of interest in the eighteenth century. This was because of a general rise of secret societies in Western Europe (Scottish in particular as well as German Freemasons), followed by the Romantic movement that also had a fascination with times past.[16] Today, Rosslyn Chapel is visited by Freemasons from all over the world because of its alleged Masonic and Templar architecture and the symbolism featured on the chapel walls. The chapel stands on thirteen pillars, forming an arcade of twelve pointed arches (see numerological symbolism, p. 233). Rosslyn Chapel is famous for two of its pillars: the Apprentice Pillar and the Master Pillar (standing either side of the Journeyman's Pillar) which display most elaborate stone carving. The Apprentice Pillar is interpreted by many authors as a representation of the Tree of Life.[17]

There is a young yew tree in the grounds of Rosslyn Chapel and a venerable old one at Rosslyn Castle, formerly the seat of the Sinclair family. While modern legend marvels about where in the vaults of this castle the supposed 'Templar treasure' might be hidden, there certainly is a real medieval treasure above ground: it is female and has a girth of 397cm (in 2004; figure 44.14). Given its close proximity to the castle it seems remarkable that it survived the ravages of fire, the bombardment by Cromwell's troops, and other attacks, although there appears to be evidence at about 4.5m height of a loss of major limbs in the past.

Visiting Assisi, the birthplace of St Francis in Umbria, central Italy, in April 2005, we were in for a surprise: just outside the main gate to the Franciscan friary, at the main pilgrimage route to the basilica, we discovered a shaded corner where the monks grow yew trees in pots! A hand-painted wooden sign

44.16 *Statue of Jesus Christ at the church of Mary at Rieubach-Raynaude, France.*

reflects the good spirits in which these gardeners do their work: it says *Pax et Bonum*, 'Peace and Good(ness)'. Several restaurants around town seem to pick up the message (and the pots) and have small yews at their doors or along their façades between the tables. Otherwise, this part of Umbria is certainly too hot and dry for *Taxus*, and at about 400m above sea level it is well below the natural distribution zone of the region. But then, Francis of Assisi (1181/2–1226), the holy man who talked to doves, is not only the patron saint of Italy but was also recognised in 1979 by Pope John Paul II as the patron saint of ecology. Assisi itself is an ancient town going back to Etruscan times, with a Roman temple to Minerva (who originally was the Etruscan goddess Menerva, later Menrva[18]) preserved in the town centre; the original small Romanesque church of St Francis (and his female contemporary, St Clare) was enshrined in 1569 by the Church of Santa Maria degli Angeli, 'Saint Mary of the Angels'.

44.17 Taxus *regeneration by the Franciscan monks of Assisi (April 2004).*
44.18 *May the growing of trees increase* Pax et Bonum, *'Peace and Goodness', in the world!*

45.1 Flourishing (female) yew at Mary's Church of the Immaculate Conception, with stations of the cross behind the church; Rieubach-Raynaude, France.

CHAPTER 45

'BEHOLD THE HEALTHFUL DAY'

THE VIRGIN MARY

The history and sites of the worship of the Virgin Mary (the mother of Jesus of Nazareth) also cross paths with the cultural history of yew. This begins nowhere less than at Ephesus (on the Aegean coast of western Turkey), the city allegedly founded by the 'Amazons', and from where the worship of the 'Mother of the Gods' (Leto/Kybele; note the plural in Gods) had once spread all over the Greek world and then the Roman Empire *(Mater Idaia)*. The great Ephesian temple of Artemis, however, was destroyed by invading Goths in 262 CE and was never rebuilt. A popular Christian legend maintains that the Virgin Mary came with the Apostle John to Ephesus and lived and died there.[1] The Council of Ephesus met in 431 CE in the great double church of St Mary; it was at this council that the Virgin was declared Theotokos, 'Mother of God'. This was a timely and clever move in the fierce ecclesiastical battle over whether Christ was by nature Man or God,[2] and also a measure to justify the widespread cult of Mary.

45.2 *The ancient yew at La Haye-de-Routot (Eure département, France) has become a shrine of the Virgin Mary.*

And there was certainly much to 'justify'; in a time of widespread suppression of religions other than Christianity (which included pre-Christian goddess cults), the Virgin Mary was becoming the sole inheritor of public reverence for the divine female. Even before the Roman Empire,[3] and more so during its duration, eastern Mediterranean goddess cults such as those of Isis, Venus/Aphrodite and Kybele had spread right across Europe and quietly replaced or merged with those of indigenous goddesses. Now the name was changing again, but the image remained the same: the eternal virgin mother and her divine son – Madonna and Child. Like Aphrodite *Urania* and Isis before her, Mary became the Queen of Heaven; she inherited Isis' blue coat representing the universe, and Inanna's five-pointed star which can be seen, for example, on the ceiling of Rosslyn Chapel, on the walls of Mary's church at Rieubach-Raynaude, and arranged in a double circle forming Mary's crown at Loreto and Le Puy (the circle of twelve golden stars also appears as the emblem of modern Europe). And like her predecessors, Mary, too, came to protect and guide not only in the light, but also in the darkness: divinities such as the black Isis, the black Kybele, Aphrodite *Melaenis* (the 'black one'), in many locations changed into the Black Madonna.[4] Later, during the twelfth and thirteenth centuries, the veneration of the Virgin Mary (and female saints) increased significantly.[5] Mary was also the patroness of the Cistercian Order, the Teutonic Order and the Knights Templar, a fact that contributed further to her popularity. This was the age when the Gothic cathedrals dedicated to Our Lady were built.

Notre Dame

In contrast to Wales, southern England and northern Spain (Asturia, Cantabria[6]) the yew is not widespread in the churchyards of France. In fact, the

45.3 *The Black Madonna in the crypt of Chartres Cathedral.*

It is believed to have been modelled directly after a pre-Christian statue, and is called Notre Dame de Sous-Terre, 'Our Lady of the Under-World'. Chartres, incidentally, has nine towers and nine doors (Walchensteiner 2006).

45.4 *The statue of the Virgin Mary inside the hollow ancient yew of St Rémy sur Orne, Normandy, France.*

whole country appears rather yew-less,[7] apart from a very few (famous and ancient) yews in churchyards in Normandy, and, of course, Ste Baume (see below). It is all the more surprising to find yews at sacred sites, and many of these are dedicated to the Virgin Mary, also in her guise as the Black Madonna. Six yews, for example, have been planted only a few centuries ago at the entrance of the abbey church of St Yved and Notre Dame ('Our Lady') in Braine[8] (Picardie, Arr. de Soissons). Another church dedicated to Mary is the Eglise Elevée de l'Immaculée Conception (Elevated Church of the Immaculate Conception) at Rieubach-Raynaude (Midi-Pyrénnées, Arr. de Pamiers). A handsome and healthy mature yew (girth 247cm) guards the entrance to the church and its affiliated Stations of the Cross dotted over the hill-slope behind (figure 45.1). In these lowlands north of the Pyrenees (Rieubach is below 300m altitude) there are no known *Taxus* trees even in the greater vicinity, and the pollination of this richly aril-bearing female yew is almost a miracle in itself – but then, it is a site in adoration of immaculate conception, after all. Rieubach-Raynaude is located in an area rich in Neolithic dolmens and just about a mile down the road from the famous prehistoric caves of Mas d'Azil.

Le Puy en Velay (Haute-Loire, France) has been an important religious site for millennia. The city is dominated by two impressive peaks of volcanic rock: one is crowned by a Romanesque chapel to the archangel Michael, which replaced a temple to Romano-Gallic Mercury (who in turn had replaced Celtic Belenos), the other by Notre-Dame de France. Along the ridge of this second hill, a Romanesque cathedral of the Black Virgin has replaced a church that was originally built *around* a prehistoric dolmen (instead of replacing it, as was the usual way of proceeding).[9] The Roman name of the town was Anisium, 'city of Ana'. The Celtic goddess Ana in Le Puy was *dia Ana* (*déesse* Ana), 'goddess Ana', hence Diana – whose temple at the foot of the rock of Mercury/St Michael is still preserved. After Christianisation, the town became known as

45.5 *Young yews at the abbey church of St Yved and Notre Dame, Braine, France.*

45.6 Taxus *has begun to colonise Mt Anis, or Rocher Corneille ('Rock of the Crow'), at Le Puy en Velais; in the background is the statue of Notre-Dame de France.*

Podium Sanctae Mariae, in local Occitan language *podium* changed to *pog*, 'summit', and finally to French *le Puy*. The underworld aspect of Anis/Ana also relates to the Breton word *anaon* which denotes the dead (compare the Welsh Annwn as the realm of the dead).[10] The ancient female divinity of Le Puy was also called Mélusine or Mala Lucina, whose name can mean both 'mother of light' as well as 'darkness – light'.[11] Her divine son, Gargan, was a psychopomp leading the souls to the underworld (hence the Roman Mercury); he was occasionally depicted as a ram-headed serpent.[12]

During its long Christian history, Le Puy has been visited by at least six popes and fifteen kings. In 1096 the Count of Toulouse ordered perpetual light before the Black Virgin on the high altar.[13] With such a history this place seemed more than fit in 1860 to become the perpetual home of Notre-Dame de France, a huge statue of the Madonna and Child, 16m high (without the pedestal) and weighing 110 metric tonnes. The statue was made of cast iron from the 213 canons from the battle of Sebastopol in the Crimean War,[14] and raised on the very summit of the hill of Anis.

Much to the surprise of a visitor to this virtually yew-less area, there is a fine yew at the foot of Rocher Saint-Michel, St Michael's Rock, a flourishing male tree with 268cm girth, exposed to the western side of the volcanic vent. From this one, younger yews seem to have spread over the last few decades (we visited in March 2005), their crowns being visible over various (garden) walls in nearby locations, and also growing from the small crevices within walls. Three tiny yews emerge from almost 15m up the steep wall that comprises the southern edge of

45.7 *Rocher Saint-Michel ('St Michael's Rock') with the Romanesque chapel on the top and the male yew at its foot, Le Puy en Velais, France.*

the Place de For, the outdoor ceremonial area of the cathedral. In a few years they will peep over the wall's edge and join the ceremonies. Curiously enough, they are evenly spaced across the width of the wall. A few slightly older yews have begun to colonise Mt Anis and can be seen from the path leading to the statue of Notre-Dame de France.

Madonna Lauretana

Loreto is a town near Ancona on the Adriatic coast of central Italy. It is one of the most famous pilgrimage resorts in the world, because of the Santa Casa, or Holy House of the Virgin. According to tradition, the Santa Casa was threatened with destruction in its original location at Nazareth by the Turks in 1291, from where it was carried away by the ministry of angels and deposited on a hill in Dalmatia (modern Croatia). In 1294 it was similarly transported across the Adriatic to a laurel grove (*lauretum*, whence Loreto) in the hills near Recanati. Archaeology has lately confirmed the Near Eastern origin of the Casa;[15] the interference of angels, however, was rather down to the noble Angeli family who ruled Epirus at the time and who financed the transport to Italy.[16] The original rescue is believed to have been performed by 'knights of a military order that defended the holy places and relics during the Middle Ages'.[17]

What makes the Santa Casa so important as a Christian relic is the recognition that these very stones have witnessed the Annunciation (i.e. the Angel Gabriel appearing to the Virgin Mary and announcing her divine pregnancy) and the Word becoming flesh.[18] As Pope John Paul II pointed out in his Letter for the 7th Centenary of Loreto, the 'second moment of the mystery of the Incarnation' is the Yes of Mary ('be it unto me according to thy word', Luke 1: 38), which made her, in the words of John Paul II, 'the first believer in the new alliance [the forgiveness of sin by God sending his Son], she who 'advanced in the peregrination of faith'.[19] John Paul encouraged one to follow her example in one's own heart: 'In this way, without physically starting to walk you live your own spiritual pilgrimage ...'[20]

There is absolutely no *Taxus* anywhere near Loreto, which has an unsuitable climate. Yet the town is mentioned here for two reasons. For one, some of the elements of the iconography of the Black Madonna of Loreto can be traced back to pre-Christian divinities and have a history of about 6,000 years (see box next page).[21] The other reason is a single line in the Litany of Loreto, a text that goes back to the thirteenth and twelfth centuries.[22] In this prayer, the Madonna is praised as:

Virgin most powerful
Virgin most merciful
Mirror of justice
Seat of wisdom
Fountain of our joy
Vessel of spirit
Vessel of glory
Vessel of devotion
Mystical rose ...

The next two lines call her *Turris davidica*, 'Tower of David' – referring to her descent from the house of David – and *Turris eburnea*. The image of a tower might surprise us as an analogy for a tender maiden, but a tower is a symbol of ultimate strength, like a rock in the storm. The term *eburnea*, moreover, has puzzled commentators for generations. Today, this line is even skipped in many popular versions of the prayer, as the arguments for 'ivory' were never quite convincing – the only biblical references for ivory refer either to a distinctly erotic context (the smooth skin of lovers in the Song of Solomon, 5: 14 and 7: 4), or to inlay work, for instance in Solomon's throne (I Kings 10: 18), nothing remotely the size (or meaning) of a tower of strength. In medieval Italy, however, *ebura* would still have been a word recognised (by some) as 'yew' (see p. 140).[23] Although having lost much of its meaning during and after the Middle Ages, this tree was still a pan-European graveyard tree, and associated with long life and immortality. With the churchyard yew being the symbolic tree of the Resurrection, or the rebirth of Jesus Christ, what better symbol could have been chosen for Mary the

***45.8** Symbolism in the design elements of the Black Madonna of Loreto.*

Long before the Romanesque period, the round arch above a female divinity represented the universe and her as the Queen of Heaven. These examples from Greek copper coins (from the time of the Roman Empire) show Aphrodite (a), a goddess with child, possibly Semele and Dionysos (b), Nemesis (c), and Isis (d) beneath an arch.

The design of the 'Dalmatian robe' that has become an integral part of the appearance of the Virgin of Loreto (f), might have been influenced by the cicada talismans that have been found at Ephesus, Mycenae, north of the Black Sea and across Greece and the Balkans. This piece (e) from the late Roman and early medieval period was found in Hungary.

Mother of God who gave birth to him in the first place? For the personal life of the believer (local or pilgrim to Loreto), a 'tower' of yew would symbolise a firm rock, and one that is not affected by time. The association of tower and goddess is reminiscent of the huge towering statue of Artemis at Ephesus – a deity doubtlessly connected with *Taxus* (see p. 167) – and the many smaller replicas which generally show her with a crown comprising a single or double tower, topped by a temple.[24] In her 'Dalmatian robe'[25] the statue of the Virgin of Loreto (remade[26] of blackened cedar wood) does rather look like a tower. The robe is embroidered with a vertical sequence of seven crescent-shaped bowls or vessels,[27] symbolising the spiritual importance of receiving, of being an open vessel like the Virgin. It hardly needs an explanation why this sequence starts at the Virgin's feet and ends at her heart. Contemplating the heart of the Madonna, one can live one's own 'spiritual pilgrimage'.

MARY MAGDALENE

The French forest authorities are in agreement that the most significant yew population of France is to be found at Ste Baume, Provence, in southern France. Over an area of 138 hectares *Taxus* is part of a mixed forest on the slopes beneath the limestone precipice containing the sacred cave of Mary Magdalene. The forest is hailed as one of the most ancient woodlands of France, going back 2,000 years or more. The oldest trees therein are clearly the yews, although none of them shares the actual age of the forest as such: many of the oldest yews are estimated at about 500 years, with a single veteran of about 800. None of the leaf trees at Ste Baume is older than two centuries, since the last major felling operations took place in

45.9 *Post-medieval etching of Mary Magdalene's ascension at Ste Baume.*

including a wooden statue of the many-breasted Artemis. A temple for the goddess was built on the headland above the harbour, and Aristarche became high priestess.[30]

How a cult of Mary Magdalene first arose in Provence is not clear,[31] but according to French tradition Mary, her brother Lazarus and a few other companions fled from persecution in the Holy Land, traversed the Mediterranean in a frail boat and landed at Saintes Maries-de-la Mer near Arles. Mary Magdalene then came to Marseilles where she spread the gospel and afterwards retired to the limestone cave of Ste Baume (*baumo*, 'holy cave' in Provençal) for the remaining thirty years of her life. A post-medieval etching shows the ascension of

45.10 *The young yew at the entrance to Mary Magdalene's cave church, Ste Baume, France.*

1789 during the French Revolution (which implies that the revolutionaries left the remaining yews standing (!), for reasons unknown today). There is a healthy natural regeneration of *Taxus* here, and in the past some foresters have worried about yew regeneration even inhibiting that of beech. The current warden, Christian Vacquié, however, is very aware of the extraordinary status of yew, both botanically and culturally, and has done much to disperse unsympathetic feelings towards this tree. France has begun to take pride in the yews of Ste Baume.

Goddess worship at Ste Baume is considered to go back to the Neolithic. When Phocaean Greeks (not to be confused with Phoenicians from the Levant) founded nearby Massilia in about 600 BCE,[28] they dedicated Ste Baume to the goddess Artemis.[29] Their hometown, Phocaea, was a Greek colony on the Aegean coast of Asia Minor, just about 100km along the coast from Ephesus. According to the foundation myth of Massalia, the Phocaeans, before leaving home were told by an oracle that they should take with them on their journey a guide provided by the great temple of Artemis at Ephesus. They sailed down to Ephesus and took on board Aristarche, 'a woman held in very high honour' (Strabo), as well as a number of sacred images

45.11 *Detail of stained-glass window at Ste Baume, showing Mary Magdalene.*

Mary Magdalene, and the rather low habit of the trees in the drawing may be interpreted as the woodland at the time having been dominated by young to mature yew trees. Monastic life at Ste Baume began with St Cassien in the fifth century, and there has been a monastery there ever since.[32] A wall was built at the cave entrance, giving it the appearance of a church from the outside, but the inside is 'natural'. The life of the Magdalene is represented in a series of seven stunning stained-glass windows made by Pierre Petit between 1976 and 1981.[33] To the right of the entrance door, the monks have recently planted a yew tree.

The true identity of Mary Magdalene is not known as the Bible does not make it clear whether the Magdalene, Mary of Bethany and the woman who anointed Jesus' feet are actually one and the same person. Her (suggested) identification with Mary of Bethany, however, also led to her being identified with 'the woman who was living an immoral life' (Luke 7: 37). Church fathers of the third and fourth centuries considered her 'unchaste'. Pope Gregory I then called her a sinner *(peccatrix)* who repented, but he did not call her a prostitute *(meretrix)* – that was a folk belief that arose later and gained much popularity among writers and artists until the twentieth century. According to the Gospel of Mary (Magdalene) in the Nag Hammadi codices (found in 1945), she was rather the opposite, more of a chosen one, as it were. The apostle Levi asks 'if the Saviour made her worthy, who are you indeed to reject her? Surely, the Saviour knows her very well. That is why he loved her more than us.' And she teaches the twelve apostles 'what is hidden' from them, namely what they could not grasp the first time when Christ had said it.[34] Mary the Mother was the first to believe, Mary Magdalene was the first to understand the gospel.

However, the Magdalene became a symbol of repentance for the vanities of the world – and which true Christian cannot identify in heartfelt sympathy with the woman who had washed Christ's feet with her tears, wiped them dry with her own hair, and anointed them with precious myrrh-infused olive oil?[35] It is this very humbleness and love, the spirit symbolised by the dove, that still today can be felt immediately upon entering Magdalene's cave at Ste Baume – for over two millennia pilgrims have come here and opened to it, their souls being nourished by what this place of devotion holds.

KUAN YIN

In Asia, the deity of compassion is Kuan Yin, in full Kuan Shih Yin, 'She who hears the cries of the world', and she is the most popular and best-loved deity of the Chinese world. She is the same goddess after whom the yew tree was named in south-western China, *Kuan Yin sha* (see p. 142).

The figure of Kuan Yin can be traced back to the *Saddharma Pundarika Sutra*, or Lotus Sutra, a key text of Mahayana Buddhism written in Sanskrit in about the first century CE. It tells of a Bodhisattva named Avalokitesvara, 'The Lord who regards the cries of the world', who pours out his compassion on those who seek release from the wheel of suffering. The Lotus Sutra was one of the first Buddhist texts to be translated into Chinese, in which *Avalokitesvara* translates into *Kuan Shih Yin*. When statues of the Bodhisattva began to appear in China in the fifth century CE, Kuan Shi Yin was still male, albeit very graceful and sometimes even androgynous, but during the eighth and ninth centuries something revolutionary happened: the transformation of a male Bodhisattva into a goddess of mercy.[36] This was because of the complex religious landscape of China at the time, and the need of large parts of the population for a female divinity after Confucianism had dismissed and excluded the feminine and shamanistic elements from religion and culture in the fifth century BCE. Kuan Yin turned female in north-western China during the T'ang Dynasty (618–907 CE), in the areas along the Silk Road, a cultural melting pot where Buddhism and Taoism, shamanism and Bon (the indigenous Tibetan religion), Christianity, Manichaeism, Islam and Zoroastrianism rubbed shoulders. Particularly influential on the development of the female Kuan Yin was Taoism and its notion of a 'Queen Mother of the West'[37] – the west for the Chinese is 'the direction of paradise, of the true mystics and of mystery itself'.[38] Another influence came with the Nestorian Christians from Persia, and a later Christian impulse followed much later when Portuguese and Spanish Jesuits in the late sixteenth century brought Renaissance statues of the Madonna to China. These immediately influenced the Chinese artists and porcelain-makers, and have done so ever since. Kuan Yin spread eastwards across China, reached Hong Kong and Taiwan, and also Korea and Japan, where she is known as Kannon.[39] She can be found in Buddhist, Taoist and Shinto sacred sites (and homes) alike.[40]

Behold I am come to take pity on thy fortune and tribulation; behold I am present to favour and aid thee; leave off thy weeping and lamentation, put away all thy sorrow, for behold the healthful day which is ordained by my providence.

The goddess Isis, in Lucius Apuleius, *Metamorphoses* (second century CE)

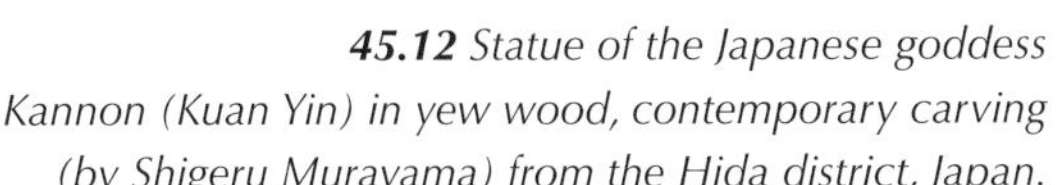

45.12 *Statue of the Japanese goddess Kannon (Kuan Yin) in yew wood, contemporary carving (by Shigeru Murayama) from the Hida district, Japan.*

Yew tree at Wakehurst Place, Sussex, where the Millennium Seed Project of Kew Gardens is located.

Appendices

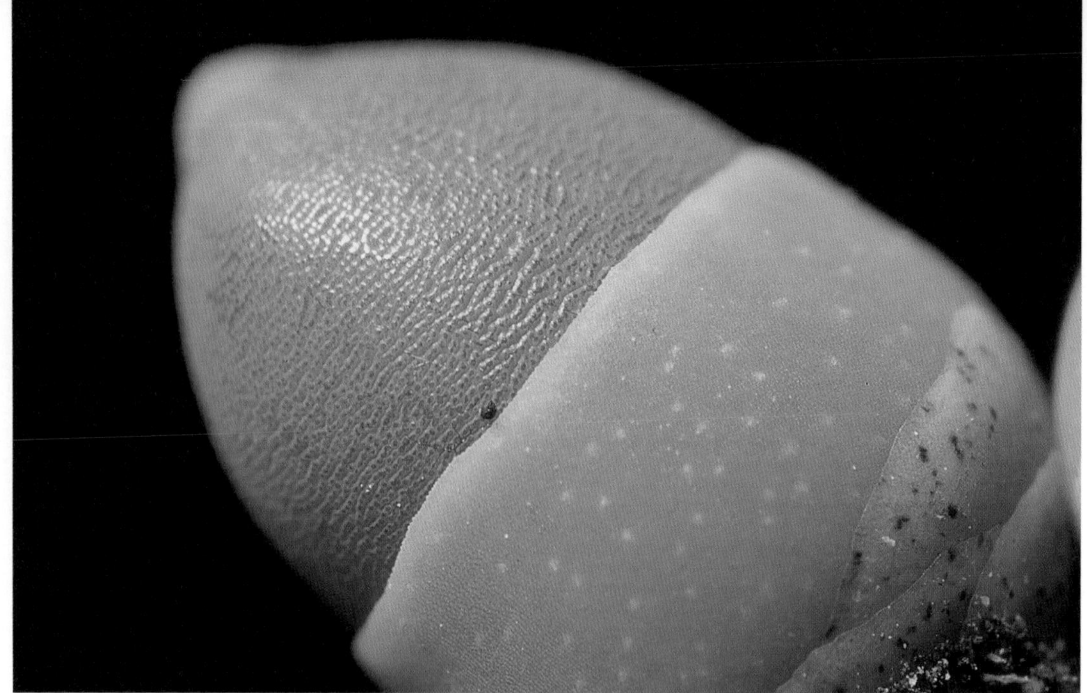

APPENDIX I

BOTANICAL GLOSSARY

abaxial facing away from the axis, dorsal; e.g. the lower surface of a leaf

adventitious growth (also known as epicormic) the phenomenon of green shoots that originate in the cambium breaking through the bark anywhere on the tree

angiosperm vascular seed plant with a highly specialised organ of reproduction (flower), in which the ovule (egg) is fertilised and (inside an ovary) develops into a seed

apex the tip, the distal end

apical at the tip

appressed, adpressed closely and flatly pressed against

aril usually a fleshy tissue on the stalk by which an ovule is attached to the placenta in the ovary, sometimes partially or completely covering the seed

axil the upper angle of a lateral organ (e.g. leaf, seed scale) and the axis or stem that bears it

branch layering process in which a branch grows towards the ground, then roots and grows up as a new plant

cambium layer of actively dividing cells that is responsible for the secondary growth of stems and roots. Cambial cells produce secondary xylem cells towards the inside and secondary phloem cells towards the outside

conifer a cone-bearing tree

cuttings new plants that are propagated from cut twigs of parent plants of those species which are able to regenerate missing parts

cytoplasm the entire contents of a plant cell (within the cell membrane) except the nucleus

deciduous not persistent, falling at end of a functional period (petals), or seasonal (leaves on deciduous trees)

decussate arranged in pairs alternately at right angles, resulting in four rows along the axis

dichotomous forking regularly and repeatedly, the two branches of each fork usually essentially equal

digitate resembling a hand, i.e. with leaves spreading from a centre

dioecious having the male and female flowers on different plants

endosperm the starch- and oil-containing tissue of many seeds

fastigiate with branches erect and more or less appressed

flexuous zigzag, bent or curved alternately in opposite directions

fluted of a trunk with rounded grooves running vertically

fruit the ripened ovary, the seed-bearing organ

globose more or less globular, irregularly globular

gymnosperm ('naked seeds') vascular plant whose ovules (seeds) are exposed on the surface of cone scales

heartwood wood consisting of xylem cells that have ceased to transport water and have been converted to supporting tissue, hence it is mechanically strong and resistant to decay

interior root in *Taxus* a root produced by the cambium (usually in the upper part of the trunk) that proceeds through the sapwood and down the decaying heartwood of a hollowing trunk

lanceolate lance-shaped, widest below the middle, tapering with convex sides towards the apex

layering *see* branch layering

monoecious having the male and female flowers separate, but on the same plant

morphological regarding the shape of a plant and its organs

mycorrhiza mutually beneficial root symbiosis of tree and fungi

obconic, obconical more or less conic

oblong longer than broad, with the sides nearly parallel most of their length

osmosis a process of physical law by which higher plants maintain their inner water pressure

ovule the body which after fertilisation becomes the seed

parenchyma plant tissue made up of undifferentiated and usually unspecialised cells

pendent, pendulous drooping, hanging downwards

petiole the stalk of a leaf

phloem conducting tissue that transports sugar and other organic substances

photosynthesis the process by which all green plants use solar energy to convert water and carbon dioxide into energy-rich organic compounds (carbohydrates)

pinnate constructed similar to a feather, with the parts (e.g. branches, lobes, veins) arranged along both sides of an axis

plane flat

phytosociological regarding the relationships between plants

raceme an unbranched, indeterminate inflorescence with flowers on individual stalks

racemose having raceme-like inflorescences that may or may not be true racemes

rays tissue that facilitates the radial and horizontal water transport in sapwood

reflexed abruptly recurved or bent downwards or backwards

sapwood layer of xylem cells that transports water and minerals

scandent climbing

seed a ripened (and fertilised) ovule, containing the embryonic plant

spiral thickenings anatomical feature of the tracheids, particularly strongly developed in *Taxus*, contributing significantly to the remarkable elasticity of its wood

stoma (pl. stomata) a minute pore in the epidermis of a leaf (or stem) through which gases are exchanged

temperate forest the naturally dominant vegetation type between approximately latitudes 25° and 50° in both hemispheres

temperate zone the climatical zone between the subpolar and the subtropical. In the temperate regions of both hemispheres the great deciduous and coniferous forests can be found

terminal at the tip, apical

tracheids the water- and mineral-conducting cells of conifers

vascular system the specialised conducting tissue, usually comprising xylem and phloem cells

xylem conducting tissue that transports water and minerals

A P P E N D I X I I

CLASSIFICATION OF *TAXUS* (TAXACEAE) SPECIES AND VARIETIES

INCLUDING KEY CHARACTERISTICS AND NATURAL GEOGRAPHICAL DISTRIBUTION

By Richard Spjut,
World Botanical Associates, Bakersfield,
CA 93380-1145, 21 July 2006

<table>
<tr><th colspan="4">I. Wallichiana Group
stomata bands scarcely differentiated from marginal and midrib cells, the abaxial leaf surface conspicuously papillose*; leaf epidermal cells usually red in dried specimens; leaf mesophyll with bone-like spongy parenchyma cells.</th></tr>
<tr><td rowspan="8">Wallichiana Subgroup
leaf epidermal cells quadrangular to tall rectangular;
second-year branchlets usually reddish purple to reddish orange;
stomata usually periclinal**; bud-scales usually persistent and conspicuous at base of branchlets.</td><td rowspan="3">stomata: 12 or more rows per band</td><td rowspan="2">T. wallichiana Zuccarini
scales at base of branchlets less than 2mm long
Nepal to SW China</td><td>var. wallichiana
abaxial leaf epidermal cells not enlarged towards margin, papillae marginal
NE India, Nepal, Bhutan, Myanmar, Sichuan, Hubei</td></tr>
<tr><td>var. yunnanensis (W.C. Cheng & L.K. Fu) C.T. Kuan
abaxial leaf epidermal cells enlarged towards margin, papillae medial
NE India, Nepal, Tibet, Myanmar, Yunnan</td></tr>
<tr><td colspan="2">T. suffnessii Spjut ineditus
scales at base of branchlets relatively large, 2–3mm long;
Myanmar</td></tr>
<tr><td rowspan="5">stomata: less than 12 rows per band</td><td rowspan="2">T. globosa Schlechtendal
abaxial leaf margin scarcely differentiated
SE United States (Florida) to El Salvador</td><td>var. globosa abaxial marginal and midrib cells sinuous
Mexico to El Salvador</td></tr>
<tr><td>var. floridana (Nuttall ex Chapman) Spjut ineditus
abaxial marginal and midrib cells relatively straight
SE United States (Florida), NW Mexico</td></tr>
<tr><td rowspan="2">T. brevifolia Nuttall
abaxial leaf margin with enlarged cells; stomata in 4–7 (9) rows
Pacific NW</td><td>var. polychaeta Spjut ineditus
cones appear stipitate, wormlike. CA, WA</td></tr>
<tr><td>var. reptaneta Spjut ineditus
scandent layering shrubs</td></tr>
<tr><td colspan="2">T. florinii Spjut ineditus
abaxial leaf margin with enlarged cells; stomata in 7–10 (12) rows
Yunnan, Sichuan</td></tr>
</table>

(I. Wallichiana Group continued)

Subgroup	Leaves	Species
***Chinensis* Subgroup** leaf epidermal cells wider than tall, elliptical to wide rectangular; second-year branchlets usually yellowish green; stomata often anticlinal**; bud-scales variable.	*Leaves linear to oblong*	***T. obscura*** Spjut ineditus flexuous branchlets, leaves drying dark metallic green above, dull rusty greenish orange below; leaves and branchlets crowded, overlapping Myanmar, China (Fujian, Taiwan), Philippines, Indonesia (Sumatra, Sulawesi)
		T. phytonii Spjut ineditus rigid branchlets; leaves drying dark metallic green above, abaxial surface with yellowish-green stomata bands bordered by reddish midrib and margins; branchlets and leaves not or scarcely overlapping. Nepal, NE India, Thailand, China (Yunnan, Taiwan), Philippines (Luzon)
		T. rehderiana Spjut ineditus rigid branchlets; leaves drying glossy green to olive green above, duller green on abaxial surface, becoming increasingly more curled from upper mid region to apex Vietnam, Taiwan, Sulawesi
	Leaves oblong or broad linear and slightly elliptical	***T. chinensis*** (Pilger) Rehder scales lacking or minute at base of branchlets Vietnam, China (Guangxi, Gansu, Yunnan, Sichuan, Guizhou, Hubei, Anhui, Zhejiang)
		T. ocreata Spjut ineditus scales numerous and tightly adhering to base of branchlets, tooth-like Yunnan, Sichuan
		T. scutata Spjut ineditus scales many and loosely attached at base of branchlets, overlapping like pointed terracotta roof tiles, loosely attached, spreading and falling off. Yunnan, Sichuan, Hubei

II. *Sumatrana* Group
stomata bands sharply differentiated from marginal and midrib cells, the abaxial leaf surface often drying reddish along margins and on midrib, lacking papillae along margins 8–36 cells across; leaf epidermal cells usually red or orange in dried specimens; leaf mesophyll with bone-like spongy parenchyma cells
T. celebica (Warburg) H.L. Li large lanceolate, plane green leaves. Bhutan, NE India, Vietnam, China (Tibet, Yunnan, Sichuan, Ningxia Huizu, Guizhou, Fujian, Zhejiang), Indonesia (Sulawesi)
T. kingstonii Spjut ineditus rusty orange leaves and branchlets India (Khasi Hills), Myanmar, China (Tibet, Gansu, Shaanxi, Sichuan, Yunnan, Taiwan)

* **mammillose/papillose**. Sharp *et al.* (1994) define mammillose or mammillate as 'prominently convex-bulging'; thus, this is more like the whole breast, not just the nipple, in contrast to papilla which they define as 'a cell ornamentation, a microscopic protuberance of various forms', which I compare to a pimple. Mammilla in *Taxus* develops as a single rounded lens (breast) like protuberance usually on the upper (adaxial) leaf epidermal cells and along the leaf margins (easily visible with a hand lens), but mammillae also appear in a different form (bulging cell cuticle or thin lens-like protuberance) on the lower (abaxial) epidermal midrib cells of *T. mairei* (only evident under the microscope, 250–400 x). Papillae in *Taxus* are numerous pimple-like projections, visible only under a microscope, often aligned in two rows on each cell. The papillae may be so close together that they become partially fused (concrescent), or in some species, e.g., *T. celebica*, papillae on a cell appear quite small and distinctly separate.

** **anticlinal** (vs. **periclinal**). Oriented perpendicular to the general line of development. Epidermal cells and stomata rows all run lengthwise, and usually the axis of the stomata as well, but in the *Chinensis* Subgroup, stomata often appear anticlinal, i.e., the stomata (a donut-like hole appearing squeezed along two opposing sides), have their longer axis (of the pore) pointed towards the leaf margins.

(II. Sumatrana Group continued)

T. mairei (Lemée & Léveillé) S. Y. Hu ex T. S. Liu enlarged epidermal cells along leaf abaxial midrib, appearing mammillose leaves elliptical to oblong	var. *mairei* isodichotomous* branching China (Sichuan, Yunnan, Anhui, Guizhou, Guangxi, Jiangxi, Fujian, Hunan, Zhejiang, Taiwan)
	var. ***speciosa*** (Florin) Spjut ineditus anisodichotomous* branching India (Khasia Hills), China (S Shaanxi, Sichuan, NE Yunnan, Guizhou, N Guangxi, Hunan, Guangdong, W Hubei, Jiangxi, Zhejiang)
T. sumatrana (Miquel) De Laubenfels leaves pucker [becoming contracted and wrinkled] on drying	var. *sumatrana* leaves drying dark brownish green above Thailand, China (Taiwan), Philippines, Indonesia (Sulawesi, Sumatera)
	var. ***atrovirens*** Spjut ineditus leaves green when dried Nepal, India (Khasia), China (Zhejiang)
	var. ***concolorata*** Spjut ineditus leaves papillose across midrib and marginal cells to 6 cells from margins Philippines, Indonesia (Sumatera)

III. Baccata Group

stomata bands partially differentiated from marginal and midrib cells; the dried abaxial leaf surface often green along margins and on midrib; papillae variable in development; leaf epidermal cells usually clear in dried specimens; leaf mesophyll usually with spherical to ellipsoidal spongy parenchyma cells.

***Baccata* Alliance** abaxial surface of leaves mostly papillose between margins and stomata bands, usually lacking across 8 or fewer cells from margins Euro-Mediterranean	***T. baccata*** Linnaeus leaves along one side of a branchlet mostly parallel, overlapping more than criss-crossing, straight or bending upwards	
	1 — Baccata Complex branchlets often dividing at less than 70° angle, reddish orange to purplish; leaves mostly linear, spreading outwards from branchlets along two sides in one plane; male cones in loose aggregates, or solitary	var. *baccata* leaves slightly overlapping along one side of a branchlet, generally not spreading more than 60° from branchlets, uniformly plane to slightly recurved along margins Euro-Mediterranean
		var. ***washingtonii*** (hortus ex Richard Smith) Beissner leaves closely parallel but not overlapping along one side of a branchlet, often spreading 45–90°, usually more strongly recurved along margins in upper third (*T. recurvata*, which is sometimes similar, is distinguished by the leaves sharply bent downwards near apex) Europe, SW Asia (Syria, Turkey)

* **anisodichotomous/isodichotomous.** *Aniso*, Greek, generally a prefix, in regard to unequal, in contrast to *iso* meaning equal. Dichotomous means having two divisions or dividing into two. In Spjut 1996 isodichotomous is defined as 'branches' that 'divide more or less equally into secondary branches', and anisodichotomous as branches showing a distinct dominance with secondary smaller divisions.

(III. Baccata Group continued)

Alliance	Complex	Variety
***Baccata* Alliance** abaxial surface of leaves mostly papillose between margins and stomata bands, usually lacking across 8 or fewer cells from margin Euro-Mediterranean	***2 — Elegantissima Complex*** branchlets spreading to recurved, yellowish green to yellowish orange; leaves bending upwards, male cones often reflexed or pendulous	var. ***elegantissima*** hortus ex C. Lawson branchlets not wide spreading; leaves turned upwards near ends of branchlets, in hair-like tufts Euro-Mediterranean
		var. ***jacksonii*** (Paul) Gordon with regularly pinnately arranged branchlets, the branchlets often wide spreading, flexuous; leaves strongly convex across adaxial surface Euro-Mediterranean
		var. ***variegata*** Weston branchlets wide spreading and rigid; leaves spreading upwards Europe
	3 — Ericoides Complex branchlets mostly ascending to erect, yellowish orange; leaves usually not in one plane but appearing radial, overlapping; male cones in terminal subglobose aggregates	var. ***erecta*** Loudon branches and leaves stiffly ascending, leaves crowded, somewhat spreading radially, mostly green Portugal, Spain
		var. ***ericoides*** Carrière branchlets appearing pendulous, fastigiate, with radial spreading oblong leaves Morocco
		var. ***pyramidalis*** C. Lawson with isodichotomous* branching; leaves spreading somewhat radially, mostly straight, yellowish green, often falling off by the third year Euro-Mediterranean
		var. ***subpyramidalis*** Jacques ex Carrière seed wider than long, obconic or four-lobed Iran
	4 — Glauca Complex branchlets variable in orientation, often reddish orange to yellowish orange; leaves overlapping and bending upwards; male cones in loose aggregates, or reflexed and racemose; arillocarpia** often in pairs	var. ***dovastoniana*** Leighton branchlets crowded or spreading digitately near apex of main branch, often drooping; uppermost leaves often spreading upwards or towards the end of branchlet Euro-Mediterranean
		var. ***fructo-lutea*** Loudon aril yellow SW Asia (Caucasus), Europe (Ireland)?, Japan?
		var. ***glauca*** Carrière branchlets recurved, drying yellowish orange as also evident on abaxial surface of leaves Euro-Mediterranean

** **arillocarpia** (*arillus*, fleshy outgrowth around seed; *karpos*, fruit). Defined in Spjut, 'A systematic treatment of fruit types' (1994, *Memoirs of the New York Botanical Garden*, 70), 'A fruit of the Pinopsida characterized by a seed being covered by a fleshy aril as in species of Cephalotaxaceae and Taxaceae. Typical: *Taxus baccata*.'

(III. Baccata Group continued)

Alliance	Complex	Species	Variety
***Baccata* Alliance** abaxial surface of leaves mostly papillose between margins and stomata bands, usually lacking across 8 or fewer cells from margins. Euro-Mediterranean		***T. contorta*** Griffith dried leaves with reddish, eggshell-like parenchyma cells	var. *contorta* leaves usually long linear, narrow to wide spreading Afghanistan, Pakistan, India, W Nepal, China (SW Tibet)
			var. ***mucronata*** Spjut ineditus leaves oblong to short linear and wide spreading Nepal, Bhutan, China (SW Tibet) (?)
		T. fastigiata Lindley leaves imbricate, spreading radially, recurved	var. *fastigiata* tall plants, narrow and cylindrical in outline. Ireland, Scotland, England (?)
			var. ***nana*** (Carrière) Spjut ineditus small shrubs up to 60 cm high; leaves dark green. Europe
			var. ***sparsifolia*** (Loudon) Spjut ineditus branchlets spreading to ascending, leaves yellowish to olive green; tall shrubs or trees. Scotland, England
		T. recurvata hortus ex C. Lawson leaves crisscrossing more than overlapping, often curved along the blade downwards, and often sharply bent downwards near apex	var. *recurvata* branchlets recurved; leaves up to 12 times longer than wide; Europe, SW Asia (Caucasus)
			var. ***intermedia*** (Carrière) Spjut ineditus branchlets straight to flexuous; leaves up to 12 times longer than wide Euro-Mediterranean
			var. ***linearis*** (Carrière) Spjut ineditus leaves 12 to 20 times longer than wide. Europe (Madeira, England, Germany, Hungary, Bulgaria, Austria), SW Asia (Syria, Turkey)
***Cuspidata* Alliance** abaxial surface of leaves mostly without papillae between margin and stomata band, 8–24 cells across; temperate E Asia except *T. canadensis*.	***1—Canadensis* Complex** leaves spreading mostly in two ranks	***T. canadensis*** Marshall leaves spreading along petiole near branchlet Atlantic North America, Euro-Mediterranean	var. *canadensis* branching nearly pinnate, terminal branchlets often isodichotomous; leaves linear
			var. ***adpressa*** (Carrière) Spjut ineditus low spreading shrubs with oblong leaves and occasional alternate branchlets
			var. ***minor*** (Michaux) Spjut ineditus leaves overlapping, strongly curved, not distinctly in two ranks; branchlets often recurved
		T. cuspidata Siebold & Zuccarini leaves sharply bent along petiole near blade; branching mostly pinnate; branchlets thick, leaves relatively thick, uniformly recurved along margins South Korea, Japan (Hokkaido)	
		T. biternata Spjut ineditus leaves sharply bent near petiole, branching mostly ternate; branchlets thin; leaves thin, appearing pinched inwards in upper third when dry. Widespread in temperate E Asia; China (NE Manchuria), Russia (SE region), North Korea, South Korea, Japan	

(III. Baccata Group continued)

***Cuspidata* Alliance** abaxial surface of leaves mostly without papillae between margin and stomata band, 8–24 cells across; temperate E Asia except *T. canadensis.*	***2 — Unbraculifera Complex*** leaves spreading radially on erect to ascending branchlets, at least near apex	***T. caespitosa*** Nakai leaves mostly imbricate, usually bending upwards along petiole	var. ***angustifolia*** Spjut ineditus branches prostrate with erect branchlets; leaves linear, relatively narrow, up to 2mm wide Japan, Korea
			var. ***caespitosa*** branchlets dense, short, ascending to erect; leaves oblong Japan
			var. ***latifolia*** (Pilger) Spjut ineditus branchlets wide spreading or trailing, not densely compacted; leaves oblong to broad linear Korea, Japan, SE Russia, NE China
		T. umbraculifera (Siebold ex Endlicher) C. Lawson leaves more or less decussate on ascending to erect branchlets, or appearing two-ranked on horizontal branchlets, sharply reflexed	var. ***umbraculifera*** diffusely branched shrubs or trees Japan, Manchuria
			var. ***hicksii*** (Rehder) Spjut ineditus columnar; branchlets crowded, ascending to erect Japan
			var. ***microcarpa*** (Trautvetter) Spjut ineditus typically a low shrub with branches spreading radially from a central point, all flat against the ground, or shrubs with branches spreading above ground but very much divided leaves generally not or only slightly overlapping in two ranks, except near apex Korea, Manchuria, Japan
			var. ***nana*** (hortus ex Rehder) Spjut ineditus low densely branched shrub with radially spreading leaves China (Shanxi), Japan

Yew trees in the mountains of Yusmarg, Kashmir, India (far left), El Cielo in Tamaulipas, Mexico's northernmost cloud forest (left), and the Hyrcanian Forest, Iran (above).

Regional names of the Himalayan yew
after Narayan 2002

Gurung salin
Nepali barma salla, bham salla, bung, luinth, pate salla, silangi, thingre salla
Newari la swan
Sherpa chyangsing
Tamang sigi

APPENDIX III

IMPORTANT OCCURRENCES OF EUROPEAN YEW

The European yew occurs in all of Europe. Its northern limit extends from the British Isles to Norway (*c.* 63° N), Sweden and Finland (61° N), the eastern border runs from the Riga bay (Latvia) through Bialowiecza (Belarus–Polish border) along the 23° meridian to the eastern Carpathians and the Black Sea where *Taxus* occurs in the Crimean peninsula and across northern Turkey. The southern limit includes Portugal and the Mediterranean countries of Europe, but also Madeira, the Atlas Mountains (Algeria, *c.* 33° N), the northern Pontus, the Taurus and Amanus Mountains (southern Turkey, northern Syria), the entire Caucasus, and the Elburs Mountains in northern Iran. Within this extensive range, *Taxus* is missing in the regions influenced by continental climate (i.e. Eastern Europe, the Anatolian Highland, the Hungarian Lowland) as well as in the higher mountains (central Alps, central Carpathians) (see also Chapter 4). Ecologically, eleven types of different plant associations can be distinguished (Table 10).

Table 10 Types of yew *(T. baccata)* forest[1]

1	Bakony yew forests
2	Carpathian yew forests
3	Yew forests of the German–Bohemian highland
4	Yew forest in the margins of Alps
5	Croatian yew forests
6	Yew forests of Greece (Balkan)
7	Yew forests of Turkey
8	Yew forests of Caucasus, Crimea and Iran
9	Iberian and Italian yew forests
10	Algerian yew forests
11	Yew forests of the North European lowlands

In a number of countries, *Taxus baccata* is included in the Red Books (e.g. in the Czech Republic, Slovakia, Bulgaria, Romania, Russia, Iran), and in several countries it is subject to nature conservation (e.g. Germany, Poland, Czech Republic, Slovakia, Romania, Russia), albeit to varying degrees. In Italy, for example, only single monumental trees are protected on the national level but in most administrative regions the species is submitted to special protection management on the regional level; and following the indications of the Natura 2000 network, beech forests with *Taxus* and *Ilex* are protected as Sites of Regional Importance (SIC = Siti di Importanza Comunitaria).[2] In Switzerland, the tree is not protected either but the management of *Taxus* regeneration and timber production is handled by the single cantons in a sustainable way and with great care. Single old trees are protected, and in two cantons – Basel Land and Schaffhausen – the species is protected entirely.[3] In the Caucasus, some forests within the nature reserves are threatened by ethnical/political unrest and by military operations against separatist or resistance groups that hide in the mountains. Even more worrying is the situation in Iran where the Hyrcanian Forest is threatened by large-scale logging, forest fires and the traditional practice of wood pasture: between 1970 and 2000, at least 27 per cent of the Hyrcanian Forest was destroyed.[4] The deforestation continues despite *Taxus* being legally protected and the efforts of environmentalists.

For more locations (worldwide) see
http://www.worldbotanical.com/taxus_baccata.htm#baccata
For more information on single countries see
www.ancient-yew.org/baccata-stands.shtml
For extensive information on the ancient yews of Britain see
www.ancient-yew.org

Table 11 Important occurrences of European yew in individual countries[5]

country	*locality*	*area (ha)*	*number of individuals*	*comments*
Ireland	Reenadinna Forest, Killarney Nat. Park			25ha of rainforest; high density of mature *Taxus* trees rooting on very thin soil or even penetrating into fissures in the bare limestone
Britain	Castle Eden Dene, Co. Durham	22	...	steep Magnesian limestone hillsides with ash, hazel and high density of mature yew trees. The name *(Œden)* derives from Saxon *yoden*, 'valley of yew'.
	Kingley Vale, West Sussex	150	30,000	National Nature Reserve; *c.* 70 ancient yews; soil is clay with flints over upper chalk. The name derives from the bronze age burial mounds on the top of the hill.
	Druids Grove, Norbury Park, Surrey			More than 20 large yews with girths of up to 7m, on chalk slope
	Newlands Corner, Surrey			23 ancient yews (girth between 4m and 7m) scattered over *c.* 50ha; soil is clay with flints on chalk
	(private) yew woodland, Wiltshire	56	...	almost pure yew woodland with many old to ancient specimens (see photo pp. 238–9)
Spain[6]	Mt Sueve, Asturias	200	8,000	currently under threat
	Sierra Tejeda (Umbría de la Maroma), Málaga		...	181 individuals recorded in 1997
	Sierra de Guara, Huesca province	...	...	occupying small areas in steep riverbanks; some individuals reach 18m[7]
	Misaclós, Montagut, Girona	4–5	...	mixed forest with Pinus sylvestris and Quercus ilex in 330–350m a.s.l.; highest density of yew: 400 individuals on 0.5ha; oldest trees have 40cm diameter
France	Ste Baume, Provence	138	...	Mixed forest on north-facing slope; in 1882, 4,000 yew trees were counted of which 2,700 had 30–45cm trunk diameter; today, largest tree (80cm trunk diameter) estimated at *c.* 800 years, the others at *c.* 500; protected since 1838, now NATURA 2000 Reserve.[8]
Corsica	(various)	...	900	mature and old trees in various locations; *Taxus* is concentrated to the south of Corte, to the south-east of Calvi and to the west of Porto-Vecchio, also in the north (Cap Corse), south (Montagne de Cagna), east (San Giovani di Moriani) and west (Piana) of the island
Italy and Sardinia	*Taxus* occurs as a forest tree along the entire length of Italy, mostly in mountainous settings; a national inventory of stands is under way. Apart from the British Isles, Italy and Sardinia probably have the highest number of old and ancient single specimens in (Western) Europe, among them are Ucca 'e Grille, Sos Niberos, Bono (Sardinia): girth 705cm, height 11m; Nattari, Urzulei, Nuoro (Sardinia): girth 530cm, height 22m; Valle Naforte, Sezze LT, Lazio: girth 500cm, height 15m; Fonte Avellana, Serra Sant'Abbondio PU, Marche: girth 475cm, height 15m (see figures 28.8, 41.2).[9]			
Switzerland	Switzerland has a relatively large number of young yew trees: the overall estimate amounts to *c.* 700,000 trees with a girth over 10cm, 50,000 of which are located in the Hörnli region (St Gallen/Thurgau/Zürich). Around Zürich there are *c.* 70,000 young yews with a girth over 4cm.[10]			
Germany	Paterzell	90	1,600	30ha of which comprise the public nature reserve
	Rudolstadt	128	6,000	mostly young trees
	Lengenberg	23	4,500	mostly young trees
	Wasserberg Gössweinstein	10	4,100	mostly young trees

<table>
<tr><th></th><th>locality</th><th>(ha)</th><th>individuals</th><th colspan="2">comments</th></tr>
<tr><td>Austria</td><td>Pichlwald, Vöcklabruck, Upper Austria</td><td colspan="4">2.6ha of mixed yew forest in 480–530m a.s.l.; lakeside slope</td></tr>
<tr><td></td><td>Stiwollgraben, Graz region, Steiermark</td><td colspan="4">17.0ha of mixed yew forest in 580–700m a.s.l.</td></tr>
<tr><td></td><td>Hinterstein, Kufstein, Tyrol</td><td colspan="4">28.4ha of mixed yew forest in 900–1,050m a.s.l.[11]</td></tr>
<tr><td>Czech Republic</td><td>Krivoklát</td><td>...</td><td>5,000</td><td colspan="2">mostly young trees</td></tr>
<tr><td></td><td>Moravian Karst</td><td>...</td><td>2,000</td><td colspan="2">mostly young trees</td></tr>
<tr><td></td><td>Stechovice</td><td>...</td><td>418</td><td colspan="2">mostly young trees</td></tr>
<tr><td></td><td>Kanice, Domazlice</td><td>...</td><td>200</td><td colspan="2">yews in mixed woodland around the summit of Mt Netreb</td></tr>
<tr><td>Poland</td><td>Wierzchlas</td><td>18</td><td>3,500</td><td rowspan="13">mostly young trees</td><td rowspan="13">* These figures are older data and in the meantime the numbers have decreased dramatically, e.g. in Harmanec due to red deer browsing, natural mortality, etc. Forest statistics are a double-edged sword anyway, as forests are not static but in constant flux.</td></tr>
<tr><td>Slovakia</td><td>Harmanec</td><td>860</td><td>(160,000)*</td></tr>
<tr><td></td><td>Gader</td><td>513</td><td>(17,000)*</td></tr>
<tr><td></td><td>Plavno</td><td>27</td><td>(9,000)*</td></tr>
<tr><td></td><td>Slovensky raj</td><td>230</td><td>1,560</td></tr>
<tr><td></td><td>Lucivná</td><td>...</td><td>1,000</td></tr>
<tr><td>Hungary</td><td>Bakony</td><td>287</td><td>(120,000)*</td></tr>
<tr><td>Romania</td><td>Forest Tudora</td><td>125</td><td>1,095</td></tr>
<tr><td></td><td>Forest Comarnic</td><td>4</td><td>1,025</td></tr>
<tr><td></td><td>Dosu Stoglui</td><td>...</td><td>763</td></tr>
<tr><td></td><td>Cenaru</td><td>383</td><td>741</td></tr>
<tr><td></td><td>Cartsoara</td><td>25</td><td>600</td></tr>
<tr><td>Bulgaria</td><td>Vitosha</td><td>...</td><td>276</td></tr>
<tr><td>Ukraine</td><td>Knyazhdvir (Kolomyja)</td><td>30</td><td>22,000</td><td colspan="2">yew trees up to 30cm diameter (in 1976) as understorey of mixed woodland dominated by beech (80%) and fir (A. alba, 20%) not older than 100 years; steep sandstone slopes[12]</td></tr>
<tr><td></td><td>Ugolka</td><td>208</td><td>10,000</td><td colspan="2">mostly young trees</td></tr>
<tr><td>Greece</td><td colspan="5">The first national Taxus inventory in Greece (1995) shows that the species is in decline, but still present in 173 yew stands (usually not exceeding 50 individuals) in 117 forests across the country, mostly in areas of central and northern Greece, with small natural stands in the Peloponnese and on the island of Evia; woods are dominated by beech, fir, black pine and oak, yew occurs most often with juniper and holly; 80% of yew populations occur in ravines (mostly 500–1,500m a.s.l.), especially along the Pindos mountain range, Mt Olympus, Mt Rodopi and Mt Cholomontas in Halkidiki.[13]</td></tr>
<tr><td>Crimea</td><td colspan="5">57 yew stands are known; mixed forest with oak, beech, hornbeam and juniper[14]</td></tr>
<tr><td>Turkey</td><td colspan="5">Usually on slopes between 1,000 and 1,900m altitude. Black Sea region: Alapli; Yenice; Düzce; Rize forest; Kürtün (Gümüshane); Ayancik (Sinop); Yedigöl above Devrek (1,000m a.s.l.), Bolu; Demirköy. In the south: Hatay (1,800–1,900m a.s.l.), Amanus Mountains; Canakkale province in the Kaz Mountains; Denizli province on Mt Akdag (1,800m a.s.l.); Icel, Cilician Gates (Gülek Bogaz).[15]</td></tr>
<tr><td></td><td>Alapli (near Eregli), Zonguldak</td><td>200</td><td>...</td><td colspan="2">200ha with yews in nature reserve of 11,000ha; valley with ancient yews (see figures p. 215) in open beech-oak-yew forest</td></tr>
</table>

	locality	(ha)	individuals	comments
	Yenice, Karakuk			ancient yews scattered among oak, beech (*Fagus orientalis*), fir (*A. bormulleriana*), pine (*P. nigra* and *P.sylvestris*); nature reserve also rich in old box (*Buxus sempervirens*)
Caucasus	There are 130 (!) known yew stands in the Western Caucasus alone of which the Batzara Reserve (also called Batzvara) in Kakhetia (*c.* 80km north-north-west of Tiblisi in Georgia) is the largest. Batzara has been protected since Queen Tamara in the twelfth century (see p. 142), it was sacred to the local population before that, and now is a European Biosphere Reserve.[16]			
	Batzara	237	220,000	yew-beech-wood in 900–1,500m a.s.l. is part of the Bazari Canyon that covers 3,000ha; yew dominates 11ha with about 80% of the woody mass; only *c.* 13,000 trees are older than 100 years, single trees are 1,500–2,000 years old[17]
	Khosta	190	...	mixed forest of yew, beech, laurel, oak, ash, hornbeam, lime and maple, with an understorey of box, usually on limestone slopes with eastern or north-eastern exposition; 15ha of the area have 50–90% yew, another 36 have 10–40%; majority of yews *c.* 600–*c.* 1,000 years old; sparse regeneration[18]
	Sochi	301	...	largest yew *c.* 2m trunk diameter and 30m height[19]
Iran	The Hyrcanian Forest covers the northern expositions of the Elburs Mountains, facing the Caspian Sea which supports a mild and moist climate (600–2,000mm annually).[20] It is huge – its original size is *c.* 1.3 million hectares – and it is an extremely unique ecotope because without being affected by glaciation during the last Ice Age it had 10 million years to develop a rich diversity of tree species. It consists mainly of deciduous trees – Oriental beech, hornbeam, Caspian alder *(A. subcordata)*, Caspian oak *(Q. castaneifolia)* and velvet maple *(A. velutinum)* – with an understorey of box and yew (between 900 and 2,000m a.s.l.).[21]			
	d'Afra-Takhté, eastern Elburs Mts	150	...	high density of yew, especially on *c.* 28ha at Punearam (over 75% of the trees are yew); average trunk diameter over 60cm,[22] age of yews 500–800 years; sparse regeneration[23]
	Arasbaran (near Kallaleh), Azarbaijan[24]	...	...	

APPENDIX IV

THEOPHRASTUS ON YEW

Excerpts from *Enquiry into Plants*

III. X. 2: The yew [*milos*] has also but one kind, is straight-growing, grows readily, and is like the silver-fir [*elate*], except that it is not so tall and is more branched. Its leaf is also like that of the silver-fir, but glossier and less stiff. As to the wood, in the Arcadian yew it is black or red, in that of Ida bright yellow and like prickly cedar; wherefore they say that dealers practise deceit, selling it for that wood: for that it is all heart, when the bark is stripped off; its bark also resembles that of prickly cedar in roughness and colour, its roots are few slender and shallow. The tree is rare about Ida, but common in Macedonia and Arcadia; it bears a round fruit a little larger than a bean, which is red in colour and soft; and they say that, if beasts of burden eat of the leaves, they die, while ruminants take no hurt. Even men sometimes eat the fruit, which is sweet and harmless.

III. IV. 5: As for silver-fir and yew, they flower a little before the solstice ... they bear their fruit after the setting of the Pleiad. Fir and Aleppo pine are a little earlier in budding, about fifteen days, but produce their fruit after the setting of the Pleiad, though proportionately earlier than silver-fir and yew.

APPENDIX V

A NOTE ON FRAZER'S *GOLDEN BOUGH*

There is hardly another tree or plant in the world with such a rich body of evidence in religious history, mythology and folklore as the yew. So why has it been largely ignored in almost all publications in these fields?

The 'natural' decline of acknowledging this special tree in the course of early history, mainly because of general developments in culture and religion, has been documented in Part II. Another reason is that even in post-medieval Iceland, people no longer understood the kennings of Skaldic poetry, and the references to the World Tree in the *Poetic Edda* and other Nordic texts – a mix of 'yew' and 'ash' (see Chapter 41) – became solely interpreted as 'ash'. Texts and illustrations produced during the subsequent centuries perpetuated the view of an assumed 'World Ash', and still today many scholars stick to it adamantly. The almost complete absence of *Taxus* in general mythology and religious studies of the twentieth century, however, is largely due to one man and his most influential book – James Frazer and *The Golden Bough*, which does not mention *Taxus*.

Written between 1886 and 1889, *The Golden Bough* was first published in 1890. It appeared in a much expanded twelve-volume edition between 1907 and 1915, and an abridged version in 1922. With this study of the religions of classical antiquity, Sir James George Frazer (1854–1941) came to be regarded as one of the founders of modern anthropology. His book laid the foundations for countless studies in anthropology, ethnology, sociology, biblical studies, mythology, and related fields. It also provided twentieth-century literature with its raw material: William Butler Yeats, T.S. Eliot, D.H. Lawrence, James Joyce, to name but a few, were vastly influenced by it.[1] However, *The Golden Bough* has very serious flaws, and 'was *never* accepted by most historians and theologians, the specialists in its field'.[2] And when early anthropology adopted essential rules – meticulous personal fieldwork, due regard to the context of each observation and avoidance of a patronising attitude towards tribal societies – it showed that Frazer broke all of them. Without sufficiently verifying his (wild mix of) sources,[3] he gathered and compared elements of nineteenth-century European folklore of the 'crude peasantry' with the rituals of the 'savage mind' of contemporary tribal peoples and with the 'primitive' consciousness of early and prehistoric religion. In view of the abundance of yew material from all ages, the fact that this tree does not feature in any meaningful way in Frazer's massive study is most curious, and not easy to explain. Did he really not know? Or did he choose not to tell?

For one, Frazer selected his material in keeping with a special mission he intended his book to fulfil – to show that Christianity had the same crude and 'primitive' roots as the pagan religions – and he presented the kind of material that helped his cause. Secondly, he was preoccupied with mistletoe and oak.[4] For a botanical identification of the mythical golden bough, Frazer set his mind on the mistletoe, with only three – entirely disconnected – leads to evidence a religious significance for this plant: the Norse deity Balder being slain with a mistletoe branch, Pliny's (unverified and fanciful) account of the Druidic golden sickle in Gaul (see p. 149), and Virgil's note in the *Aeneid* (see p. 210). As Frazer's biographer, Robert Ackerman, says, 'despite the fact that Virgil does not identify the Golden Bough with the mistletoe but only compares the two, Frazer claims that the Golden Bough was the mistletoe nevertheless: "... seen through the haze of poetry or of popular superstition."'[5] Frazer, however, knew how feeble his mistletoe argument really was: 'The whitish-yellow of the mistletoe berries is hardly enough to account for the name [golden bough] ... Perhaps the name may be derived from the rich golden yellow which a bough of mistletoe assumes when it has been cut and kept for some months.'[6] But Virgil's point was that it was golden *on* the tree ...

Above: *The Fraser yew above Loch Ness, Scotland.*
Right: *One of the trunks of the Fraser yew (the parent one has a girth of 458cm at ground level).*

Frazer was also an enthusiastic perpetuator of the 'sacred oak' thesis. The sacred trees of Greece *(drus)* that over time *on paper* had become 'oaks' *(drys)*, the sacred trees of the Bible that had been translated into 'oaks' (see Chapter 30), Pliny's 'oak' of the Gaulic druids, all merged nicely in Frazer's *opus grande* to become the sacred oak as the carrier of the sacred mistletoe. He claims that the Indo-Europeans 'commonly' kindled their sacred fires with oak, and because the perpetual holy fire at Nemi, too, was said to have been fed with this wood the sacred tree at Nemi must have been the evergreen oak, 'the sacred tree of Jupiter'.[7] This is doubly illogical because (1) the tree species whose wood is ceremonially burnt is not necessarily the one that is the most sacred; [8] the concept of a sacred tree usually implies a taboo to touch it, hence it would not be burnt away for a perpetual fire in its honour (particularly if it was a single ancient yew in a mixed oak woodland), and (2) the sacred tree of the sky-god (Jupiter) is certainly not the tree of the Anatolian mother of the gods from the Taurus Mountains (Frazer's 'Tauric Diana')[9] to whom the grove in question was dedicated. Furthermore, Frazer (like Robert Graves half a century later) did not account for the different species of evergreen oaks across the Mediterranean (see discussion pp. 146–9).

Apart from this, Frazer could not know what we know today about *Taxus*. The entire body of Hittite evidence, for example, was still locked in undeciphered cuneiform tablets (until Hrozny's break-through in 1915, see p. 144), and, on top of that, the duration of his work on *The Golden Bough* coincides exactly with the period in which the dendrological discussion of the age of yews drastically reduced their estimated ages and ridiculed any other opinion; it was the time of a demystification of *Taxus*, and no rational Victorian gentleman would have opposed it.

On the other hand, it is hard to believe that Sir James knew nothing about the cultural significance of *Taxus* since it features prominently in his own family history: yew is the traditional clan badge of the Fraser clan, and on the Fraser land stands a magnificent grove that was the clan's meeting place for generations.[10] It consists of about thirty trees that have grown from the roots of a single central tree. Would someone as interested in old customs as J.G. Frazer never look into his own family history (and then stumble across Christian monastic yews, Templar yews, Scottish Freemasonry and so on)? Whatever the reasons, the absence of yew in Frazer's book accounts largely for its absence in the important works of authors such as A.B. Cook and other historians and anthropologists of the Cambridge-based Myth and Ritual School inspired by Frazer,[11] and later Joseph Campbell, Mircea Eliade and many other influential writers of the twentieth century.

Despite its flaws and antiquity, *The Golden Bough* remains one of the very few books on religious history referred to in botanical and dendrological papers even of very recent years.

USEFUL ADDRESSES

The Ancient Yew Group was founded to raise awareness of Britain's unique ancient yew heritage and to highlight the risks that still threaten many of these trees. Its website, sponsored by the Tree Register and the Conservation Foundation, was launched in June 2005 and has attracted considerable interest and helped raise the yew's profile in Britain. The increase in the number of people seeking help to save threatened yews makes all the more urgent the need to inform and educate those who make decisions about the fate of individual trees. We need to be aware that more than 80 per cent of our oldest yews grow in churchyards belonging to the Church of England and the Church in Wales. In 2006 the Ancient Yew Group forged links with Eibenfreunde (Friends of the Yew, founded in Germany in 1994), organising their annual conference and field trip to visit the yew sites of southern Britain. For more information go to www.ancient-yew.org or write to The Ancient Yew Group, Vine Cottage, 3 Ham Green, Pill, North Somerset, BS20 0EY.

The Friends of the Trees was founded in March 2003 and received charitable status in November that same year. Its two main objects are to raise awareness about the 'non-material' values of trees, i.e. their right to exist independent of their usability for human convenience or profit, and secondly to provide places where people of all religions (or none) can come together. It is the Friends of the Trees' long-term objective to create special places of contemplation, peace and mutual friendship in nature, where people can find inspiration, relaxation and healing. The charity has also set up a separate bank account, the Friends of the Trees Yew Fund, which supports the Ancient Yew Group, and receives half of the author's royalties for this book. For more information go to www.friendsofthetrees.org.uk or write to Friends of the Trees, The Secretary, 269 Melbourne Court, Battlefield, Newcastle on Tyne, NE1 2AU.

The Tree Register of the British Isles was co-founded in 1988 by Alan Mitchell, author of *A Field Guide to the Trees of Britain and Northern Europe*, who compiled records of 100,000 trees on a handwritten card index system. This was the beginning of the modern computerised register that now contains details of more than 150,000 of Britain and Ireland's most notable trees. Rare, unusual and historically significant trees are recorded, and a definitive record is provided of Britain's champion trees – the tallest and largest girthed of each species. The Tree Register recognised the significance and scope of the work being carried out by a handful of yew enthusiasts and helped create the first website dedicated to a single tree species www.ancient-yew.org. For more information go to www.tree-register.org or write to The Tree Register, 77a Hall End, Wootton, Bedford, MK43 9HL.

The Conservation Foundation was founded by David Bellamy and David Shreeve. It was responsible for setting up The Yew Tree Campaign in 1986 to encourage the protection of Britain's ancient yews and actively supports the work of the Ancient Yew Group. It has campaigned to raise awareness of the yew's antiquity and the role it has played in British history and has been able to provide finance to help protect yews. To celebrate the Millennium it organised the collection and distribution of thousands of cuttings taken from Britain's most ancient yews, and these are now growing in churchyards and gardens throughout the country. The latest initiative is a UK Yew Guardianship Project, which seeks to establish a code of care for all ancient, veteran and significant yews. For more information go to www.conservationfoundation.co.uk or write to The Conservation Foundation, 1 Kensington Gore, London, SW7 2AR.

ACKNOWLEDGEMENTS

Most heartfelt thanks, first of all, to my partner Elaine Vijaya; without her, writing this book would not have been possible.

Many thanks to all who came before me and whose patient and passionate work has made it possible to carry on from where they left off. This includes pioneers in yew research like John Lowe and Hugo Conwentz, and, more contemporary, Dr Thomas Scheeder, Hugo Rössner and Christian Wolf; in Britain, the members of the AYG, namely Andy McGeeney, Tim Hills, Paul Greenwood and Toby Hindson – without their years of work that predate this book I would never have had the courage to undertake such an enormous task.

For their professional support I deeply acknowledge, most of all, Dr Arthur Brande, Ecological Institute at the Technical University of Berlin, and Prof. Ladislav Paule, Faculty of Forestry, Zvolen (Slovakia), for their comments on my botanical sections. Many thanks to Dr Ulrich Pietzarka, Forest Botanical Garden at Tharandt (Germany), and Prof. Mikhail Pridnya, University of Sochi (Russia), for additional revision and inspiration. Deepest thanks to Prof. Ronald Hutton, Department of History at Bristol University, to Dr Robert J. Wallis, Richmond the American International University in London, and to Graham Harvey, Open University, for their comments on my history and religious history; and to Dr Helen Nicholson, Cardiff University, for her advice on ecclesiastical history. A great many thanks to Dr Frank Depoix, University of Mannheim, for the electron microscope photography, and to Dr Necmi Aksoy for organising our trip through Turkey, namely the forestry centres at Düzce, Alapli and Yenice, and arranging for us to meet their staff, and also the staff of the forestry departments of the universities of Düzce and Istanbul. Thanks to Prof. Bartomoleo Schirone, University of Tuscía, Italy, Prof. Mauro Ballero, University of Cagliari, Sardinia, Franz Tod and Dr Berthold Heinze in Austria, and Urs Leuzinger and Prof. Christoph Leuthold in Switzerland. In Spain I would like to thank Bosco Imbert, Ignacio Abella, Enrique Garcia Gomáriz, and Dr Sven Muetke Regneri (University of Madrid). It was also a great pleasure to communicate overseas with those associated with the Native Yew Conservation Council, namely Dr Stanley Scher, Shimon Schwartzschild, Hal Hartzell and Dr David Pilz. Regarding Japan, I am most grateful to Chris Worrall for being my right hand (*and* left, actually) in the land of the rising sun, and for being the excellent photographer he is. Further thanks go to Prof. Gunnar D. Hansson in Göteborg, Sweden, Dr Peta Hayes at the Natural History Museum in London and to Hiroyuki Tanouchi and Miwa Murata of the Japanese forestry commission (FFPRI). The credit is theirs for what I have managed to get right; if there are errors, they are mine.

For regional studies and/or 'simple' translations I would like to thank Dr Haraldur Erlendsson, Caitlin Matthews, Cornelia Wunsch, Pierette Housdon, Jehanne Mehta, Laura Ridolfi, Gabriel Millar and Yvan Rioux, Martin Trilk and Christine Konrad, Norbert Drews, Ceridwen Lentz and Varda Zisman; and for Chinese studies, Zhihui Holzhäuser, Liting Guo, Jian Sheng and Jörg Wenzel.

Many thanks also for all the other kind people who helped along the way and cannot all be mentioned here. And last but not least, to Claudine Bloomfield for reading and editing the typescript before submission.

Very special thanks to my 'guest authors', Dr Margaret Redfern, Andy K. Moir, and Russell Ball, and, of course, to Prof. David Bellamy and Robert Hardy for all their support. Also to all the photographers and artists who contributed stunning work to this unique book, to Jaz Media in Cheltenham for optimum high-end scanning, to the whole editing and production teams, and to you, the reader.

Fred Hageneder,
Stroud, February 2007

PICTURE CREDITS

PHOTOGRAPHY

Abbreviations

AAT Amt für Archäologie des Kantons Thurgau, Switzerland
BM British Museum, London
HM Hull and East Riding Museum, Humberside
MC Museo Archeologico Nazionale di Cagliari, Cagliari, Sardinia
MI Museum of the Ancient Orient, Istanbul, Turkey
ML Musée de Lodève, Lodève, France
MP Museo Archeologico Nazionale dell'Umbria, Perugia, Italy
MS Ittireddu, Civico Museo Archeologico ed Etnografico, Sassari, Sardinia
SMB Ägyptisches Museum und Papyrussammlung, Staatliche Museen zu Berlin, Germany

I am grateful to these organisations for permission to reproduce material in their collections.

Ignacio Abella fig. 30.3–4. **Necmi Aksoy** figs 40.3, 40.8. **Dan Barton and Tony Tickle** fig. 26.9. **Bridgeman Art Library** fig. 33.14 (HM). **Margarete Büsing** fig. 39.3 (SMB). **Canterbury Archaeological Trust** figs 22.4–6. **Christopher Cornwell** pp. 1, 10–11; figs 7.1, 26.10–12, 41.1, 41.8. **Frank Depoix** p. 6, first left; figs 6.6, 6.7, 7.2–4, 7.5–6, 9.2, 10.3, 18.9–10. **Dragon Design UK** figs 1.3, 24.4 (satellite images by Reto Stockli, NASA Earth Observatory), 40.2 (satellite image by GeoEye). **Enrique Garcia Gomariz** p. 8; fig. 4.1. **Chris Gomersall** fig. 14.1 (with kind permission of rspb-images.com). **Paul Greenwood** p. 7, far right; figs 5.6, 6.3, 10.4, 16.1, 30.10, 41.17, 44.14. **Fred Hageneder** pp. 6, third and fourth from left, 7, second from right (BM), figs 3.1, 4.2, 5.1, 5.3, 5.4, 5.8, 6.1, 6.2, 6.4, 8.7, 13.3–4, 13.9, 15.9, 16.3, 16.9, 19.2–5, 19.8–12, 21.6, 21.12–13, 23.4, 24.6, 30.2, 30.7–8 (BM), 30.11–12, 31.1–2, 31.3 (MP), 32.1 (BM), 32.3h (BM), 33.6 (BM), 33.8 (BM), 34.2 (MC), 34.4 (MS), 35.3, 37.1–3, 37.21, 37.24 (BM), 37.25, 38.6 (MC), 38.9, 39.11, 40.4–5, 40.6 (MS), 40.11 (BM), 41.2, 41.4, 41.9–10, 41.13, 42.5 (BM), 42.6 (BM), 42.7 (MI), 43.8, 44.3–4, 44.8, 44.10–12, 44.13 (ML), 44.15, 44.16–18, 45.1, 45.5–7, 45.8f, 45.10–11. **Cliff Hansford** fig. 16.5. **Hal Hartzell** fig. 21.1–2. **Jürg Hassler** fig. 24.9. **Berthold Heinze** p. 265, right. **Tim Hills** figs 3.2, 6.5, 18.5–6, 19.6, 20.6, 21.5, 21.7–10, 21.15–16, 28.2, 29.2, 29.4, 34.3, 37.16, 44.1–2, 45.2, 45.4. **Toby Hindson** figs 21.17, 24.3. **Historisches Museum der Pfalz, Speyer (Germany)** fig. 39.1. **Josef Hsalek** fig. 14.11. **Lubomir Hsalek** figs 14.2–3. **Jagiellonian Library, Cracow** fig. 23.5. **Robin Jones** fig. 12.2. **Thomas Kellner** figs 21.3–4. **Doug Larson** fig. 20.2. **Gerry McCann** fig. 29.15 (www.scran.ac.uk). **Andy McGeeney** p. 2, 6, fifth from left, 7, third from right, 96–7, 271, left and right; figs 1.1, 4.3, 5.5, 5.9a–c, 14.4–5, 14.10, 15.2a, 16.2, 16.4, 17.1–2, 18.3, 18.8, 19.7, 19.13–14, 19.16, 21.11, 21.14, 26.2, 26.4, 26.8, 27.4–5, 29.1, 29.3, 29.5, 30.13, 30.16, 32.6, 33.1, 33.3–5, 33.19, 34.6–8, 35.1, 36.1, 36.3, 37.26, 38.1, 38.10, 39.6–10, 41.3, 42.1, 44.5–7. **Jehanne Mehta** fig. 34.12. **Archie Miles** figs 3.3, 34.1. **Reinhard Mosandl** p. 265, left. **Natural History Museum** fig. 2.2. **Rob Nicholson** p. 265, centre. **Hugh Palmer** fig. 26.7 (with kind permission by Andrew Russell, Gravetye Manor Hotel). **Edward Parker** pp. 4, 256–7; fig. 28.5. **photos.com** figs 13.5–6, 13.8, 13.12–13, 30.15. **Ulrich Pietzarka** figs 11.4–5. **Vladimír Rajda** fig. 17.3. **Margaret Redfern** fig. 15.2b, f, g. **Yvan Rioux** fig. 24.7. **Hubert Rössner** figs 11.3, 13.2, 19.1, 20.5, 24.4–5, 30.9, 33.16–17, 38.7–8. **Blazej Stanislawski** fig. 41.6. **Ulrich Stehli** figs 23.2–3. **Franz Steiner** fig. 45.3. **John Sunderland** figs 42.2–3 (with kind permission of Margaret Gowen and Co. Ltd). **Swedish Museum of Natural History** fig. 2.1. **Franz Tod** fig. 24.10. **Christian Wolf** pp. 5, 6, second left, 7, fourth from right, 258; figs 5.2, 5.7, 8.1–6, 8.8, 9.1, 10.1, 11.1, 12.1, 13.1, 13.7, 13.10–11, 14.6–9, 14.12–13, 16.6–8, 18.1–2, 18.4, 18.7, 20.3–4, 24.12, 26.3, 30.1, 38.11–12, 39.5, 43.12. **Chris Worrall** p. 9; figs 10.2, 15.1, 15.5–8, 20.1, 20.7, 26.1, 29.6–14, 31.4, 33.12, 33.18, 39.12–15, 41.14–16, 45.12.

The only photographs that have been (colour)manipulated are: p. 4; figs 24.3, 41.9–10, 42.5.

ILLUSTRATIONS AND DIAGRAMS

Hans Diebschlag figs 27.1, 27.3 (Hans Diebschlag, 3 West View Gardens, East Grinstead, RH19 4EH; www.diebschlag.com). **Dragon Design UK** diagrams 1 after Pietzarka 2005, 2–4 after Scheeder 1994, 5–6 after Rajda 1992 and 2005, 7 after Hindson 2000, 8 after Meredith n.d., 9 after Fred Hageneder, 11 after Thomas and Polwart 2003, 12 after Kaya 1998, 38.3, 13 after Sanofi Aventis material, 14–16 after Fred Hageneder; figs 33.7 after Gimbutas 1995, 37.6 after Schliemann 1880, 41.12 after Cook 1914. **Jan Fry** fig. 27.6. **Seán Goddard** figs 41.7. **Fred Hageneder** figs 32.3g after Goff 1963, 36.2, 36.4–5 after Goff 1963, 37.9–13, 39.2 after Gurney 1952, 39.4 after Schefold 1989. **Nicole Melis** figs 25.1–4. **Andy K. Moir** diagrams 10a–b. **Sidney Rust** fig. 29.16. **Elaine Vijaya** figs 11.2 after Thomas and Polwart 2003, 33.9–11 after Gimbutas 1995, 34.9, 34.10–11 after Gimbutas 1995, 37.5 after Butterworth 1970, 37.7 after Wirth 1979, 38.4 after Butterworth 1970, 40.6 after Gimbutas 1995, 41.5 after Waddington 1997.

Any omissions should be notified to the publishers for correction in future editions.

NOTES

Abbreviations
KUB = identification standard for Hittite cuneiform tablets from Boghazköy *(Keilschrift-Urkunden aus Boghazköy)*
KBo = identification standard for Hittite cuneiform texts from Boghazköy *(Keilschrift-Texte aus Boghazköy)*
n.d. = no date
subsp. = subspecies
var. = variety

Quotations from works in foreign languages have been translated by the author.

PART I

Chapter 1. Baccata – 'berry-bearing'

1. E.g. *Strasburger Lehrbuch der Botanik*, 35th edn, 2002.
2. Krüssmann (1985), for example, divides *Taxus* into eight species proper: (1) *T. baccata* in Europe, (2) *T. brevifolia* Nutt. on the west coast of North America, (3) *T. canadensis* Marsh. in the north-east of North America, (4) *T. celebica* (Warburg) L. in China (and otherwise known as *T. chinensis* Rehd.), (5) *T. cuspidata* Sieb. and Zucc. in Japan, (6) *T. floridana*, (7) *T. globosa* in Central America, and (8) *T. wallichiana* in the Himalayas (and one variety, *T. wallichiana* var. *chinensis*, and one form, *T. cuspidata* f. *latifolia*).

For Pilger (1916) these five species are simply subspecies, to which he adds *T. baccata* subsp. *floridana* Nutt. in Florida and *T. baccata* subsp. *globosa* Schlechtd. in Mexico. Krüssmann's *T. cuspidata* and *T. celebica* are Pilger's *T. baccata* subsp. *cuspidata* var. *latifolia* and *T. baccata* subsp. *cuspidata* var. *chinensis* respectively. And time partly proved Pilger right, as most contemporary authors see *cuspidata* in Japan and eastern China (and Korea), while the south-western Chinese trees are considered to belong to the Himalayan yew, *T. wallichiana* Zucc. (or *T. baccata* subsp. *wallichiana* Pilg.).

The author in Engler (1960) interprets the non-European species as varieties of *T. baccata*, and Hess (in Leuthold 1980, Dumitru 1992, pp. 8–9) sees one species with seven geographical taxa. Voliotis (1986), too, sees variants of one basic species.

3. Bolsinger and Jaramillo 1990, p. 17.
4. Thomas and Polwart 2003.
5. Spjut was involved in the procurement of large-scale samples of *Taxus brevifolia* by the USDA Agricultural Research Service for the NCI isolation and development of taxol as new anticancer drug.
6. www.worldbotanical.com/Nomenclature.htm#nomenclature; www.worldbotanical.com/Introduction.htm#introduction; www.worldbotanical.com/TAXNA.HTM.

Chapter 2. Evolution and Climate History

1. Website of the University of San Francisco (http://bss.sfsu.edu, January 2004).
2. Emberger 1968, p. 583.
3. Brande 2001, p. 24, after May 1995.
4. Harris 1961.
5. *Ibid.*, pp. 112–14. 'The leaf cuticle of *Marskea* resembles *Taxus* in its stomatal ramparts and papillose stomatal bands, and also in the ridges outside these bands, but differs in its monocyclic stomata. Its cell walls when wavy differ from the normal form of *Taxus* and are more like those of the Triassic fossil *Palaeotaxus rediviva* Nathorst (see Florin 1944) though in other respects the cuticle of *Palaeotaxus* is very different.' (Harris 1961, p. 113).
6. Thomas and Polwart 2003, p. 513.
7. Brande 2001, p. 25.
8. Brande 2003, pp. 58–9.
9. Personal communication with members of the Forestry Faculty at Düzce, Turkey.
10. Brande 2001, p. 26.
11. Thomas and Polwart 2003, p. 514.
12. Brande 2001, p. 26.

Chapter 3. The 'tree archetype'

1. Pridnya 2000, p. 28.
2. Kelly 1981.
3. Leuthold 1998, p. 364.
4. *Ibid.*, pp. 362–3.
5. 'der Ur-Baum Europas', *ibid.*

Chapter 4. Climate and Altitude

1. Carruthers 1998.
2. Tittensor 1980, pp. 244–5.
3. Weather station Hohenpeissenberg.
4. Yew woods on northern and north-eastern exposures in the Szentgál mountain group of the Bakony Mountains, northern Hungary; personal communication with N. Frank PhD, Westhungarian University.
5. Pridnya 2002, p. 147.
6. *Ibid.*
7. *Ibid.*
8. Korori *et al.* 2001.
9. Measured at Manavgat; Mayer *et al.* 1986, pp. 182–3.
10. *Ibid.*
11. *Ibid.*
12. Pietzarka 2005, 3.2.2.
13. Moir 1999.
14. Pietzarka 2005, 4.1.10.
15. Thomas and Polwart 2003, p. 499.
16. Pietzarka 2005, 2.3.
17. *Ibid.*, 3.3.2.
18. Thomas and Polwart 2003, pp. 498–9.
19. Carruthers 1998.
20. Tittensor 1980, pp. 244–5.
21. Weather station Hohenpeissenberg.
22. See note 4 above.
23. Pridnya 2002, p. 147.

24. *Ibid.*
25. *Ibid.*
26. Korori *et al.* 2001.
27. Measured at Manavgat; Mayer *et al.* 1986, pp. 182–3.
28. *Ibid.*
29. Average from Mukteswar and Shimla meteorological stations, north India; Yadav and Singh 2002.
30. Pisek *et al.* 1967.
31. Szaniawski 1978, p. 61.
32. Lange in Dumitru 1992, p. 74.
33. Núñez-Regueira *et al.* 1997.
34. Groves and Rackham 2001, pp. 218, 236.
35. Thomas and Polwart 2003, p. 490; Pridnya 2000b, p. 28.
36. My own observations; compare Chapter 40, note 18, also Leuthold 1980, 1998.
37. Observations by senior forester Hubert Rössner, Bavaria.
38. Williamson 1978.
39. Thomas and Polwart 2003, p. 490; Voliotis 1986.

Chapter 5. Plant Communities

1. Thomas and Polwart 2003, pp. 492–3.
2. Heinze 2004. This has also been shown for yew stands in Slovakia (Jaloviar 1998; Korpel 1996; Saniga 1996).
3. Pietzarka 2005, 4.1.1, after Heinze 2004. Also compare Scheeder 1996, Stahr 1982, Willerding 1968.
4. Leuthold 1980.
5. Pietzarka 2005, 4.1.
6. *Ibid.*, 4.2.
7. *Ibid.*, 3.6.2.
8. Larcher 2001.
9. Pietzarka 2005, 4.1.5, 4.2.1.
10. *Ibid.*
11. *Ibid.*, 4.2.2.
12. Kanngiesser 1906; Roloff 1998; Scheeder 1994.
13. Pietzarka 2005, 4.1.11.
14. *Ibid.*, 4.2.1.
15. 'This is true for the individual tree and its various investments into securing its own survival, as well as for the population and its various mechanisms to maintain a high genetic variety' (Pietzarka 2005, 6).
16. *Ibid.*
17. *Ibid.*, 4.1.7.
18. Thomas and Polwart 2003, p. 492.
19. Dumitru 1992, pp. 37–8.
20. Newbould 1960.
21. Thomas and Polwart 2003, p. 494.
22. In the Sierra Nevada mountains of Spain rather Spanish barberry *(Berberis hispanica)* and wild rose bushes (*Rosa* spp.) but most of all junipers *(J. communis, J. sabina)*. Garcia *et al.* 2000.
23. Watt in Thomas and Polwart 2003, p. 493.
24. Kelly 1981.
25. Including *Vaccinium myrtillus, Dryopteris dilatata* and *Oxalis acetosella*.
26. Pilkington *et al.* 1994.
27. Wilks in Thomas and Polwart 2003, p. 513.
28. Pilkington in Thomas and Polwart 2003.
29. Tittensor 1980, p. 260.

Chapter 6. The Roots

1. Dumitru 1992, p. 60; Thomas and Polwart 2003.
2. Kelly 1981.
3. Thomas and Polwart 2003, p. 492.
4. Tittensor 1980, p. 244.
5. Dumitru 1992, pp. 60–1.
6. Ballero *et al.* 2003.
7. Attems-Heiligenkreuz in Dumitru 1992, p. 62.
8. Szaniawski 1978, p. 59.
9. Howard *et al.* 1998.
10. Larson 1999.
11. Löblein 1995.
12. 'Frischbier 2001, for example, shows that the correlation between light exposure and root growth is clearly stronger than that between light exposure and height increase. Studies of the girth increase of yew also suggest a positive effect of light exposure on this parameter (Difazio *et al.* 1997; Hofmann 1989; Scheeder 1994; Worbes *et al.* 1992)' (Pietzarka 2005, 3.8.2, also 4.1.3).
13. Pietzarka 2005, 4.1.10.
14. *Ibid.*, 4.1.6.
15. Löblein (1995) examined yews in urban locations.
16. For a deeper treatment of root anatomy see Hejnowicz 1978, p. 43–4, or Scheeder 1994, pp. 23–4.
17. Prof. J. Gaper, mycologist, personal communication with Prof. L. Paule, Faculty of Forestry, Zvolen, Slovakia.
18. Korori *et al.* 2001.

Chapter 7. The Foliage

1. Photosynthesis is also performed below the compensation point, but its gain is less than the respiration of the living tissues.
2. Szaniawski 1978, p. 57.
3. Muhle (1978) gives the lengths of young yew shoots under different light conditions: 2.5cm in deep shade, 3cm at 10–20 per cent relative light intensity, 6cm in 60 per cent relative light intensity, but only 4.5cm in full light. This led many researchers to believe that a half-shade of about 60 per cent of the light exposure in the open would result in optimum growth. However, Muhle took nursery seedlings *used to full sunlight* and planted them into woodland locations with differing but unmonitored shade intensity, which renders his results questionable.
4. Pietzarka 2005.
5. *Ibid.*, 3.2.2.
6. *Ibid.*, 3.6.1.1, after Larcher 2001.
7. Pietzarka 2005, 3.6.2.
8. *Ibid.*, 3.6.2.
9. *Ibid.*, 3.7.1.1, 3.7.2.
10. The tip of the European yew is short, while the Japanese yew has a longer tip. The non-stinging characteristic refers to the softness of all *Taxus* leaves.
11. In general, needles on older shoots are not replaced, unless these shoots have produced buds, which is, however, quite common in yew.
12. Szaniawski 1978.
13. Hejnowicz 1978, p. 41, Scheeder 1994, p. 22.
14. Apart from yew the only other conifers lacking a sclerenchymatical hypoderm are Eastern hemlock *(Tsuga canadensis)*, Pacific silver fir *(Abies amabilis)*, bald cypress *(Taxodium distichum)* and Chinese swamp cypress *(Glyptostrobus heterophyllus)* (Kirchner *et al.* 1908); also dawn redwood *(Metasequoia glyptostroboides)*.
15. Pietzarka 2005, 3.8.2; Larcher 2001.
16. Pietzarka 2005, 4.1.9.

17. 'After Kozlowski *et al.* (1990), shade tolerance is only the visible result of a whole sequence of physiological adaptations, particularly on the biochemical level (lower contents of rubisco, ATP-synthetase and other elements of the electron transport chain), which, among other things, lead to decreased capacities in the electron transport and hence the fixation of CO_2. The successes of these adaptations are found in lower respiration losses and a lower light compensation point … Hence, their [the shade plants'] adaptability regarding a more effective usage of this resource [the light] is small, but what they achieve is that higher light intensities cause them less harm (Larcher, 2001)' (Pietzarka 2005, 4.1.9).
18. Hejnowicz 1978, p. 41; Scheeder 1994, p. 22.
19. Scheeder 1994, p. 35; Pietzarka 2005.
20. Di Sapio *et al.*, Dempsey and Hook, Salisbury in Thomas and Polwart 2003, respectively.
21. Dempsey and Hook 2000.
22. Mitchell 1998.
23. Kartusch and Richter 1984.
24. Parker 1971.
25. 'During these favourable periods, the photosynthesis performance is even higher than that of the actual growth season' (Pietzarka 2005, 4.1.7).
26. Pietzarka 2005, 3.8.1.4.
27. Pietzarka 2005, 3.8.2, after Strauss-Debenedetti and Bazzaz 1991.
28. Pisek *et al.* 1967, 1968, 1969. Taxus' net photosynthesis was measured in the Austrian Alps, near Innsbruck, at an elevation of 550m a.s.l. The optimum was measured at 10,000 lux = 0.1 cal/cm^2/min).
29. Szaniawski 1978.

Chapter 8. The Flowers

1. Thomas and Polwart 2003, p. 503.
2. Leonhardt *et al.* 1998.
3. Thomas and Polwart 2003, p. 503.
4. Frank 2003.
5. This notion comes from Dr Pietzarka, Forest-botanical Garden of the University of Dresden (personal communication). In his test nurseries, Pietzarka identified up to 9 per cent monoecious yews. For climatical data for Tharandt see Chapter 11, note 8.
6. Kirchner *et al.* 1908; Rössner 1996.
7. Svenning and Magård 1999.
8. Pietzarka 2005, 3.5.2.
9. *Ibid.*, 4.1.5, after Rohde 1987.
10. Cao *et al.* 2004; Lewandowski *et al.* 1995; Thoma 1992, 1995.
11. Lange *et al.* 2001.
12. Compare the full treatment in Pietzarka 2005, 4.1.2.
13. Thomas and Polwart 2003, p. 503.
14. Described in Kirchner *et al.* 1908.
15. Pietzarka 2005, 4.1.5, after Rohde 1987.
16. Hejnowicz 1978, pp. 44–5.

Chapter 9. Pollination and Fertilisation

1. Hejnowicz 1978, p. 48.
2. *Ibid.*, p. 48; Thomas and Polwart 2003, p. 501.
3. Hejnowicz 1978, pp. 45–51.
4. Dark 1932; Sax and Sax 1933; Moore 1982.
5. Sax and Sax 1933. Details of the meiotic stages of *T. baccata* are given by Dark (1932).
6. Hejnowicz 1978, pp. 50–1.

Chapter 10. The Seed

1. Thomas and Polwart 2003, p. 503.
2. Smal and Fairley 1980b.
3. Suszka in Thomas and Polwart 2003, p. 501.
4. Herrera in Thomas and Polwart 2003.
5. *Ibid.*
6. Hulme, also Smal and Fairley in Thomas and Polwart 2003, p. 502.
7. For seed oils and seed storage proteins of *T. baccata* see Wolff *et al.* 1996; Allona *et al.* 1994.
8. Hulme and Borelli, Melzack and Watts in Thomas and Polwart 2003.
9. Detz and Kemperman, Herrera, Suszka, Brzeziecki and Kienast in Thomas and Polwart 2003.
10. Personal communication with Prof. L. Paule, Faculty of Forestry, Zvolen, Slovakia.

Chapter 11. Natural Regeneration

1. Thomas and Polwart 2003, p. 504.
2. Williamson in Thomas and Polwart 2003, p. 505.
3. USDA 1948; Suszka 1978.
4. Thomas and Polwart 2003, p. 505.
5. Pietzarka 2005, 3.1.1.2.
6. Probably one of the causes of the natural regeneration problems reported by Prof. Pridnya for certain Caucasian yew stands.
7. Pietzarka 2005, 3.1.1.2 and 3.3.1.3, after Gregorius and Degen 1994; Rajewski *et al.* 2000.
8. All following data refers to the natural regeneration of *Taxus baccata* in open areas of the Forest Botanical Garden at Tharandt (near Dresden, Germany). Climate data: height above sea level: 250–280m. Long-term average of air temperature 7.9°C. Average annual precipitation 736mm. January temp.: 9.8–16.6°C, July temp.: 26.4–31.5°C.
9. Pietzarka 2005, 3.2.2.
10. Also in silver fir: for the first three to five years priority is on the growth of the tap root and the side shoots, not on height increase.
11. Pietzarka 2005, 3.4.1.1.
12. *Ibid.*, 3.5.
13. *Ibid.*
14. Roloff 1989; Roloff and Pietzarka 2001.
15. Roloff 1989; Roloff *et al.* 2001.
16. In Pietzarka's test areas at Tharandt, Pietzarka 2005, 3.4.1.3.
17. *Ibid.*, 3.4.2.
18. *Ibid.*
19. Korpel 1996.

Chapter 12. A Potent Poison

1. The main active substances of the taxoid group in yew are Taxin A, B and C and also taxan-tetraol in a pseudo-alcaloid compound. The molecular formula of taxine is $C_{37}H_{51}O_{10}N$. The partial formula $C_6H_6{\cdot}CH(NMe_2){\cdot}CH_2{\cdot}CO{\cdot}O{\cdot}C_{24}H_{34}O_6{\cdot}O{\cdot}CO{\cdot}CH_3$ suggests taxine to be related to the cardiac poisons of the *Digitalis, Strophanthus* and toad-poison groups (bufotoxin) (Callow *et al.* 1931). Also Dumitru 1992, p. 98.
2. Personal communication with Dr H. Osthoff.
3. Erdtman and Tsuno 1969.

4. Elsohly *et al.* in Thomas and Polwart 2003.
5. Hence the report states that 'systematic studies on the accumulation of bioactive taxanes in cultivated yew trees failed to demonstrate any general correlation between botanical (horticultural) features and the accumulation of specific taxoids ... suggesting that the production of taxanes is the result of a complex and still poorly understood interplay of genetic and environmental factors' (Ballero *et al.* 2003, p. 38).
6. Thomas and Polwart 2003, p. 506. Also compare Waldemar Zank at the dendrology symposium 29 October 1994 in Potsdam, summary by Dr O. Hermann. *Der Eibenfreund*, 2 1996, p. 82.
7. Krenzelok *et al.* 1998.
8. Van Ingen *et al.* 1992.
9. Dumitru 1992, p. 104.
10. Jensen in Dumitru 1992, p. 109.
11. Dumitru 1992, pp. 104, 108.
12. Thomas and Polwart 2003, p. 506, referring to a French investigation by Charles Cornevin in 1892. Cornevin's horse entry (2g) modified after Dumitru 1992, p. 103.
13. *Ibid.*
14. 'Foliage is reputed to be even more poisonous when wilted or dried (Elwes and Henry 1906; Williamson 1978) but according to Cooper and Johnson (1984) they [the needles] are "as toxic as the fresh plant".' Thomas and Polwart 2003, p. 506.
15. Thomas and Polwart 2003; Williamson 1978; Osthoff (2001).
16. Osthoff 2001, p. 71.
17. Moir 2004, p. 5, referring to W. Johnson 1908.
18. Personal communication with Prof. L. Paule, Faculty of Forestry, Zvolen, Slovakia.
19. Williamson 1978, p. 43.
20. Theophrastus, *Enquiry into Plants*, III, X, 2, in Hort (tr.) 1916.
21. Virgil, *Eclogues*, IX, 30, in Lewis (tr.) 1983. This remark is also an indication that Corsica must have had considerable *Taxus* populations in the first century BCE.
22. Pliny XVI, X, in Hartzell (tr.) 1991, p. 15.
23. Dioscurides cited in Voliotis 1986, p. 47.
24. Lowe 1897, p. 136, referring to Plutarch.
25. Dumitru 1992, p. 107.
26. Gulland *et al.* 1931.
27. Kukowka in Dumitru 1992, p. 96. Also Graeter 1994.
28. Personal communications.

Chapter 13. Mammals

1. Tittensor 1980.
2. Dumitru 1992, p. 81; Thomas and Polwart 2003, after Mysterud and Ostbye 1995.
3. Thomas and Polwart 2003, after Lowe 1897, Watt 1926, Williamson 1978.
4. Watt 1926.
5. Williamson 1978, p. 63.
6. *Ibid.*, p. 84.
7. *Ibid.*, p. 83.
8. Bartkowiak 1978.
9. Williamson 1978, p. 83.
10. Thomas and Polwart 2003, after Bartkowiak 1978, also Williamson 1978.
11. Hassler 2003.
12. Thomas and Polwart 2003, after Mehlman 1988.
13. Tittensor 1980; Smal and Fairley 1980a.
14. Hulme 1996; Hulme and Borelli 1999.
15. Hulme 1997.
16. Thomas and Polwart 2003, after Hulme 1997.
17. Rössner 2001.
18. Rössner, in personal communications.
19. Williamson 1978, pp. 28ff.
20. Thomas and Polwart 2003; Rössner 2001.
21. Williamson 1978, pp. 68–9, 77.
22. Schroeder, in Stern 1979, p. 256.
23. Hassler 2003.

Chapter 14. Birds

1. Williamson (1978) reports as many as 2,000 redwings and fieldfares alone being drawn to the autumnal Kingley Vale reserve by the 'berry crops of yew and hawthorn (and to a lesser extent of spindle, privet and dogwood)', p. 126.
2. Thomas and Polwart 2003, after Bartkowiak 1978; Fuller 1982; Snow and Snow 1988; Hulme 1996.
3. According to USDA (1948); Suszka (1978); and Namvar and Spethmann (1986), germination is improved if *Taxus* seeds have passed through the digestive tracts of birds. Treatments with hot water or sulphuric acid to break down the seed coat do not result in higher germination (Suszka 1978). Heit (1969), however, states that it is not the bird's digestive juices that help germination, but the simple removal of the pulp.
4. Larson 2000, p. 227, after Vogler 1904.
5. Williamson 1978; Bartkowiak 1978; Snow and Snow 1988.
6. Snow and Snow 1988.
7. Bartkowiak 1978.
8. Williamson cited in Thomas and Polwart 2003, p. 507.
9. Snow and Snow 1988.
10. 1952/53 in the Pillnitz area near Dresden; Snow and Snow 1988, referring to Creutz 1952.
11. Dumitru 1992, p.83, referring to Schönichen 1933 and Namvar and Spethmann 1986.
12. Snow and Snow 1988, after Creutz 1952.
13. Hulme and Borelli 1999.
14. Rössner 2001.
15. Williamson 1978, pp. 28, 126–7. And Pridnya 2002 mentions migrating bird populations in yew stands of the Caucasus.
16. Snow and Snow 1988.
17. Thomas and Polwart 2003, p. 507.
18. Snow and Snow 1988.
19. After Snow and Snow 1988, who took their records in the Chilterns, England.
20. *Ibid*; Barnea *et al.* 1993.
21. Williamson 1978, p. 128.
22. *Ibid.*, pp. 77, 131.
23. *Ibid.*, p. 134, referring here to Simms 1971.
24. Scher 1998; Wolf 2002; Huf 2002.
25. In the Plesswald forest near Bovenden, Germany. Dumitru 1992, p. 80.
26. Wolf 2002, p. 172.
27. Scher 1998.
28. Dr S. Scher, Dept. of Environmental Science, Policy, and Management, University of California, Berkeley, in Scher 1998.

29. Scher 1998, p. 414; Wolf 2002, p. 175.
30. Huf 2002, referring to an ornithological article by Prof. E. Martini, Kronberg, Germany.
31. Wolf 2002, p. 173.
32. *Ibid.*

Chapter 15. Invertebrates

1. Daniewski *et al.* in Thomas and Polwart 2003, p. 508.
2. This stimulation will certainly be chemical, using chemicals/plant hormones produced naturally by the plant and/or the insect, but the mechanisms involved are unknown (in this and all other insect galls).
3. This, however, is not proven in *T. taxi,* and there is no damage to the cell walls visible under the microscope. (Personal communication with Dr M. Redfern.)
4. Heath 1961, in Redfern 1975, p. 530.
5. 'the most dangerous parasite of *Taxus baccata* in Europe' (Skuhravá 1965), cited in Siwecki 1978.
6. Redfern and Hunter 2005, p. 86.
7. Of course, it does not actually *control* the host in the sense of the word – it seems to just *follow* the numbers of its host whether it is abundant or rare.
8. Redfern 1998, p. 25.
9. *Ibid.*, p. 26.
10. Thomas and Polwart 2003, p. 508. See also Coutin 2003.
11. Rössner, personal communication.
12. Hassler (2003) in Switzerland; and for the Andalusian highlands in Spain, Hulme (1997) specified two species of aril-eating ants: *Cataglyphis velox* Sants. and *Aphaenogaster iberica* Em.
13. Kirchner *et al.* 1908.

Chapter 16. Parasites

1. Swift *et al.* 1976, in de Vries and Kuyper 1990.
2. Lewandowski *et al.* 1995.
3. The yew population spreads over 8.5ha in the northern part of the Fürstenwald, 600–700m above sea level, in a mixed community of beech and pine, with some spruce and larch.
4. Hassler *et al.* 2004.
5. Strouts and Winter in Thomas and Polwart 2003.
6. Thomas and Polwart 2003, p. 508.
7. Strouts and Winter 1994.
8. *Ibid.*
9. Strouts 1993; Strouts and Winter 1994.
10. De Vries and Kuyper 1990.
11. Among them, two that usually specify in juniper: *Amylostereum laevigatum* and *Kavinia alborividis*. Juniper and yew not only often share a similar habitat (poor soils), but also a relatively high pH value of the wood, *Juniperus* 5.15 and *Taxus* 5.65 (the other conifers are usually below 5). But such pH measurements of the wood have to be taken with care, as the metabolic activity of the fungi can change the acidity.
12. This fungus has rarely been found on leaf plants (Rosaceae, Ericaceae) as well, but the specimens on *Taxus* show different micro-characteristics and might constitute their own taxon, according to de Vries and Kuyper.
13. British Mycological Society Fungal Records Database (BMSFRD); in Thomas and Polwart 2003, p. 508.
14. De Vries and Kuyper 1990.
15. *C. psilaspis* Nalepa: Eriophyidae.
16. Duncan *et al.* 1997.
17. Siwecki 2002.
18. Skorupski and Luxton 1998.

Chapter 17. Vitality

1. By the Swiss electro-technician and electrotherapist Eugen Konrad Müller (1853–1948); Hageneder 2000, p. 32.
2. Long-term studies of the bio-electrical fields of an elm and a sycamore at the University of Yale, conducted between 1943 and 1966 by Harold Saxton Burr; *ibid.*
3. Cf. Hageneder 2000, pp. 31–2.
4. Vladimír Rajda, Elektrodiagnostika, Smetanova 947, CZ-69701 Kyjov, Czech Republic.
5. After Rajda 2004: average plant vitality in Central Europe (1980–2002) 68 per cent; Germany (1993–99) 68 per cent (with oaks in Nordrhein-Westfalia (1999) 86 per cent, poplars in Hessia (1993) 95 per cent, beech in Hessia (1993) 45 per cent); Czechia average 68 per cent (with stands facing industrial sites 29–37 per cent); oaks in Austria (Weinviertel 1998) 59 per cent, pines in Austria 42 per cent; Mexico (1995): pines 93 per cent, oaks, firs and cedars 73–9 per cent, apple and apricot trees 100 per cent.
6. Rajda 1992, 1995, p. 351; cf. Hageneder 1999, 2004.
7. By H.S. Burr, see note 2 above.
8. Burr, too, had found individual characteristics for each species; Hageneder 2004, p. 45.
9. Rajda 1992, 1995, 2004, 2005.
10. Rajda in a personal letter to the author in early summer 2004. It was this result that moved Mr Rajda to prepare the twelve-month study of *Taxus*.
11. Rajda 2005.
12. Personal communication of the author with Mr Rajda, September/October 2005.

Chapter 18. The Wood

1. Kucera 1998, p. 330.
2. In more continental climates, e.g. Bavaria, two months later.
3. Moir 1999.
4. Dumitru 1992, p. 50.
5. *Taxus* value from Thomas and Polwart 2003, p. 500; Lincoln 1986 gives 670kg/m^3. The redwood mentioned is *Sequoia sempervirens*, the pine *P. sylvestris*, beech and oak from north-western Europe (Desch 1975; Lincoln 1986).
6. Pietzarka 2005, 4.1.3, after Korpel and Paule 1976.
7. According to Alan Mitchell (1972), all yews over 4.57m (15ft) girth begin to hollow. Field data over the last thirty years, however, show clearly that this process can already begin in trunks with much smaller girths. It is impossible to pinpoint an *age* for the beginning of the hollowing stage. The hollowing can occur in trees *c.* 300 years of age (Moir, personal communication), in others not before 1,000 years (Pridnya, see next chapter) – *but there is no rule.*
8. Tabbush and White 1996.
9. The adventitious growth imprints the wood with dark spots, which are highly valued as 'pepper' or 'cat's paw' in the veneer trade.
10. Hejnowicz 1978, pp. 35–6.
11. Pietzarka 2005, 3.8.2.
12. Kucera 1998, p. 332, table 1.
13. Hejnowicz 1978, p. 40; Scheeder 1994, p. 22.

Chapter 19. Regeneration Ability

1. This is the principle of the theoretical model. Forest trees often have an additional decrease in ring width because the close stand with neighbouring trees reduces their crown size and hence their assimilation, which also affects girth increase.
2. White 1998, p. 2.
3. Chetan and Brueton 1994, App. 3.
4. The promotion of growth by gaining additional nutrients through the roots of layered branches is not in doubt but not fully examined either. It is not known at what timescale daughter and mother tree become independent of each other (and it is unlikely anyway that the versatile yew would fit into any scheme). As long as substances are exchanged, the 'bridge' between the two will keep growing as well. There is room for future research here.
5. White 1998, p. 2. The most regular reason, however, simply is the disappearance of shading trees in the upper tier of the forest.
6. Hindson combines his empirical data and remeasurements of historical measurements (a comprehensive remeasuring exercise on twenty-one ancient Hampshire yews, most of them not included in Chetan and Brueton 1994, but in Lowe 1897 etc.) with known planting dates for trees up to about 800 years of age, and with the inclusion of data from Meredith's gazetteer, and thus creates an extended base to relate age to girth measurements. See Hindson 2000.
7. Scheeder 2000, p. 68.
8. Interviewed by John Craven for a *Countryfile* programme on the BBC in 1991.
9. Toby Hindson, 'Death by pollarding', unpublished essay, available at the AYG website (www.ancient-yew.org).
10. See note 4 above.
11. According to Tabbush and White 1996, p. 198.
12. Green 2003.
13. Interior roots were first mentioned in writing by Reverend William Bree in his 1833 article in the *Natural History Magazine*. He also contributed to Loudon's *Arboretum et Fructicetum Britannicum* in 1842, where his observation of the vegetative regeneration of yew led him to conclude, 'In cases where this process takes place, the existence of a yew tree on a particular spot might continue as long as the world endures.' Later, interior roots as well as the great longevity of yew were almost entirely forgotten until these phenomena were rediscovered by Allen Meredith in 1974. For a full account see Chetan and Brueton 1994, pp. 23–4.

Chapter 20. Dating Old Yews

1. For his figures on *Taxus*, de Candolle himself measured no more than three young yews (of 71, 150 and 280 years respectively) and announced (1831) the growth rate of *Taxus* to be 'a little more than one line annually in diameter in the first 150 years, and a little less from 150 to 250 years' (Bowman 1837, p. 29). For very old yews he assumed an average of one line annually in diameter (one line being one-twelfth of an inch or 2.117mm). Half a line in radius equals about 1.05mm ring width. However, de Candolle assumed his rule of thumb would tend to underestimate the age of some trees.
2. Encouraged by de Candolle's remark on underestimation of the age of yew trees, J.E. Bowman (in publications 1835–37) verified de Candolle's formula because it 'makes old yews to be younger than they are', and 'gives too great an age to those of more recent growth' (Bowman 1837, p. 35). Bowman allowed two lines diameter (*c.* 2mm ring width) for young trees, even 3 lines for trees in rich soil, until they had attained a diameter of about 60cm (2ft). He suggested 'an additional allowance for a probable intermediate rate of increase between the age of 150 and 250 years, rather than pass at once from the vigorous growth of youth to the slower progress of more advanced periods' (*ibid.* p. 32f). And his formula for old trees suggested higher ages than de Candolle's. For example, the Darley Dale tree in Derbyshire, at the time with a circumference of 9.4m (27ft) at the base, would have been 1, 356 years of age according to de Candolle, and 2006 years after Bowman's formula.
3. Lowe (1897) speaks of the 'erroneous views' (p. 35) of de Candolle, and discards Bowman entirely (p. 57: 'of no utility whatever'). Lowe attested to de Candolle having used 'stunted and ill-grown' (*ibid.* p. 46) trees, found himself some samples with fast growth rates, objected to the idea that yews in old age would slow down in growth speed at all and claimed the exact opposite: 'on the contrary it increases, on the whole, with a rapidity as great as, and in many instances much greater than, that of young trees' (p. 61 f). His average rate of increase is 1ft (30.48cm) of diameter in sixty to seventy years, which equals an average ring width from 2.18–2.54mm. To stick with the example in the note above, the Darley Dale yew, according to Lowe, would have been no older than 612 years.
4. Although Prof. H. Eddelbüttel (1937, pp. 149–50) knew of a yew stump with more than 1,000 annual rings within 21.6in (55cm) (average ring width below 0.55mm), and a cross-sectional disk with 55 rings in 12.5cm (6in) (average ring width 0.23mm) (Krause 1884, Trojan 1903, in Eddelbüttel p. 150), his foremost goal remained the avoidance of overestimating young trees. Not trusting the data of low growth rates of old trees he generalised the fast growth rates of young trees.
5. Paul Tabbush and John White of the Forestry Commission in Scotland have shown that the distortion of age estimates caused by trunk fusion is minor and can be discarded (providing the original circle of 'trees' was not much larger in diameter than 1m). They illustrated this with an example given by Richard Williamson, warden of Kingley Vale from 1963 to 1978, who described a felled yew with a trunk diameter of 37cm and seven centres ranging from 21 to 53cm in circumference, each 46 years old. Following White's formula for a *single-core* trunk, a 37cm-diameter would have been attained after *c.* sixty-nine years, an estimate that proves 50 per cent too high for this young tree! But when it comes to the question of whether the hollow trunk of a large ancient yew was once multi-stemmed in the core, the twenty-three years of our example would hardly make a difference to the vast overall age. Also, the actual period of bark and cambium fusion would cause a growth *delay* not accounted for in the above example.
6. Kirchner *et al.* 1908, p. 65.
7. *Encyclopedia Britannica* 1984, p. 915.
8. This *opinion* of yews 'faking' a great age even spread to Cambridge and Oxford. Only in 1994 could this view for the first time be doubted again, when Chetan and Brueton published the data and arguments of Allen Meredith.

9. Pilcher *et al.* 1995.
10. See bibliography: Chetan and Brueton 1994. Meredith's gazetteer has since been incorporated into the extended yew database of the AYG, the Ancient Yew Group, publicly accessible at www.ancient-yew.org.
11. White 1994.
12. White 1998, p. 2.
13. Tabbush and White 1996, p. 202.
14. Moir 1999.
15. Tabbush 1997.
16. Moir 1999.
17. Hindson, Toby 2003, *A Longitudinal Study of Monnington Walk* (unpublished manuscript).
18. Measurements by Paul Greenwood, AYG.
19. Location: Bartin, Ulus Ilcesi, Kumluca Orman Isletme Sefligi, Kumluca Serisi, Yenisencay, Turkey.
20. Larson 1999.
21. Larson 2000, p. 91f and personal correspondence 2003.
22. Larson, personal correspondence with the author, August 2004.
23. Pridnya 2002, p. 152.
24. Personal correspondence with the author, January 2004.
25. Lowe 1897, p. 45.

Chapter 21. Green Monuments

1. Thomas and Polwart 2003, p. 515.
2. Svenning and Magård 1999.
3. In Bavaria, for example, see Ludwig and Bauer 2000.
4. Scheeder 1994, p. 106.
5. Thomas and Polwart 2003, p. 516.
6. Goodman and Walsh 2001, p. 56. See also Videnseek *et al.* 1990.
7. Goodman and Walsh 2001, pp. 51, 54–61, 131, 208.
8. 1991: over 850,000 pounds from Federal lands, a further 225,000 from state lands, and 525,000 from private lands. In 1992, it was split almost 50/50 between public and private land. Goodman and Walsh, pp. 229–30.
9. In the US, these were the National Environmental Policy Act of 1969, the Endangered Species Act of 1973, the National Forest Management Act of 1976, and the Federal Land Policy and Management Act of 1976. Goodman and Walsh 2001, p. 88.
10. Franklin *et al.* 1981.
11. *Ibid.*, p. 40, in Goodman and Walsh 2001, p. 103.
12. Probably Carl Oskar Drude (1852–1933), the first director of the Botanical Garden of the Technical University of Dresden.
13. Personal communication from Richard Williamson.
14. The Staatliche Stelle für Naturdenkmalpflege in Preussen was founded in 1906 at Gdansk, and moved to Berlin in 1911 (Scheeder and Brande 1997).
15. Hartzell 1991, in Goodman and Walsh 2001, p. 87.
16. Furthermore, some small companies still occasionally harvest yew bark since Bristol-Myers Squibb's exclusive access to the yew resources on federal lands expired. For instance, the Roseburg District of the BLM (Bureau of Land Management) allowed yew harvesting in protected riparian zones in 2002, prompting criticism by the Oregon Natural Resources Council, a statewide environmental advocacy organisation (personal communication from Dr David Pilz, Oregon State University, in September 2005).
17. Goodman and Walsh 2001, p. 1.
18. Shemluck *et al.* 2003.
19. Scher 2000, after World Conservation and Monitoring Centre (WCMC) 1999: Tree Conservation Database. http://wcmc.org.uk/cgi-bin/SaCGI.cgi/trees.exe.
20. *Ibid.*, after WWF 1998, and Walter and Gillitt 1997.
21. *Ibid.*
22. *Ibid.*
23. *Ibid.*
24. Scher 2005a. On genetic contamination see also Scher 2005b and 2005c.
25. About 85 per cent of the old yews in the United Kingdom are found in churchyards and are owned by the Church of England or the Church in Wales. Neither organisation, however, has national policies or specific guidance on the treatment and care of old yews in their grounds. 'Instead, decision-making is delegated to the local level. Parochial church councils (PCC) are supposed to seek permission from their diocese for any significant work they want to do to their church or churchyard. But neither the PCC nor the diocese is likely to have a yew expert within their ranks. Often those taking the decisions do not realise how important their yew is. There is no central register of experts to call upon for advice on how to treat an old yew, and there are no central funds available to help towards the cost of such expert advice and treatment. And everyone is unnecessarily afraid of health and safety regulations' (Anderson 2005).
26. Simmonds 1979.
27. In his Alan Mitchell Memorial Lecture 2000.

PART II

Chapter 22. The Art of Survival

1. Tittensor 1980, p. 249, refers to the English South Downs, but I think this statement is fairly valid for north-western Europe in general.
2. Milner 1992, p. 42.
3. Beckhoff 1963 and Wetzel 1966 mention five of the seven sites: Holmegaard IV (a Danish site after which this particular type of bow is named), Ochsenmoor/Dümmer, Satrup/Förstermoor, Aamosen/Muldberg I, and Barleben/Kr. Wolmirstedt. Contacting a good number of north German museums in 2004 did not enable me to find out whether more bows have been found since 1966.
4. Beckhoff 1963.
5. Wetzel 1966.
6. See also Adler 1915; Reinerth 1926.
7. Earwood 1993; Dumitru 1992, pp. 113–14.
8. Thomas and Polwart 2003, p. 514.
9. Maiden Castle in Dorset: Neolithic, *c.* 2500–2000 BCE, and early Iron Age, *c.* 400–200 BCE (Thomas and Polwart 2003, p. 514); Whitehawk Camp, Brighton: Neolithic; Holdenhurst, Hampshire: Bronze Age (Tittensor 1980, p. 249).
10. Tittensor 1980, p. 249.
11. Thomas and Polwart 2003, p. 515, after Mitchell 1990.
12. Niederwil/Gachnang TG Egelsee; Robenhausen, Pfäffikersee; Seeberg BE Burgäschisee Süd; Horgen ZH Scheller, Zürichsee.

13. Scheeder 1994, p. 50.
14. Milner 1992, p. 43. However, this was not the common technique. Unfortunately Milner does not give his source, or more details.
15. The dates as released by the Hull and East Riding Museum in March 2001 are: F1 (boat no.1): 1880–1680 BCE, F2: 1940–1720 BCE, F3: 2030–1780 BCE. See homepage of the Ferriby Boats charity, www.ferribyboats.co.uk. See also Wright and Churchill 1965.
16. Meiggs 1982, p. 408. The royal barge is now housed in its own building, the Solar Boat Museum just in front (south) of the Cheops Pyramid.
17. See www.canterburytrust.co.uk.
18. See Hageneder 2001.
19. Excavation sites at Wurt Elisenhof (eighth to eleventh century), at the River Eider estuary, Schleswig-Holstein, Germany, and at Wollin (tenth to twelfth century), near Stettin harbour, Pomerania, Poland (Dumitru 1992, p. 121).

Conwentz (1898; 1921) reported eighteen yew vessels (late Roman to Viking) in the Museum of Oslo, twenty-six items (buckets, knives, bows) in the National Museum of Denmark in Kopenhagen, and others in the museums of Stockholm and Lund (Sweden), and Kiel, Berlin and Hannover (Germany). In 2005 I contacted about ten museums in Germany (about a century after Conwentz) that had held yew artefacts in the past (according to Dumitru), but all replies were negative.
20. Scheeder 1994, p. 51.
21. Bows made of Pacific yew are attested for the following tribes: Costanoan, Flathead, Hanaksiala, Hoh, Karok, Klamath, Mendocino, Montana, Nitinaht, Okanagan-Colville, Oweekeno, Paiute, Pomo, Kashaya, Coast Salish, Shasta, Thompson and Yurok (Moerman 1998, p. 552).
22. Yew bows and arrows or arrowheads are to be found among the Bella Coola, Chehalis, Haishais, Kitasoo, Klallam, Makah, Quileute, Quinault, Samish, Snohomish, and Swinomish (*ibid.*).
23. Harpoons or harpoon shafts: Hoh, Makah, Nitinaht, Quinault, Coast Salish, Samish, Swinomish (*ibid.*).
24. Moerman 1998, pp. 551–2. See also Hartzell 1991, pp. 132–49.
25. Hartzell 1991, p. 136. The use of yew wood in deer traps in ancient Greece is described by Xenophon (*c.* 427–355 BCE) in *The Sportsmann*, IX, 18, in Dakyns (tr.) n.d.

Chapter 23. The Longbow

1. Hardy 1992, pp. 11–14.
2. The Yurok, Hupa, Karok, Shasta, Maidu, Wintu and Yahi (Hartzell 1991, p. 145).
3. Hartzell 1991, pp. 142, 143.
4. Hardy 1992, p. 13. 'There are four Egyptian longbows in the British Museum [...]. The bows date from between 2,300 BC and 1,400 BC; three of them appear to be made from acacia wood.' (p. 22).
5. Hardy 1992, p. 17.
6. The bow of the 'Ice Man' was still being made and was left unfinished. Its owner also carried a bronze axe with a yew wood handle, which he probably used to work on the bow. The wood of the bow contains twenty-eight annual rings of slow growth (Spindler 2004).
7. Dated 4040–3640 BCE by the Oxford University Radiocarbon Accelerator Unit (Sheridan, n.d.).
8. Hardy 1992, p. 17.
9. *Ibid.*, pp. 19–20.
10. Sheridan, n.d.
11. Scheeder 2000, p. 68.
12. Found at Lupfen in Württemburg (Hardy 1992, p. 30).
13. Hardy 1992, p. 26.
14. Some 138 yew longbows and 6,000 Tudor arrows were found in various stages of preservation in the *Mary Rose*. Only the horn 'nocks' at the ends of the bows (where the string was hooked in) had perished in the salty silt (except one which was protected in an accretion of metallic residue). Some of the bows are on display at the Portsmouth Historic Dockyard (Hardy 2003). See www.maryrose.org.
15. Scheeder 1994, p. 41.
16. 'Once, in AD 354, they [the Romans] were prevented from crossing the Rhine by showers of arrows from the Alamans, and 34 years later the Roman attack on Neuss was repulsed by a hail of arrows "falling as thick as if thrown by arcubalistae".' (Hardy 1992, p. 21).
17. Hardy 2003, p. 18. The whole dramatic story of the dominance of the longbow is excellently told in Hardy 1992.
18. Lowe 1897, pp. 118–19.
19. Hardy 1992, pp. 49–50.
20. Quoted in Hartzell 1991, p. 39.
21. Lowe 1897, p. 120.
22. *Encyclopedia Britannica* 2004. The question of the numbers has recently been discussed in great depth in Curry 2005.
23. Scheeder 2000, p. 72. Currency: until 1971, £1 sterling equalled 20*s*, which equalled 240*d*. In medieval Latin documents the words *libra, solidus* and *denarius* were used to denote the pound, shilling, and penny, hence the symbols £, *s* and *d*. For prices see also Strickland and Hardy 2005, pp. 42–3.
24. Scheeder 2000, p. 72. For the life and work methods of bowyers see Strickland and Hardy 2005, pp. 20–5.
25. Hardy 1992, p. 129.
26. *Ibid.*, p. 83.

Chapter 24. The Catastrophe

1. This phrasing is Robert Hardy's, regarding the reign of Edward III (1992, p. 53), but there is no doubt that this also applies to a much longer period than that.
2. Strickland and Hardy 2005, p. 42.
3. The extent of the continental yew stave trade seems to be entirely unknown to English authors but has been reasonably well documented in German and Polish works since Hilf 1926.
4. The English cherished the tight-grained wood from the Alps, and also the timber from the Baltic harbours, the quality of which was explained 'by reason of the coldness of those countries' (a contemporary writer, quoted by Hardy 1992). It was the Italian yews, however, that were hailed as the 'principal finest and steadfastest woods by reason of the heat of the sun'.
5. Scheeder 1994, p. 43; Scheeder 2000, p. 72.
6. Lowe 1897, pp. 104–5.
7. This order was reinforced by Henry VIII in 1511 (Scheeder 1994, p. 43).
8. Chetan and Brueton 1994, p. 80.
9. Such cross-cultural deliveries did occasionally happen, according to the trade archives of Vienna. Additionally, the

Dutch probably sold yew staves to the Turks in 1612, after the Dutch–Turkish peace treaty (Scheeder 1994, p. 45).
10. The Rhineland guilder or florin (fl.) is not identical with the Dutch guilder or gold florin of the time. The great number of small states each with their own currency in sixteenth-century Europe makes it impossible to reliably relate this currency to the pound sterling at the time, or to give another clue about its value without a major detour.
11. Scheeder 1994, p. 45.
12. The Hatfield Papers from 1572 list four main sources of yew staves: Germany and Austria (via the diocese of Salzburg), Switzerland, the eastern Baltic (Gdansk and Reval) and Italy (Venice) (Scheeder 2000, p. 73).
13. Scheeder 1994, p. 48.
14. Apart from their 'small side business', Stockhammer was the Secretary of Charles V, Fürer was Royal Adviser (Küchli 1987).
15. Scheeder 1994, pp. 45–7; Scheeder 2000, pp. 72–3.
16. Scheeder 2000, p. 72.
17. Strickland and Hardy 2005, p. 25.
18. Scheeder 2000, p. 72.
19. Scheeder 1994, p. 46.
20. *Ibid.*, p. 48.
21. *Ibid.*, pp. 49–50, after Mutschlechner and Kostenzer 1973, pp. 277–8.
22. Roger Asham, *Toxophilus*, 1545, quoted in Lowe 1897, p. 131.
23. Scheeder 2000, p. 76.
24. *Ibid.*, p. 75.
25. Williamson 1978, p. 39.
26. Quoted in Lowe 1897, p. 125; also Williamson 1978, pp. 49–50.
27. However, according to another source an archery contingent took part in the siege of Rey in 1627. And for 1601 and 1602 a last yew import from Hamburg-Stade is documented (Scheeder 2000, p. 80).

The longbow flared up once more in history, as Charles I in 1629 enforced its use at the beginning of the Civil War. The last record of its military use is from the siege of Devizes under Cromwell (Williamson 1978, p. 40).

Chapter 25. A Potent Medicine

1. European yew and Pacific yew have recently been tested as new remedies in classical homoeopathy (see J. Sherr in Bibliography). However, there are too few practical experiences yet as to generate a discussion in this publication.
2. Moerman 1998, pp. 552–3. Cf. Hartzell 1991, p. 136.
3. *Ibid.*
4. Moerman 1998, p. 551. Cf. Hartzell 1991, pp. 136, 139.
5. Williamson 1978, p. 45.
6. Dumitru 1992, p. 105.
7. Osthoff 2001.
8. Dumitru 1992, p. 105.
9. *Ibid.*, p. 105.
10. Manandhar 2002.
11. Osthoff 2001, p. 72.
12. Manandhar 2002; Brandis 1874, in Lowe 1897, p. 139.
13. Wujastyk 2003, p. 195.
14. Hoernle 1893–1912.
15. In May 2005; personal communication with the group product manager for Taxotere at Sanofi-Aventis, Guildford, Surrey.
16. Guchelaar *et al.* 1994, quoted by Maria Castello and Kelly Kellmel, website of Wilkes University.
17. Both Sanofi-Aventis (Taxotere) and Bristol-Myers Squibb (Taxol) produce well-informed leaflets for the practitioner and the patient. See their websites www.taxotere.com and www.bms.com. But see also Chapter 21.
18. The registered charity Friends of the Trees, for example, receives such requests – and usually refers to the AYG's website, www.ancient-yew.org, which can be a helpful travel planner too.
19. See Hageneder 2000, p. 25, fig. 3.
20. The French chemist Pierre Potier (22 August 1934–3 February 2006) was, in the mid-1980s, the first to propose the possibility of a semi-synthesis of paclitaxel, and a few years later he discovered the precursor molecule, DAB-III, and how to transform it in a very few steps into a semi-synthetic compound more active than paclitaxel itself. Potier's work saved thousands of trees and even more human lives.
21. Gradishar *et al.* 2005.

Chapter 26. For the Senses

1. Jablonski 2001; Sommer 1998.
2. Llewellyn 1997, p. 61.
3. Gravetye Manor is now a luxury hotel. The gardens are not open to the public but 'perimeter' walks are permitted every Tuesday and Friday. See www.gravetyemanor.co.uk.
4. Jablonski 2001.
5. Hermann 2000.
6. Jablonski 2001.
7. *Ibid.*, after Brande 2001.
8. Dumitru 1992, p. 124.
9. *Ibid.*, p. 119.
10. Other woods were cypress, juniper and oak (Baumann 1999, p. 37, after Pausanias).
11. Scheeder 1994, p. 50 f; Dumitru 1992, pp. 122–5.

Chapter 27. Poetry

1. *The White Devil.* All the following are quoted from the excellent overview in Hartzell 1991, pp. 243–79.
2. *The Complaint.*
3. *Endymion* I, 731–3.
4. *Macbeth*, IV. i. 28–9.
5. *Romeo and Juliet*, V, iii, 3–7.
6. *Voices from things growing in a churchyard*, in Hartzell 1991, p. 262.
7. Ackroyd 2006.
8. Coleridge, *Reflections*.
9. Holmes 1982, p. 556.
10. 'Burnt Norton', 'Wasteland', 'Dry Salvage', 'Little Gidding'.
11. 'Ash Wednesday', section IV.
12. *Poems by William Wordsworth,* vol. 1, pp. 303–4, in *Der Eibenfreund*, 12: 189–90.
13. *In Memoriam*, 2, 1–16.
14. 'The Moon and the Yew Tree'.
15. Previously unpublished poem, February 2000.

Chapter 28. Sympathy

1. Chetan and Brueton 1994, pp. 243–4.
2. Quoted from volume 3 (1873) of Fontane's four-volume account of his travels in the March of Brandenburg published between 1862 and 1882. In *Der Eibenfreund*, 8: 36–41.

3. Dr Arthur Brande in his notes on Fontane 1873, *Der Eibenfreund*, 8: 41–3.
4. Das Herrenhaus, Leipziger Strasse 3; now the location of the Deutsche Bundesrat with an old plane tree, two robinias, and a number of maples and horse chestnuts, but no yews.
5. Quoted in *Der Eibenfreund*, 8: 36; translation by the author.
6. *Ibid.*, p. 40.
7. Brande 2001, p. 33.
8. Scheeder 1994, pp. 53–65.
9. According to an article about the yew research of the Senckenberg charity published in the *Frankfurter Nachrichten* on 9 June 1907, during the Frankfurt yew tree relocation.
10. Dumitru 1992, p. 133, after Quantz 1937.
11. Dr Arthur Brande in his notes on Fontane 1873, *Der Eibenfreund*, 8: 41–3.
12. Fontane 1873, *Der Eibenfreund*, 8: 36–41.
13. Brande 2001, quoting the Bundesrat President K. Biedenkopf who in July 2000 said about the location, 'a memorial plaque to the composer which had been removed by the Nazis shall be replaced and a yew tree shall be planted in the western part of the garden. Under such trees the premiere of *Midsummer Night's Dream* was performed in 1826 in front of the palais.' Source: Lippold, F. E.: 'König Kurt will noch eine Eibe pflanzen lassen. Bundesratspräsident Biedenkopf inspizierte das neue Domizil des Bundesrates in Berlin.' *Berliner Morgenpost*, 18 July 2000, Berlin.
14. 'To Minnie', *A Child's Garden of Verses*, 1885; quoted in Rodger *et al.* 2003.
15. 'The Manse', *Memories and Portraits*, 1887; quoted in Rodger *et al.* 2003.
16. Quoted in Chetan and Brueton 1994, p. 171.
17. Personal communication with A. Meredith. See also Chapman and Young 1979, p. 127; Dunbar 1970, pp. 64, 168.
18. The publication year of his *On the Origin of Species*.
19. In May 1848, Darwin responded to J.D. Hooker in regard to their controversy about the origin of coal, 'I shall never rest easy in Down church-yard [if it is not solved] before I die' (quoted in Desmond and Moore 1991, p. 357).
20. Darwin in a letter to Hooker dated 15 June 1881 (Croft 1989, p. 108).
21. Desmond and Moore 1991, p. 664.
22. Chetan and Brueton 1994, p. 194.
23. The 'oldest' oak, a 588-year-old tree, stood in the Spessart and was felled in 1957. Most ancient oaks are hollow and cannot be ring-counted. Siegler in Bach 2004; also Dieterle 1999.
24. Chetan and Brueton 1994, p. 191, after Holt 1982.
25. Chetan and Brueton 1994, p. 193, after Wilks 1972.
26. Chetan and Brueton (1994, pp. 190–4) give a good summary of the Robin Hood material.

Chapter 29. Sanctuaries

1. Hansard 1841.
2. Quoted in Chetan and Brueton 1994, p. 77.
3. Hutton 1991, p. 271, and personal communication with the author.
4. Morris 1989, p. 79.
5. Some examples: the Council of Tours in 567 deplored pagan festivities at New Year, and also illicit practices at rocks, trees and springs; a meeting at Auxerre in the 580s urged the abrogation of vows made beside thorn bushes, holly trees or springs; at Toledo in 665 the pagan New Year customs were again condemned; Eligius, Bishop of Noyon, in about 640, discouraged placing lights at temples, stones, springs or trees; etc. General Catholic trends were usually soon echoed by national or regional law (Morris 1989, p. 60).
6. Personal communication with Ronald Hutton, 2006.
7. Lowe 1897, p. 99, after John Evelyn's *Sylva* (1664), Nichol's *Extracts from Church-warden's Accounts* (1797), Brady's *Clavis Calendaria*.
8. Kent: Lowe 1897, pp. 99–100. Wales: personal communication with A. Meredith.
9. *Encyclopedia Britannica* 2004.
10. Hageneder 2000, p. 142, after Davies 1911, p. 55.
11. John Mason Neale's was not one of the many Protestant voices that opposed Catholic ritual. He was a great hymnodist and an explorer of the riches of the ancient roots of the Catholic tradition. He founded a religious community of Anglican women and was roundly condemned and reviled for this and for his explorations of ancient Greek liturgy by Protestants who opposed what they regarded as 'popish' practices.
12. Quoted in Cornish 1946, p. 42.
13. Dumitru 1992, p. 94.
14. Coleridge in Holmes 1982, p. 214.
15. One unverified piece of information comes from central or southern Germany where the conifer traditionally employed to decorate the apex of a newly built house for the finishing celebration used to be yew. Personal communication with A. Meredith, 2004.
16. Personal communication with A. Meredith.
17. *Encyclopedia Britannica* 2004.
18. When Buddhism came to Japan in the sixth century CE, it did not attempt to undermine or supplant Shinto but simply founded its temples next to the Shinto shrines, proclaiming there was no fundamental conflict between the two religions. Since then, Shinto has absorbed many Buddhist deities into its 'pantheon' of *kami* (Littleton 2002).
19. *Ibid.*
20. *Ibid.*
21. Personal communication with Chris Worrall.
22. Hageneder 2004, chapter 'Cypress'.
23. Thanks to the excellent fieldwork of C. Worrall in Japan.
24. I am grateful to Chris Worrall for this information.
25. Wilson 1929, p. 219. In Wilson's day, some of these monasteries contained more than 2,000 priests, and many thousand pilgrims came each year from all parts of China, Tibet and even Nepal.
26. In the higher regions (1,800–3,000m) the temples were entirely constructed of the abundant local silver fir *(Abies delavayi)*. The only other trees that could possibly grace the temple gardens at those altitudes were hemlock spruce *(Tsuga yunnanensis)*, dwarf juniper *(Juniperus squamata)* and Chinese yew. *Ibid.*, p. 225.
27. Personal communication with yew researcher Paul Greenwood, AYG.
28. Moerman 1998, p. 552.

Chapter 30. The Secrets of Names

1. In Dumitru 1992, p. 100.
2. Virgil, *Georg.* 2, 448, in Fallon (tr.) 2004; Pliny: XVI, 10, 51, in Jones (tr.) 1960.

3. In Dumitru 1992, p. 100.
4. Quoted in Lowe 1897, p. 137.
5. Henslow 1906, p. 51 n. 2.
6. Grimm 2004.
7. Alessio 1957; Chetan and Brueton 1994; Cortés *et al.* 2000; de Vries 1977; Dumitru 1992; Hageneder 2000; Hassler 1999; Scheeder 1994; Schirone 2001.
8. Ivanov *(Problemy indoevropejskogo jazykoznanija)* in Puhvel 1984.
9. Pytheas, Diodorus Siculus, Strabo, in Cunliffe 2002, pp. 94, 106.
10. Reaney 1964, p. 24.
11. 'How many yews grow in Novum Eboracum?', in 'To the Editor', the *New York Times*, 11 April 2006.
12. Waddington 1997, p. 221.
13. Mawer 1920.
14. The castle at Chateau d'If on the Taxiana insula near Marseilles is of a much younger date but the name might relate to deeper and older local traditions.
15. Bertoldi 1928; Alessio 1957.
16. For example, Ewden ('yew pasture'), Ewe Down, Ewel ('yew well'), Ewen, Ewetree, Ewhurst, Ide (from *ida*), Ideford, Ifold ('yew valley'), Ivegill ('yew valley'), Iwode, Uley, Youlton (all three 'yew wood') in England; Ystrad Yw ('valley of the yew') in Wales; Co. Mayo ('plain of the yew', from Maigh Eo) and Iveragh peninsula, Co. Kerry in Ireland; Udale ('yew valley', from Norse *ydalr*) in Scotland; Jåtten (from *Ja*, 'yew', and *-túna*, 'yew grove') in Norway; Eibenberg ('yew mountain'), Eibenhorst ('yew grove'), Eibensbach ('yew brook'), Eibenstock ('yew branch'), Eibsee ('yew lake') in Germany; Eibengraben ('yew ditch'), Ibach ('yew brook'), Iberg, ('yew mountain'), Eyholz ('yew grove'), Eiwald ('yew forest'), Jona, Yverdon ('yew hill'), Yvonand ('yew valley') in Switzerland (Alessio 1957; Chetan and Brueton 1994; de Vries 1977; Hassler 1999; Scheeder 1994).
17. Referring to Dioscurides, *De materia medica*, 'the yew growing in Narvonia [Spain]...', and Strabo IV, 202 in Dumitru 1992.
18. Julius Caesar, *De bello Gallico,* V, 24, in Deissmann 1980.
19. Bertoldi 1928. There is no evidence whatsoever whether this tribe was directly named after the tree or rather after a (legendary) leader or (divine) ancestor – whose name Lemos, 'Elm', however, might indicate an elm cult anyway.
20. Julius Caesar, *De bello Gallico,* VI, 13, in Deissmann 1980.
21. There is no archaeological or historical hard evidence for Chartres being in the area of the Forest of the Carnutes mentioned by Caesar. Compare, however, the case of Le Puy, pp. 249–51.
22. Julius Caesar, *De bello Gallico*, VI, 31, in Deissmann 1980.
23. Dumitru 1992, p. 106.
24. *Encyclopedia Britannica* 2004.
25. Personal communication with Prof. M. Pridnya, September 2005.
26. Pande 1965, p. 39.
27. Lowe 1897, p. 98, after Brandis 1874.
28. *Sha* as a general term for conifer is an entirely different word to the *sha* known as 'negative energy' in Feng Shui.
29. Elwes and Henry 1906, p. 108.
30. Kawase 1975.
31. Originally, Japanese was a spoken language only, but later the Japanese borrowed the Chinese written characters to create their own written language. Because Chinese characters with a Chinese pronunciation were used for Japanese words with a Japanese pronunciation, the result is that each character can be pronounced in *at least* two ways.

Kunungi, by the way, is also the more common name for sawtooth oak (*Q. acutissima*) – the confusion of yew and oak does seem to be a worldwide phenomenon!
32. www.oncotherapy.co.jp/corporate/onco.html.
33. Hoffner 1998, p. 11.
34. Meiggs 1982, p. 73. Mari is modern Tall al-Hariri in Syria. Yew and boxwood were combined in fine furniture inlays.
35. The over 200 fragmented cuneiform tablets found in Uruk, Nimrud, Ashur and particularly Nineveh are all copies from a version of the epic which did not change from the ninth to the middle of the third century BCE (Schrott 2004, p. 16).
36. *Almug* = 'lmg, *elammaku* = lmk. Modern Hebrew for yew is *tekesus*, clearly a derivative from the Latin *Taxus*.
37. 1 Kings 10: 11–12.
38. Hebrew has no word for timber, the original text simply mentions all trees in plural: *almugim*, 'yew trees'. Because the scribe of 1 Kings put down *algumim* instead of *almugim*, until recently scholars have mistaken this wood for sandalwood from India. It has been shown since that the Sanskrit word *algummim* for sandalwood which had encouraged this interpretation is in fact of later origin (J. G. Greenfield and M. Mayhofer, 'The algummim/almuggim problem re-examined', Suppl. to *Vetus Test.* 6 (1967), pp. 83–9). It is also very doubtful, according to Meiggs 1982, whether there was any trade between the Near East and India at this early time.
39. About *Taxus* on Mt Lebanon see pp. 200–1.
40. 1 Kings 6: 29–35; see also discussion in Hageneder 2001, p. 61.
41. This occasion is mentioned twice in the Bible. In 2 Chronicles 2:8 (quoted here) it is the legendary Queen of Sheba who presents him with these gifts. While in 2 Chronicles 9:8 it rather sounds like it is the servants of Hiram who bring the goods. In both cases the cargo is not related to the seaport of Tyre (Phoenicia) but to Ophir on the Red Sea, which obviously accommodates the tale of Sheba (located in south-western Arabia) – and was understood by Meiggs as supporting the theory of a *southern* wood import (see note 38), although Henslow (1906, p. 48) had already warned that this does not necessarily imply that the algum wood also came from Ophir. Indeed, the Queen of Sheba, coming to Jerusalem by caravan from Eloth (modern Elat at the head of the Gulf of Aqaba), could have easily sent a negotiator to Tyre in time to buy yew wood and then meet up with the northern caravan before the gates of Jerusalem. After all, the timber cargo could not have been *that* large as it was to make a number of stools and musical instruments. The accounts from Ugarit, too, refer to this wood as being used for harps and lutes, hence Meiggs adds that for the identity of almug 'one should look for a precious hardwood used for furnishings rather than a building-timber. The rarity of this wood is emphasized in both accounts ' (pp. 486–7.) But even Meiggs did not suggest yew because he was still looking south for this rare and precious wood.
42. Josephus, *Antiq.*, VIII, 7, 1, in Henslow 1906, p. 48.

43. Meiggs 1982, p. 105.
44. Examination by Dr A. H. Layard, in Henslow 1906, p. 49.
45. Sayce 1893. The creation of palace gardens and royal arboretums was popular among the kings of Babylonia, Assyria and Persia, their palaces were not as bleak as their ruins suggest today. After all, the Hanging Gardens of Babylon (constructed eighth to sixth century BCE) were one of the Seven Wonders of the World. The ancient Orient also invented the avenue (Demandt 2002, p. 49).
46. Meiggs 1982, p. 486.
47. The Oxford Study Bible translates correctly, 'For a sword devours all around you' (Jer. 46: 14).
48. Henslow 1906, p. 52.
49. Theophrastus' writings are older than the Greek Bible.
50. Variations include *smilos, mylos, thymalos* (Dioscurides). Occasionally, Roman writers used Greek *smilax*, and, vice versa, some Greek documents contain the term *taxos* (Dumitru 1992; also Koch 1879).
51. Henslow 1906, pp. 50–1.
52. *Ibid.*
53. H. Rössner speaking to Greek foresters.
54. Hence in his footnote he points to a term used elsewhere: 'cedar blood' for resin. Hrozny 1924, p. 18.
55. KUB XII 20. 9; also a 'pear-tree', VII 44 Vs. 13. Suggestion by Brandenstein in Puhvel 1984.
56. Otten, Goetze, Güterbock and H.A. Hoffner in Puhvel 1984.
57. Szabó in Puhvel 1984.
58. In Puhvel 1984.
59. KUB XXIX 1 IV 18, in Puhvel 1984.
60. Ivanov (1969, 1973); P. Friedrich (1970a, 1970b) in Puhvel 1984.
61. Puhvel 1984, referring to Puhvel (1980), *Kratylos*, 25: 136–7.
62. By far the best source on this subject is Meiggs 1982.
63. See Hageneder 2005, 'Cedar', 'Cypress', 'Juniper'.
64. *J. foetidissima, excelsa* and, more rarely, *drupacea*.
65. A similar problem exists regarding the differentiation of cypress, juniper and pine in ancient documents (Meiggs 1982, App. 3).
66. See Appendix IV.
67. The timber trade routes were 'international', the resources for 'cedar' always the same. For example, King Esarhaddon (680–669 BCE) built at Nineveh with 'cedar' from Mt Lebanon and possibly also from Mt Hermon; King Tiglath-Pileser III built his palace at Nimrud with cedar beams from Amanus, Lebanon, and possibly also from Mt Hermon (Meiggs 1982, pp. 77–8).
68. In the first millennium BCE, the Babylonian and the Assyrian Tree of Life symbology distinctly shifted to the nourishing date palm (Hageneder 2001, pp. 43–4; Hageneder 2005, 'Date palm').
69. Brosse 1994, p. 231.
70. According to Demandt (2002, p. 78), Greek *drys*, 'oak' (also Meiggs 1982, p. 45), derived from Indo-European *deru* which originally denoted any tree. The root of the word meaning something firm, solid. It is also to be found in the classical term *dendron*, 'tree', which is related to Old Indian *dru*, Old Persian *dauru*, Gothic *triu*, Celtic *derva*, Irish *dair* and English *tree*.

In Old Greek, *drys* was often used for the entire oak family, 'but clearly it was also regarded as a separate species' (Meiggs 1982). We just do not know which one. An oak spirit, by the way, was called *dryad* (not *druad*).
71. *Ibid.*, p. 420.
72. *Ibid.*, p. 421.
73. Melville (tr.) 1986. Ovid, *Metamorphoses*, VIII, 743–5, 758 and 760–2: *cuius ut in trunco fecit manus inpia vulnus, haud aliter fluxit discusso cortice sanguis.*
74. White 1912.
75. Pliny, *Nat. Hist.* 16, 8 in Jones (tr.) 1960.
76. Beech does not extend south of Thessaly (Meiggs 1982, p. 454).
77. Groves and Rackham 2001, pp. 48, 156; Meiggs 1982, pp. 45, 109. The prickly-oak (*Q. coccifera*) is an extremely versatile tree which can take any form from a shrub a few centimetres high to a tree of 20m, with a trunk of 4m girth. However, it rarely grows higher than about 6m (not quite big enough to fix the 2m-diameter brass gong of Zeus into its branches). It is gregarious, growing in large numbers together rather than single individuals. It is not killed by burning, felling, or browsing (Groves and Rackham 2001, p. 48). In the eastern Mediterranean, according to Meiggs 1982 (p. 45), holm oak and prickly-oak are superseded by another evergreen oak, *Quercus calliprinos*, which is very similar to prickly-oak but larger in girth and growing taller when it survives the grazing of goats. Valonia oak is a shallow-rooted, semi-evergreen oak with huge acorn-cups (Groves and Rackham 2001, pp. 54, 151–2).
78. The oldest pottery shards found at Dodona belong to the Early Bronze Age (*c.* 2500–2100/1900 BCE), stone and bronze axes and double axes (labrys type) go back to *c.* 1650 BCE and the oldest architectural remains date from the thirteenth to tenth century BCE. Archaeology cannot prove or disprove, however, if Dodona was a religious site from the very beginning. The interpretation of the Dodonean bronze daggers and (double) axes as religious artefacts is controversial; definite religious objects only begin in the sixth century BCE. On the other hand, religious activity does not necessarily always leave material traces: Dodona never seemed to have had an offering pit or even an altar. The function of Bronze-Age Dodona as a mere settlement is even more disputable, particularly since there are no graves in or around the site (Dieterle 1999, V).
79. Not before the late fifth century BCE was a small stone temple to Zeus added to the site, and grandly rebuilt a century later by King Pyrrhos who also built a wall around the oracle itself and the sacred tree, added temples to Heracles and Dione, built other buildings, founded a festival featuring athletic games and musical contests, and a theatre that is more capacious than any in England today. In 219 BCE, the Aetolians invaded and burned the temple to the ground. Dodona was rebuilt by King Philip V of Macedon but never recovered fully. In 167 BCE, Dodona was once again destroyed and later (31 BCE) rebuilt by Emperor Augustus. When Pausanias (I, xviii, in Levi (tr.) 1979) visited in the second century CE, the sacred grove had been reduced to a single tree. Pilgrims still consulted the oracle into the late fourth century, when all ancient cults were forbidden and their temples destroyed under the Christian emperor Theodosius I. The last tree of the Dodonean grove was cut down by an Illyrian in 391. Dodona remained a religious centre, however, and became a bishop seat which was centuries later moved to Ioannina (Dieterle 1999, I.3.).

80. Dieterle 1999, V, after Petersmann 1986, pp. 74–82.
81. Herodotus, *Histories* 2, 54–7, in Waterfield 1998.
82. Cook 1925, pp. 214–5. However dated, Cook's gigantic work, *Zeus*, is a treasure trove of information, and I believe that the classical texts or archaeological artefacts that Cook refers to do not 'date' as such anyway, although their translation or dating may undergo some corrections.
83. Cook 1914, pp. 362ff, 524.
84. Hammond 1967, p. 18.
85. Demandt 2002, p. 78.
86. Dieterle 1999, VI, after Herzhoff 1990.
87. See Meiggs 1982, p. 25.
88. Homer (*Iliad*, XVI, 768) mentions *phagos* growing with cornel cherry (*Cornus mas*) and flowering ash (*Fraxinus ornus*), hence specifying an ecotope which does not accommodate Valonia oak, but Macedonian oak (*Quercus trojana*) – and *Taxus baccata*.

In the open, *Quercus trojana* can grow into a stately tree about 18m high and with a broad crown. Its leaves are entire and somewhat reminiscent of beech. It is not evergreen but holds its dried, copper-coloured foliage well into spring (Dieterle 1999, VI, after Herzhoff 1990).
89. The replanting of a sacred oak tree at Dodona is fully discussed in Dieterle 1999, V. The probable planting times would have fallen into the periods 1400–1200 BCE, 800–600 BCE and *c.* 200 BCE. Interestingly, the second date coincides with the arrival of the Selloi and the conversion of the site to Zeus worship.
90. Meiggs 1982, pp. 19–28.
91. *Ibid.*, pp. 421–2.
92. *Ibid.*, pp. 24–5.
93. Pliny, *Nat. Hist.*, XVI, 249, in Matthews 1996, p. 21.
94. On the absurdities of generalising Pliny's account for the entire Celtic world, and of his deriving the Celtic word *druid* from a Greek word, see Roux 1996, pp. 580–1 (also Hageneder 2001, p. 133). Ironically, it is a *Roman* commander – Pliny commanded a cavalry squadron on the Rhine, became an imperial agent *(procurator)* in provinces such as Spain, and returned to Rome for the position of commander of the main naval base – who has misled the beliefs of many modern followers of the 'Celtic tradition'.
95. See Hageneder 2001, pp. 58, 81; Hageneder 2005, 'Terebinth'.
96. Zohary 1982, p. 108.
97. Cook 1925, p. 682.
98. See Hageneder 2005, 'Apple'.
99. Irish legend: *Dindshenchas*; *The Settling of the Manor of Tara;* cf. p. 238. Also in the foundation legend of St Baglan recounted in the Welsh *Mabinogion* as well as in an eighteenth-century manuscript in the Bodleian Library, 'The Response of Anthony Thomas, Incumbent of Baglan, to Queries by Edward Lloyd, 1700', quoted in Chetan and Brueton 1994, pp. 232–3.
100. *Ibid.*, pp. 228, 230–2.
101. Demandt 2002, p. 22.
102. The 'change' of fruit tree has already been suggested by Mackenzie in 1922. See also Harris 1919.
103. Demandt 2002, p. 22.
104. Personal communication with Ronald Hutton 2006.
105. Examined by the biophysics department at Leeds University (Chetan and Brueton 1994, pp. 159–60).

Chapter 31. The Great Passage

1. See Hageneder 2005.
2. *Topog. Hibern.*, dist. iii. cap. x, ed. J.E. Dimock, London 1867; quoted in Lowe 1897, p. 96.
3. Morris 1989, p. 79; also personal communication with A. Meredith.
4. www.ancient-yew.org.
5. Personal communication with Dr Robert Brus, Dept. of Forestry and Renewable Forest Resources, University of Ljubljana. The oldest yew of Slovenia, however, does stand outside consecrated ground; see fig. 24.7.
6. *Hydriopathia* or *Urn Burial*, IV, quoted in Lowe 1897, p. 98.
7. Yew remains (leaves and wood) were found in barrow no. 85 in Amesbury, Wiltshire, *c.* 2000 BCE, by R.S. Newall, FSA.
8. Tacitus, *Germania*, 27, in Fuhrmann (tr.) 1995; also in Hageneder 2004, p. 167.
9. Gerstenberg near Gommern, Saxonia-Anhalt; Leuna, Merseburg-Querfurt, Saxonia-Anhalt; Häven/Jarchow, Parchim, Mecklenburg-Western Pomerania; Haina, Thuringia; Hassleben, Sömmerda, Thuringia; Heiligenhafen, near Oldenburg, Schleswig-Holstein; Varpelev, Seeland, Denmark; Osztrópataka and Stráze, Slovakia; Linton Heath, Cambridgeshire, England; Roundway Down, Wiltshire, England (Hellmund 2005).
10. Quoted in Lowe 1897, p. 98.
11. *Thebaid*, VIII, 9–10, quoted in Lowe 1897, p. 98: *Neclum illum (in Amphiarum) aut trunca lustraverit obvia taxo Eumenis*. Eumenis is not a personal name but singular of Eumenides, the 'Kindly Ones', another name of the Furies.
12. Dumitru 1992, p. 90. The early dominance of cypress over yew, however, can be explained by climate: most settlements (and hence graveyards) were located in the coastal lowlands where it is too hot and dry for *Taxus*.
13. Lucan (39CE–65CE), Spanish-born Roman poet; Silius Italicus (*c.* 25CE–101CE), Latin poet; Seneca the Younger (*c.* 4CE–65CE), Spanish-born Roman philosopher (in Dumitru 1992, p. 90).
14. The classical Greek sources mention four underworld rivers: the Styx, Phlegethon, Acheron and Cocytus which all converge at the centre of Hades. The sacred tree of the underworld stands in this central plain of Hades (see Chapter 38, note 85) but is also mentioned in association with Acheron. This river, not the Styx, is the boundary river in the original Greek sources, across which the ferryman Charon was believed to have transported the souls of the newly dead from this world into the next.
15. Liber quartus, 432–3: *Est via declivis, funesta nubila taxo: ducit ad infernas per muta silentia sedes.* In the Oxford World's Classics edition Melville gives 'deadly yews', a term even more prone to emotional associations than Albrecht's (1994) 'mourning yews'. However, the Latin *funesta* in this case denotes simply their ceremonial association with the rites of passage: 'of or concerned with death or mourning, funereal' (Oxford Latin Dictionary).
16. Pridnya 2000b, and personal communication.
17. *Slovo o polku Igoreve*, epic poem composed between 1185 and 1187.
18. Kayacik and Aytug 1968.
19. Pine was the main building wood of the region. Juniper was chosen for the outer burial chamber walls because it

was long-lasting and strong, pine was chosen for the finer craftwork in the inside of the chamber (Meiggs 1982).
20. The panels are now in the museum at Ankara (*ibid.*).
21. The samples originated from the remnants of five or six coffins found at an excavation site at the necropolis at Meir, near Qousieh (Kast, Cusae, Aphroditopolis) in the Egyptian province of Siout.
22. The robbers had mixed goods from tombs of the Twelfth, Eleventh and Sixth Dynasties (Beauvisage 1895, 1896).
23. Dumitru 1992, p. 118.
24. Hartzell 1991, pp. 135–6.

Chapter 32. The Tree of Life

1. Eliade 1996, chapter 8. For a fuller treatment of the Tree of Life, see Butterworth 1970, also Hageneder 2001.
2. Hageneder 2001, pp. 25–36.
3. E.g. Eliade 1958; Butterworth 1970; Cook 1992; Brosse 1994; Hageneder 2001, 2004.
4. Eliade 1996, p. 286.
5. Apart from the banyan (*Ficus bengalensis*) in India, but this is a different climate region.
6. Hathor: Hageneder 2001, pp. 46–9; Yakut: Butterworth 1970, pp. 1–2.
7. Wirth 1979, p. 422 and plate 152.

Chapter 33. Timeless Symbols

1. Symbols proper, as the founder of analytical psychology, C.G. Jung (1875–1961), once defined them, are not mere *signs* that have to be learned (e.g. like traffic signs) but rather keys that unlock or activate certain potentials within the human (subconscious) soul. They have a similar effect on every human being, regardless of creed or culture. Some of these symbols might have first appeared in the religious art of the Stone Age but work just the same for us today.
2. Inspired by Eliade 1996.
3. Green 1995, p. 169. Among its many metaphysical functions, the serpent is also the 'lord of the waters. Dwelling in the earth, among the roots of trees, frequenting springs, marshes, and water courses, it glides with a motion of waves' (Campbell 1959, p. 10). While in the skies, snakes or dragons, in many legends, govern the clouds and keep the world supplied with water (Eliade 1996, p. 170).
4. For the moon being personified as male and reptile, see Eliade 1996, p. 167. Snakes were also seen as the givers of all fertility.
5. See Hageneder 2001, p. 24; Campbell 1964, p. 259.
6. See Chapter 36. Campbell (1959, p. 388), however, says that the myth of the serpent and the maiden appeared around 7500 BCE and is likely to represent only a development from an even earlier base.
7. Hageneder 2004, p. 168.
8. The biblical serpent, too, did nothing else when it caused Adam and Eve to be expelled from the Garden of Eden before they could lay hands on the Tree of Life.
9. Campbell 1964, p. 416.
10. Eliade 1996, pp. 288, 290–1.
11. Apollo: Campbell 1964, p. 20; Asklepios: Pausanias, II. 28. I, in Levi (tr.) 1979.
12. See Green 1995, pp. 169–71.
13. '[S]ince the serpent is an epiphany of the moon, it fulfils the same function.' (Eliade 1996, p. 165).
14. Etana is sometimes identified with the historic king of that name who ruled Kish in southern Mesopotamia sometime in the first half of the third millennium BCE (Butterworth 1970, p. 149).
15. Retold in Rolleston 1993, pp. 97–8.
16. Shown in Gimbutas 1995, p. 214.
17. Other birds include waterfowl such as ducks and geese (still in China today symbols of fertility and good luck), the crane, the stork (see figure 41.11), and a variety of singing birds ('winged messengers'), and ravens and crows (commonly taken as heralds of the death goddess, particularly in Celtic tradition).
18. Butterworth 1970, p. 218.
19. Eliade 1978, p. 135. I am aware that Eliade's work has been criticised for various reasons but this, in my opinion, does not reduce the usefulness of the particular statements of his which I employ in this study.
20. Gimbutas 1995, 18.3.
21. Green 1995, p. 172.
22. Dovecotes existed in the temples of Aphrodite (Campbell 1959, p. 328), and in an early red-figured vase painting (*c.* fifth century BCE) she is holding a dove in her hand (Campbell 1964, p. 27).
23. Butterworth 1970, p. 218.
24. To most of us a dove might simply be a white pigeon. The *Encyclopedia Britannica* (2004) defines a dove as 'any of certain birds of the pigeon family, Columbidae (order Columbiformes). The names pigeon and dove are often used interchangeably. Although "dove" usually refers to the smaller, long-tailed members of the pigeon family, there are exceptions: the domestic pigeon, a rather typical pigeon, is frequently called the rock dove and is the bird portrayed and called the "dove of peace".'
25. Campbell 1959, p. 143.
26. E.g. the Hittite sun-god Tesup was worshipped as a bull (Graves 1955, 42.4). Cf. Chapter 41, note 72.
27. Green 1996, p. 32.
28. Boghazköy; Hassuna pottery, see Chapter 36.
29. Green 1995, p. 125.
30. Graves 1955, 18.6.
31. Billington and Green 1996, p. 169.
32. Graves 1955, 7.b.
33. Green 1995, p. 169.
34. *Ibid.*, p. 126.
35. Billington and Green 1996, p. 37.
36. *Ibid.*, p. 33.
37. *Ibid.*, p. 30.
38. Hesiod, *Theogony* 312, in Davidson 1998, p. 50. Similarly, the Egyptian Anubis who guides the souls of the dead, is depicted with the head of a dog.
39. Cook 1925, pp. 141–2. See exhibits at the museum of Cagliari.
40. Cook 1925, p. 44, n. 2, after Philolaos *frag.* 12.
41. *De ant. nym.*, 18, in Ransome 1937, p. 107.
42. *Ibid.*
43. As late as in medieval Welsh laws: 'The origin of Bees is from Paradise, and on account of the sin of man they came hence, and God conferred his blessing upon them, and therefore the mass cannot be said without the wax' (*Dull Gwent Code*, Book 2, Ch. 27, in Ransome 1937, p. 196). A common feature among Celtic mother-goddesses found across Gaul, Germany and Britain (Arrington, Cambridgeshire) is the beehive headdress (Green 1995, p. 110).

44. Honey as falling from the skies, e.g. in Hesiod, Aristotle, Virgil, in the Indian Vedas and in Norse myth.
45. A late Palaeolithic rock painting in the Araña Cave, Bicorp, Valencia, Spain, shows a human figure on the top of a rudimentary ladder, collecting honey, and surrounded by bees (Ransome 1937, fig. 2). In the middle of the fourth millennium BCE, the bee appears as the emblem of the King of Lower Egypt and can be traced through the entire Egyptian history until the Roman period. The Orphics studied the beehive as an ideal republic (Graves 1955, 2.2, 5.b, 5.1, 27.2). And for Virgil, bees possessed a share in divine reason and the breath of life, which originated in the ether, which is pervaded by God (*Georgics*, IV, 220–7, in Fallon (tr.) 2004).

In the Bronze Age, beeswax played an essential role in the widespread *cire perdue* technique for casting metal: the desired shape is sculpted in wax and covered in a clay form, which is then filled with liquid metal that melts away the wax (see fig. 38.6).
46. *Gilgamesh* VIII, line 216, in George 1999.
47. In Prophyrios, *De Abstin.*, II. 20, in Ransome 1937, pp. 119–20.
48. E.g. noted by Plutarch, *Banquet*, 106, in Ransome 1937, p. 119.
49. Ransome 1937, p. 120.
50. 'Rivers pour forth a stream of honey, In the land of Mannanan, son of Ler' (quoted in Ransome 1937, p. 189) – unfortunately the poet was too drunk, I believe, to mention that the honey *was* fermented.
51. Apollo, too, at his sacred grove in Epiros, had snakes (Aelian, *De Nat. Anim.*, xi. 2, in Ransome 1937 p. 128), and only the maiden priestess was allowed to approach (naked) the circular enclosure and feed the reptiles with honey-cakes – which for snakes were always made with barley, by the way. Honey-cakes were also fed to the sacred snakes of Asklepios at Epidauros, and offered at the serpent shrine at Olympia (Pausanias, VI. 20, 2, in Levi (tr.) 1979). In the *Aeneid* (VI, 420: see Lewis (tr.) 1986), the sybil soothes mighty Cerberus with a honey-cake (this one made with wheat) – probably an echo of the traditional honey libations to Persephone.
52. Campbell 1959, pp. 143, 428, referring to the cultural phases of Halaf in the upper Syro-Turkish-Iraqi piedmont and the Syro-Cilician corner of the eastern Mediterranean, which represent the earliest Neolithic settlements, from c. 6000 BCE. Cf. Chapter 36.
53. Porphyrios, *De ant. nym.*, 18, in Ransome 1937, p. 96.
54. *Ibid.*; Callimachus, *Hymn to Apollo*, 99, in Ransome 1937, p. 96.
55. Pindar, *Pythian Ode*, IV, 59, in Bowra (tr.) 1969. The insect also features on coins from Delphi, Ephesus, Crete, and from some of the Aegean islands, from the fifth to the first century BCE (Ransome 1937, p. 99 n. 1; plate vii; fig. 16).
56. No connection with the ascetic sect of that name that existed in Israel at the time of Jesus of Nazareth!
57. Ransome 1937, pp. 58, 96.
58. *Ibid.*, p. 129.
59. According to Hesiod, Mnemosyne was, by Zeus, the mother of the Muses. But this is a later idea, the Muses are older than Zeus and 'of obscure but ancient origin' (*Encyclopedia Britannica* 2004).
60. Pausanias, *Guide to Greece*, II, 11, 4, in Levi (tr.) 1979.
61. Hesiod, *Theog.*, 76; Pausanias, IX, 23.2, in Levi (tr.) 1979; Antipater of Sidon, *Greek Anthology*, VII, 13; VII, 34; XVI, 305.
62. Varro, III, 16, 7, in Ransome 1937, p. 84.
63. A third tree can be found at Rohrbach-Tobel near Kempten, Allgäu region, Bavaria (Rössner 2004).
64. See Chapter 30, note 95.
65. *Georgics* IV, in Fallon (tr.) 2004, and *Eclogues*, IX, 30, in Lewis (tr.) 1983.
66. On the soul as a butterfly or moth, Cook 1925, p. 645, n. 4, recommends O. Jahn, *Archäologische Beiträge*, Berlin 1847.
67. Eliade 1996, p. 183.
68. Most finds, however, are from burial sites. In the Aegean, exquisite butterfly designs on golden disks have been unearthed at Mycenae (figure 32.3 e), and on a bronze double axe from Phaistos, Crete (figure 33.21, Cook 1925, pp. 643–5); Evans 1921–35 compares these butterflies to Middle Minoan III seal-types from Zakro, Knossos (in Cook 1925, p. 645).

Gold pendants representing the larva or chrysalis of the cicada were found in a number of archaic graves at Mycenae, a fifth-century barrow near Temrjuk on the Sea of Azov (northern extension of the Black Sea), and in the cave of Pan and the Nymphs, Lychnospelia, on Mt Parnes. Gold brooches of cicadas have been found in the Bosporus area, and at the earliest Artemision at Ephesus (Cook 1940, pp. 252–4). The symbol also appears on third- and second-century-BCE tetradrachme coins from Athens. In German folklore too, the souls of the unborn were connected with crickets, also called *Heimchen*, a term that originally referred to bees. In another German tradition the unborn souls were conducted by Mother Perchta from her heavenly home to the earthly world, particularly in the *Perchtennacht* (Perchta's Night) on 6 January (the end of Yuletide, see p. 225) (Menzel 1870, p. 127; cf. Mannhardt 1858, p. 424).
69. Gimbutas 1991, p. 48; Musès 1991, pp. 133, 135ff.

Chapter 34. Birth

1. Baumann 1986, p. 51, after Sprengel 1971.
2. Baumann (1982, reprint 1999, p. 51, perhaps after Sprengel 1971) states that single yew trees are still clinging to the rather barren Mt Artemision. My site visit in October 2006 showed no trace of yew, the only trees there being firs, cypresses and a species of juniper with bright orange-red cones. The latter ('conifers with red berries') might have misled the informants of Baumann/Sprengel to assume the presence of yew. The area is clearly too dry and arid for *Taxus*, but well-watered yews in sanctuaries could have grown here nevertheless.
3. It is noteworthy that in a local cult in Attica, Artemis *Tauropolos* was worshipped as Bull Goddess.
4. Baumann 1986, p. 51, after Homer, *Iliad*, 24, 607, in Fitzgerald (tr.) 1974/1999.
5. Gimbutas 1995, 12.6.1. Her Celtic counterpart is Brigit (Green 1995, p. 200), her Germanic one Freyja (see p. 228).
6. See also Graves 1955, 89.b: 'on Aegina she [Artemis] is worshipped as Aphaea, because she vanished; at Sparta as Artemis, surnamed "the Lady of the Lake"; and on Cephallonia as Laphria; the Samians, however, use her true name in their invocations'.
7. Davidson 1998, p. 49.
8. During the following (and last) centuries of her earthly

career, Hekate's role became successively darker and eventually reduced to the role of roaming the earth on moonlit nights with her baying hounds, followed by the restless souls of the dead. This is similar to the 'wild hunt' of Germanic Odin and of Arawn, the Welsh lord of the dead: both have hounds, too, namely 'otherworldy' white dogs with red ears (compare figure 33.12). Hekate's image eventually declined into that of a fearsome evil witch, and finally fell into oblivion. In their aspect of guardian of the underworld, the yew and this goddess are remarkably similar: even their final, reduced image of 'death goddess', and 'tree of death', respectively, unites the tree and the deity in a parallel (d)evolution.

9. Graves (1955, 34.1) suggests that the mighty guardian dog is not only her servant but another guise of the goddess herself.

10. The sacrificial bull's 'brow rough with the foliage of yew' (Valerius Flaccus, *Argonautica*, 1.730, in Mozley (tr.) 1998). For the connection of the underworld goddess with yew, see the following chapter.

11. This has first been suggested by Cameron 1981 (p. 4 f; also in Gimbutas 1995, p. 265).

12. Walker 1988.

13. Lucan (39–65 CE), *Pharsalia*, I, 450–8, quoted in Matthews 1996, p. 21.

14. For example, in the Welsh tales of Bran *(Mabinogion)* and the Irish ones of Dagda. Classical writers were not sure whether the druids got the idea of the immortality of the soul from Pythagoras (e.g. Ammanius Marcellinus, *Works*, XV, 9, 8) or vice versa (Diogenes Laertius, *Vitae*, Introduction, I, 5: 'Some say that the study of philosophy was of barbarian origin'); both in Matthews 1996, pp. 19, 20.

15. Socrates, Plato and Aristotle. Cicero said, Socrates 'brought down philosophy from heaven to earth' (*Encyclopedia Britannica*).

16. Plato, *Phaedo*, 107E, tr. R. Hackforth, quoted in Mylonas 1961, p. 268.

17. *Ol.* II, 71, in Bowra (tr.) 1969. Pindar, who was born 518 or 522 BCE in Boeotia and died after 446 BCE, probably c. 438 BCE, was the greatest lyric poet of ancient Greece.

18. Ovid, *Metamorphoses*, I, 168–9, in Cook 1925, p. 39: *est via sublimis, caelo manifesta sereno; lactea nomen habet, candore notabilis ipso.*

19. Porphyrios (also known as Porphyry), *De antr. nymph*, 28, in Cook 1925, p. 41. Altogether, says Cook, 'three writers, steeped in neo-Platonic lore, and drawing perhaps from a single source, ascribe to Pythagoras himself the belief that the Milky Way is the road by which souls come and go' (*ibid.*).

20. *Phaidros*, 246E–247C, tr. J. Wright, quoted in Cook 1925, p. 44.

21. Macrob. comm. in somn. Scip. 1. 12. 1–3, quoted in Cook 1925, pp. 41–2. The classics are not alone in alluding to the Milky Way as the path of souls: 'The Basutos call it the "Way of the Gods"; [...] North American tribes know it as "the Path of the Master of Life," "the Path of Spirits," "the Road of Souls," where they travel to the land beyond the grave, and where their camp-fires may be seen blazing as brighter stars.' (Cook 1925, p. 38, quoting Tylor 1891. Basutoland is in South Africa.)

22. *Olympian*, II, 72–7, in Bowra (tr.) 1969.

23. On Easter Sunday 1998, the fire brigade of Kronberg in the Taunus region in Germany was called out because 'the castle was on fire'. The false alarm was caused by locals who had seen the castle surrounded by thick yellow clouds – which turned out to be the pollen clouds released by the yew trees in the castle grounds (Briehn 2001).

24. Which is remarkable because Pindar lived about seven centuries after the 'Dorian migration' had brought to Greece a distinct shift towards patriarchalism.

25. The line 'The Milky Way is the road of souls traversing the *Hades in heaven*' (my italics) is accredited to Empedotimos of Syracuse (Cook 1925, p. 43, referring to Philop. in Aristot. *meteor.*). Furthermore, in Sumerian myth – where many of the major mythological motifs of the world occur for the first time in writing – the underworld is also a place where the 'heart rejoices' and one is 'close to the gods', only he whose body lies unburied in the plain – 'his shade finds no rest in the nether world' (Kramer 1961, p. 37). For the importance of *proper* burial rites see next chapter, note 8.

Chapter 35. The Mysteries

1. The remains of Messene lie north of the modern Greek city of Messíni. Messene was probably founded in 369 BCE after the defeat of Sparta by Athens and the Boeotian League in the Battle of Leuctra in 371 BCE. It gave a new home to the descendants of formerly exiled Messenians as a fortified city-state independent of Sparta (*Encyclopedia Britannica*).

2. Pausanias, *Guide to Greece*, IV, 26, 6, in Levi (tr.) 1979.

3. *Ibid.*, 7–8. Also of interest is Peter Levi's note about the tin sheet: 'The mystery inscribed on thin metal recalls certain Orphic verse inscriptions about how to become immortal after your death, which have been found on very thin gold and were buried with the dead. There is at least one of these in the British Museum, but it is not exhibited.' (Levi 1979, p. 163) – If you should meet an immortal museum director in your afterlife you can ask why.

4. The temple was excavated and published by the Danish archaeologist Ejnar Dyggve (1948) and yielded some stunning ornamental works in painted terracotta (Levi (tr.) 1979, vol. II, p. 176 n. 143).

5. The astonishing personalisation of the tin scroll and the mysteries it contains as an 'old woman' being locked in a 'brazen chamber' (the bronze jar) can either be a poetic metaphor for the goddess and her cult being put 'on hold', as it were, or could point to a metaphysical understanding: the priests or priestesses of Artemis, when burying such an important artefact would surely have done so with appropriate ritual and spells of protection. It might have been part of this to invoke a spirit being as a guardian of the bronze jar.

6. Pausanias, IV, 27, 5, in Levi (tr.) 1979.

7. *Ibid.*, 27, 6–7.

8. As an ethnological comparison, according to the teachings of the Huichol in the mountains of Mexico (according to Eliot Cowan who was trained by traditional Huichol shamans), the *correct* burial rites are a crucial prerequisite for people being reborn into *their own* society. It makes perfect sense that any tribal group or ancient culture believing in rebirth would make sure that the souls of the most outstanding and benevolent personalities among them, for example great kings, hunters, warriors, priests or artists, would be reborn among their own people and not in any other ethnic group. This is generally overlooked when historians interpret the excessive burial rites and impressive

monuments of antiquity simply as the gigantomania of supposedly egocentric rulers. There is a reasonable chance that in some cases their society actually might have *wanted* to show their appreciation (with the hard work in most instances being done by slaves anyway), and some individual kings and queens might have even wished for less pomp – did not something similar in our time happen to Charles Darwin (see p. 127)?

9. The area appears to have been colonised in the sixteenth century BCE, the first architectural remains of the temple area are from the fifteenth century BCE (Eliade 1978, p. 293). Apart from brief interruptions (e.g. by the Persian invasion of Greece) the Mysteries were performed annually until 395 CE. The site remained deserted until the eighteenth century, when it was revived as the modern town of Eleusis (Greek Lepsina), now an industrial suburb of Athens.

10. The best summary of the Mysteries is given in Mylonas 1961, ch. 9.

11. *Ibid.*, p. 244.

12. *Ibid.*, p. 226. The state of Athens, after all, did impose severe punishments even for the accidental transgression of the secrecy, but scholars agree that it was more the respect of the divinities than jurisdiction that kept the mouths shut.

13. In the early history of Eleusis, initiation was restricted to the Eleusinians and Athenians; in the seventh century BCE when Eleusis fell under Athenian rule, the Mysteries were opened up to all Greeks; and in the Roman Empire to all Roman citizens (Mylonas 1961, p. 248, cf. Eliade 1978, p. 294).

14. Campbell 1959, p. 183 f; Graves 1955, 24; Cook 1914, pp. 229–31.

15. Diodorus, V, 48, quoted in Mylonas 1961, p. 280.

16. Also, for example, in the Indonesian Hainuwele myth, and in the Japanese myth of the death of the food goddess, Ohogetsu-hime (Naumann 1996, p. 62).

17. A brilliant point made by Eliade 1978, p. 293.

18. Mylonas 1961, p. 238. According to Graves, Hades is a 'Hellenic concept for the ineluctability of death' (1955, 31.2). If this is so, then the marriage of Hades with Persephone as queen is a perfect example of the religious syncretism in ancient Greece: the male god of the Indo-European 'invaders' unites with the goddess of the indigenous population. The incident in which Hades goes to snatch Kore illustrates that the male principle alone could not rule the underworld; after all, it is the ancient goddess of death and birth who holds the secrets of immortality.

19. Mylonas 1961, p. 276.

20. P. Foucart alone, in his *Les Mystères d'Éleusis* (1914), maintained that the enactments that were part of the initiation rites contained a simulated trip to the lower world. Mylonas gave this possibility the consideration it deserves, but ultimately was discouraged by the complete absence of any underground chambers or passages at Eleusis (Mylonas 1961, pp. 265, 268). However, Plouton's cave or the Telesterion, the Hall of Initiation, would have been a sufficient setting for a *mental* visit to the underworld, prepared and led by the secret teachings and ritual invocations, the individual experience being amplified by fasting for days (and possibly altered by the consumption of the sacrificial drink, *kykeon* (barley water with mint), which, however, to our knowledge had no psychoactive ingredients). Siberian shamans climb the World Tree to the upper and the lower worlds and leave no archaeological traces; an essential element of their trance is physical exhaustion (which appears to have been accomplished in Eleusis through the 22km pilgrimage and subsequent ecstatic dancing).

21. As Eliade (1978, p. 296) says, 'at Eleusis Demeter revealed a different religious dimension from those manifested in her [national] public cult'.

22. In Campbell 1959, p. 186. Sophocles says: 'Thrice happy are those of mortals, who having seen those rites depart for Hades; for to them alone is it granted to have true life there; to the rest all there is evil' (*Fragm.* 719 (Dindorf), quoted in Mylonas 1961, p. 285). Pindar, too, speaks of *seeing*: 'Happy is he who, having seen these rites goes below the hollow earth; for he knows the end of life and he knows its god-sent beginning' (*Fragm.*, 102, *ibid.*). Isokrates states that Demeter conferred to the Athenians a double gift: the gift of the fruits of the earth and the gift of the Mysteries (*Paneg.* 28); and even the Roman statesman, lawyer and scholar Cicero joins such praises when he writes that Athens has given nothing to the world more excellent or divine than the Eleusinian Mysteries (*De Legibus*, 2, 14, 36; both quoted in Mylonas 1961, p. 270).

23. Otherwise, the myrtle in ancient Greece was associated with Aphrodite, and myrtle wreaths were often worn in wedding ceremonies (Hageneder 2005, 'Myrtle').

24. Quoting Mylonas 1961, p. 252.

25. Willetts 1962, p. 160, after Pliny, *Nat. Hist.*, 23.159–60, 24.50, 21.126, in Jones (tr.) 1960.

26. Campbell 1959, p. 287.

27. Cook 1914, pp. 228–9.

Chapter 36. Origins

1. Larson *et al.* 2004, p. 89.

2. *Ibid.*, pp. 33, 59.

3. *Ibid.*, p. 91.

4. Even more so since the caves-plus-waterstreams sites favoured by humans (cf. *Ibid.*, figs 18, 44) significantly overlap with the moist cliffs and gorges that are hospitable to *Taxus* in warmer climates (even in Britain, yews can be found close to waterfalls).

5. Not to mention the Atlas Mountains in northern Africa.

6. The interpretation of these cave paintings as 'religious' can be doubted only in ignorance of our knowledge of early and present-day hunting communities and the role of shamanism in such cultures. See Campbell 1959, pp. 299–347.

7. *Encyclopedia Britannica* 2004, Middle East (from Stone Age): Mesolithic–Neolithic: the rise of village-farming communities.

8. Goff 1963, pp. 1–2.

9. *Ibid.*, p. 8.

10. *Ibid.*, figs 71–122.

11. The 'Neolithic revolution' also includes the other basic cultural traits of high civilisations, such as basketry, pottery, metallurgy, the wheel, the calendar, mathematics, royalty, priestcraft, taxation, book-keeping ... (Campbell 1959, p. 404).

12. *Ibid.*, pp. 142–3, 403; Halafian: also see *Encyclopedia Britannica* 2004, 'Mesopotamia'.

13. Campbell 1959, pp. 142–3, 428; for Crete also Graves 1955, 8.2.

14. Hoffner 1990, p. 11.

15. The Taurus Mountains are probably named after the

Hurrian storm- and bull-god, Taru (Haas 1977, p. 61); cf. Greek *tauros* and Roman *taurus*, 'bull'.
16. See Eliade 1978, p. 139.
17. The Babylonian name *Ishtar* (see below), however, was also used to designate the numerous local goddesses. Her Hurrian name was Shanshka (*ibid.*, p. 140).
18. See Haas 1977, pp. 133–61.
19. KUB XXV 31 Vs. 5–6. Also: 'from the altar they ... the *eya*-tree', KBo XXIII 49 IV 5, 6. Puhvel 1984.
20. KUB XXV 33 I 7–8, *ibid.*
21. KUB XIII 8 Vs. 9, *ibid.*
22. KUB XXVII 67 IV 9–10, *ibid.*
23. KBo XXII 236, 9–11, *ibid.*
24. Cf. Hageneder 2005, 'Terebinth'.
25. A similar memorial tree, for the child victims of the 2004 Iraq invasion, has been established at the National Memorial Arboretum in Staffordshire, England. http://www.friendsofnma.org.uk.
26. Personal communications with botanical and ethnobotanical observers, Istanbul, October 2004.

Chapter 37. The Mountain-mothers

1. The first Dynasty of Ur at the site of al-Ubaid, in Kramer 1963, pp. 29, 152.
2. Goff 1963, p. 224, after Thureau-Dangin 1907. Identity of Ninmah with Ninhursag, Kramer 1963, p. 122.
3. For a prehistoric Caucasian admixture in the Japanese inheritance see Blacker 1996, p. 178.
4. Callimachus, *Hymn to Artemis* 1–2, in Graves 1955, 22.b.
5. Examples are the 'throne of Nahat' in Armenia; Tuzuk-Dagh near Ikonion in Lycia; Kizil-Dagh, an isolated hill at Kara-Dagh, the 'black mountain', an outlying ridge of Taurus; the altars and thrones of Kybele on the plateau of Doghanlu, the Phrygian town of Midas; Tantalus on Mt Sipylos in Lydia, and the rock throne on the south-eastern slope of Mt Koressos at Ephesus (Cook 1914, pp. 136–8).
6. Cook 1914, p. 136.
7. The recorded mountain-cults of Zeus number nearly one hundred (Cook 1914, p. 165; Cook 1925, App. B).
8. Cook 1914, p. 141. In Phrygian times, 'Kybele appeared under the names of Artemis and Rhea in the major cities of the west coast of Asia Minor (Turkey), while the Romans called her Magna Mater ["great mother"], and the Gauls Berecynthia, among other names' (Billington and Green 1996, p. 70).
9. In Olympian theology the Muses were said to be the nine daughters of the union between Zeus and the goddess Mnemosyne ('Memory') with whom he lay for *nine* nights, but it is clear that their cult was much older than that of Hellenistic Zeus.
10. Pausanias mentions Olympia, Megalopolis, Messene and Tegea.
11. Cook 1925, p. 548; *Encyclopedia Britannica*, 'Ursa Major'.
12. Compare the shamans of the Siberian Chuckchees who say that 'at the Pole Star is a hole, through which it is possible to pass from one world to another. There are several levels or storeys of worlds, one above the other' (Butterworth 1970, p. 4).
13. Cook 1914, p. 104. The Muses always retained their strong association with music.
14. *Encyclopedia Britannica* 2004.
15. Demandt 2002, p. 89.
16. *Ibid.*, p. 90.
17. Plato, *Republic*, 14, 617, in Waterfield (tr.) 1993; names as translated by Graves 1955, 10.1.
18. In Troy III–V (corresponding to *c.* 2100–1900 BCE), for example, Heinrich Schliemann found about 18,000 terracotta spindles, most of them with symbolic incisions (the branch pattern being a recurring one within a wide range of designs; his catalogue numbers 1902, 1904, 1910 and 1933 are shown in figure 37.6). They were produced as votive offerings (Schliemann suggests to a goddess connected with spinning such as Athene *Ergané*) and deposited new, without any marks of wear. Similar spindles or whorls have been found in Italy (Villanova near Bologna; Castione and Campeggine near Parma; Modena district; significantly from the time of King Numa, see 'Diana', p. 191), Hungary (Szihaloin in Borsod county), Poland (Zywietz near Oliva/Gdansk), Germany (Neu Brandenburg, Schwerin), Switzerland (Stone-Age village at Möringen, Lake of Bienne) and the Greek island of Thera (Schliemann 1881, pp. 229–30).
19. Graves 1955, 6.3.
20. Dodds 1951, p. 18.
21. Dumitru 1992; Baumann 1986.
22. Graves 1955, 6.3.
23. Hageneder 2001, p. 120.
24. Bates 1996.
25. For a full treatment of Wyrd see Bates 1996.
26. Dodds 1951, p. 20 (note 30).
27. Quote from J. Grimm 1882, vol. 1, pp. 265 ff, in Cook 1940, p. 446. The connection of the benign Holden and the souls of the dead was widespread in Germanic culture. The evolution in the collective consciousness of a leader of the Holden – dame Holda – however, is later, probably not older than the Christian era. Her name is related to the Norse goddess of the underworld, Hel, and stems from Old High German *helan*, 'to hide, to cover', and has been related to the covering of dead bodies and the veils of mourning women (or goddesses). It could also refer to the invisibility of the 'liberated souls of the departed' (Meyer 1910, p. 114). In Christian times, however, it was distorted to the notion of 'hell' as a place of eternal punishment.
28. Munch 1926, p. 310, in Cook 1940, p. 447.
29. Willetts 1962, pp. 148–97.
30. Earliest evidence for sanctuaries on high places in Crete: 2100–1900 BCE (Eliade 1978, p. 132).
31. Willetts 1962, p. 191. Because of these attributes it comes as no surprise that some sources actually name Hekate as the mother of Britomartis, nor that the elements of this tradition became absorbed into that of Artemis.
32. Robertson-Smith (in Guthrie 1962, p. 84 n. 1) identifies Leto with the Semitic Allath (Alilat) – cf. Hebrew Elath as a title of Asherah, below.
33. Compare Pausanias (X, 12, 4, in Levi (tr.) 1979): 'overgrown places used to be called *Idai*'.
34. In 1460 BCE, according to classical sources listed in Cook 1925, p. 949.
35. Eliade 1978, p. 131; also p. 251.
36. Graves 1955, 53.2.
37. Cook 1925, p. 934, referring to Pythagoras' visit to the cave.
38. Eliade 1978, p. 251.

39. Willetts 1962, p. 197.
40. Ida: Cook 1925, p. 938; Dikte: see box pp. 186–7.
41. Cook 1925, p. 949, after his translation of *Iliad*, XIV, 286–8.
42. See Appendix IV.
43. He continues, 'and there is another small one not far off, and there are quite a number about a spring called the Lizard's Spring about twelve furlongs off. There are also some in the hill country of Ida in the same neighbourhood, in the district called Kindria and in the mountains about Praisia.' (III, 3, 4) The opening of this passage, however, is rather obscure: he calls this tree *aigeiros* (which is usually translated as black poplar) and, after stating that the Arcadians regard every mountain tree as fruit-bearing but not so the *aigeiros* – which he himself declares elsewhere that it appears to have neither fruit nor flower (III, 14, 2) – he gives a few exceptions of fruit-bearing *aigeiros* trees in the Cretan mountains, particularly in the Idaean region. Did he himself identify these trees or did he rely on reports? *Aegeiros/ aigeiros*, however, was a synonym for the underworld tree (see Chapter 38, note 85, and Chapter 39, note 58). Pliny, by the way, got it wrong again (see pp. 148–9) and called it a willow (*Nat. Hist.*, 16, 110, App. B, in Cook 1925, p. 529).
44. Sir Arthur Evans found in the Diktaean Cave the remains of a libation table with three cup-like depressions (Ransome 1937, p. 120). See Evans 1901b, p. 113.
45. Graves 1955, 7.3. According to Graves (1955, ch. 56), Io gave her name to the Ionian Sea and the Hellenistic Ionians (*ibid.*, 7.6).
46. Graves (1955, 7.3) suggests Melisseus to be a corruption of Melissa, the Goddess (or her priestess) as 'Queen bee'.
47. Diodorus Siculus, V, 70; Callimachus, *Hymn to Zeus*, 49. A folklore survival occurs in a manuscript from the second century CE or later, relating a story in which the Diktaean Cave is occupied by sacred bees (in Cook 1925, p. 928f).
48. Diodorus Siculus, V, 63, in Ransome 1937, p. 93.
49. Graves 1955, 7.3.
50. Apollo and Artemis in Crete were also called upon in various treaties, and in the oaths of citizenship of various Cretan cities (Willetts 1962, p. 191).
51. In Strabo's time, the Idaean Daktyloi were generally seen as 'wizards and attendants of the Mother of the gods' (Willetts 1962, p. 99, after Strabo 10.473).
52. *Ibid.*, p. 175.
53. Like the Minoan 'maiden' and vegetation-goddess, Ariadne, the 'very holy one' (*ibid.*, p. 194).
54. Graves 1955, 1.d.
55. Hestia, Demeter, Hera, Hades and Poseidon.
56. Virgil, *Georgics*, IV, 63, in Fallon (tr.) 2004.
57. Graves 1955, 7.b.
58. Campbell 1964, pp. 50–4; Cook 1940, pp. 403–8, after Evans 1925 and 1921–35, vol. 3.
59. Campbell 1964, pp. 50–2; in the following I largely follow Evans' interpretation as quoted in Campbell 1964.
60. For the spiritual significance of birds see Campbell 1959; Eliade 1978, p. 163 n. 2.
61. Cuneiform text dated *c.* 2050 BCE, Kramer 1956, pp. 172–3, quoted in Campbell 1964, p. 53.
62. Cook 1925, p. 929.
63. *Európe* appears also as an epithet for Demeter at Lebadeia in Boeotia and Astarte at Sidon (Graves 1955, 58.2; Cook 1925, p. 524).
64. Willetts 1962, p. 167; Cook 1914, p. 525.
65. Willetts 1962, p. 160.
66. Graves 1955, §58, cf. also 111. f.
67. In Olympic myth, *Európe* is a young woman or princess who was abducted by Zeus in the shape of a great white bull and taken from the shore of Tyre (Phoenicia) to Crete. There, himself now in the shape of an eagle, he raped her in the crown of an old tree.
68. The notion that the virgin on bull-back would have arrived on Crete's *seashore* has encouraged assumptions of a wetland tree, but willows do not like salty air. Theophrastus (I, 9, 5) received word of a plane tree (*Platanus*) in the district of Gortyna, which was said to have been the tree of *Európe* and Zeus. This sounds more like local folklore, however, as it dates from many centuries after this kind of ritual (*hieros gamos*) would have happened.
69. Not for etymological reasons, though: the goddess' name is *not* 'Yew-ropa'. Her name, *Európe,* translates as 'broad-eyed', which has been interpreted as a metaphor for the full moon by Graves (1955, 58.2) and by Cook (1925, p. 537).
70. For example, on a fifth-century-BCE coin from Sicily (Cook 1940, p. 175).
71. Graves 1955, 18.3. I use Graves's *Greek Myths* very carefully, and mostly only for his summarised versions of the tales (which are quite well-referenced to the classical texts), avoiding his interpretative sections.
72. Furthermore, the name of Anchises (Ankh-Isis) may link her to Isis (*ibid.*).
73. Cook 1940, p. 177.
74. Graves 1955, 18.4; also Cook 1940, p. 171. Cf. Germanic Freyja, p. 228.
75. Cook 1925, p. 872.
76. Pausanias (II, 10, 4, in Levi (tr.) 1979) describes a statue of Aphrodite at Sikyon, Korinth, which had a poppy in one hand and an 'apple' in the other for which various commentators (e.g. Cook 1925; Levi 1979) have linked her with Demeter and the Eleusinian Mysteries. Aphrodite, the goddess of love also being the goddess of death points to the ancient theme of *death as love*. In Greek folklore, Cook saw a survival of this in a number of folk songs which describe the beginning of the afterlife for men as an initiation in the bridal chamber of Persephone, and for women as a marriage-union with Hades himself. Death was the result of the assignment by the Fates, and the first stop-over the result of the Underworld deities claiming yet another consort, hence the saying 'Whom the gods love dies young.' (1925, pp. 1164–6).
77. And as the 'wheel of life' it played a vital role in alchemy (see Hageneder 2001, fig. 68) and appears as the tenth card of the 'major arkana' in the Tarot cards.
78. Cook 1914, pp. 268–74; Etruscan culture coming from Asia Minor: Cook 1940, p. 259.
79. Babylonian Ishtar is usually identified with Sumerian Inanna; her name, however, seems, rather, to have developed from the Sumerian mother of the gods, Ashratum, who was the bride of the primeval creator An(u) (Patai 1990, p. 37), and who still appears in Middle Bronze Age Akkadian texts. Keel and Uehlinger (1998, p. 22) mention *Ashratum* as the precursor of the name of *Asherah*. A cuneiform text from Ugarit calls the Canaanite Asherah 'Mother of the Gods' (Eliade 1978, p. 151).

80. Patai 1990, pp. 57, 302 (note 24).
81. Keel and Uehlinger 1998, pp. 19–30.
82. Patai (1978, p. 55) gives examples, 'He cries to Asherah and her children, to Elath and the band of her kindred'.
83. *Astarte* and *Anath* were mistaken by the Egyptians in the fourteenth century BCE (Amarna letters) as two separate divinities; Anath developed into the Egyptian war goddess Neith, while 'Astar of Syria' became a goddess of healing (Patai 1990, pp. 56, 61).
84. Cf. Patai 1990, section '"Church"-less Judaism', pp. 25–7.
85. As late as the second century CE, Yahweh could thus be identified (and worshipped) by Jews and non-Jews alike with Zeus-Sabazios (Lord Sabaoth), a fusion of the Greek and the Syrian celestial deities (Campbell 1964, p. 273).
86. All that the followers of Yahweh – in Hebrew referred to as *Adonai*, 'the Lord' – had to do was to replace the Canaanite Baal, 'the Lord', with his presence. Although the identical epithets surely helped in the transition, the process was not always smooth: During the reign of King Ahab (873–852 BCE) – who himself had erected a 'sacred pole' for Asherah and whose wife, Jezebel, was the daughter of the King of Sidon and high priest of the Goddess (1 Kings 16: 31–2) – the biblical prophet Elijah met 'the 450 prophets of Baal' for a rain-making contest on Mt Carmel and then had them seized, dragged down into the valley and slaughtered by the River Kishon (1 Kings 18: 19–45). The 400 followers of Asherah who were also present at the contest were left unharmed and free to go.
87. 'The worship of Asherah as the consort of Yahweh [...] was an integral element of religious life in ancient Israel prior to the reforms introduced by King Josiah in 621 BCE,' says Raphael Patai in his ground-breaking study, *The Hebrew Goddess*, and concludes that 'the one variety of religious worship in which the Israelites engaged more frequently than in any other was the cult of the Goddess Asherah, symbolised and represented by her carved wooden images.' (1978, pp. 39, 53).
88. *Ibid.*, p. 40.
89. In alternating phases of monotheistic Yahwism and pan-religious syncretism, the successive kings of Judah moved the asherah in and out of the Temple. Its final expulsion came with the reform by King Josiah in 621 BCE (*ibid.*, pp. 39–41, 46–50).
90. Jeremiah 17: 2–3; cf. 2 Kings 18: 4; 2 Chronicles 31: 1.
91. The asheras are described as 'sacred poles' or 'pillars' and are either 'made', 'set up', or even 'built' by the followers, or 'hewn down', 'broken', or 'burnt' by their enemies (see Patai 1990, p. 296 n. 9). Only once does the textual evidence mention an asherah as 'planted' (Deut. 16: 21), and only one other uses a term that can denote 'to uproot' (Micah 5: 14).
92. Already by the early 1940s, a total of no less than 300 terracotta figurines and plaques representing a nude female figure had been found (Patai 1990, p. 58).
93. Patai 1990, p. 59. The last evidence for Jewish worship of the ancient goddess (here: Astarte) is found in a letter from a Jewish military colony in Upper Egypt dating from 419 or 400 BCE (*ibid.*, p. 65 f).
94. The possibility that the herringbone pattern could even be a mere geometrical decoration can be ruled out. Generally speaking, in the religious iconography of this age – and particularly on such miniature objects with limited space but of great magical importance – every design element has meaning.
95. See Hageneder 2001; Hageneder 2005, 'Date palm'.
96. Regional agriculture of the time was generally not above 500m altitude (Meiggs 1982, p. 54).
97. This is not to say that anyone involved with the asherahs in the 'high places' of Israel would have known anything about yew trees, the tradition would simply have been a *survival* from an earlier era, with increasingly lost meaning.
98. Sycomore Fig, Persea, Acacia, Tamarisk, Terebinth (Hageneder 2001, 2005).
99. About *Taxus* on Mt Lebanon see pp. 200–1.
100. Patai 1990, p. 60, after Albright 1943, p. 139. The northern origin of these potters is assumed because Israel was the southernmost region of Canaanite/Phoenician/Syrian religion.
101. Keel and Uehlinger 1998, p. 24.

Chapter 38. Gods and Heroes

1. The Sumerian word for 'to love' is a compound verb which literally means 'to measure the earth', 'to mete out a place' (Kramer 1963, p. 250).
2. On the 'mountain of heaven and earth, the place where the sun rose', the past tense denoting it as the eternal place of origin.
3. Cook 1914, p. 3.
4. Campbell 1964, pp. 273–5.
5. The association of Zeus with yew also shines through in Aristophanes' *The Birds* (line 216, in Rogers 1979), written in 414 BCE. The hoopoe says to the nightingale that its pure sound rises through the thicket of yew *(smilakos)* to the throne of Zeus *(Dios)* where Apollo listens to it. (It was probably the presence of Apollo that caused Rogers to translate *Dios* as 'heavenly' rather than limiting its meaning to Zeus alone; Rogers also differs from the common translation by giving woodbine for *smilakos*.)

A classical Roman vase painting shows Jupiter, the Roman equivalent to Zeus, wearing a crown of yew leaves. Carrying a thunderbolt and leaning on a knotted staff he stands to the right of the god Hermes who raises his caduceus as he weighs the souls of two warriors. The painting style itself does not really allow the identification of *Taxus* but the symbolism of the scene indicates the threshold to the next world (Didron 1907, p. 180, figure 218).
6. This formed an important point of contact between the old religion and Christianity later (Cook 1925, pp. 1167–9, 1177).
7. *Ibid.*, p. 1176.
8. Plato, *Republic*, 616b, c, in Waterfield (tr.) 1993.
9. *Ibid.*, 614c–616a.
10. See Butterworth 1970, plate II, top row, third image; plate XXVI, bottom.
11. See Hageneder and Singh 2007.
12. Butterworth 1970, p. 129.
13. University of Georgia Libraries: *'Taxus',* after Sir Joseph Hooker (1854), 1: 168, 191; 2: 25. http://djvued.libs.uga.edu/QK488xE4/1f/trees_of_britain_and_ireland_vol_1.txt
14. Buddhist missionaries from India had arrived in Greece long before Alexander went to India.
15. Butterworth 1970, p. 100.
16. Hermes and Pan save Zeus – after the Three Fates have weakened the serpent with 'ephemeral fruits' (Apollodorus: i. 6. 3, in Hard (tr.) 1997).

17. Typhon is born in a cave because either the Corycian Cave was used as one of those natural places that help to induce other states of consciousness with particular ease, or because 'his fiery power was felt in the hollow cavern of the body' (Butterworth 1970, p. 99), or both. It is 'virtually certain' (*ibid.*) that trance experience was part of the cult at the Corycian Cave.

The Corycian Cave is actually not a cave but a vast chasm in a jagged high plateau of *calcareous* rock. Some of its cliffs are over 60m (200ft) deep and covered by a thick jungle of trees and shrubs which are kept fresh by a number of watercourses and the shadow of the cliffs. At its southern and deepest end, the uneven bottom merges into the cavern proper which is filled with the roar of subterranean waterways. The entrance is partly blocked by the ruins of a Byzantine church which once replaced an old temple. A second, bigger one, towered right above the western precipice (Frazer 1906, pp. 69–71).

18. Hoffner 1990, pp. 10–14. For evidence of the practice of yoga in western Asia in the third and second millennium BCE see Butterworth 1970.

19. Already the Hittites (outside the ascetic monasteries) took the serpent/dragon as a negative symbol, namely one for drought and infertility that had to be overcome by the benevolent storm- (and rain-) god.

20. Larson *et al.* 2004, p. 60.

21. Homer, *Odyssey*, XII, 68, in Ransome 1937, p. 99.

22. Pliny, *Nat. Hist.* 16.110 (in Willetts 1962, p. 168 n. 141), after Theophrastus (3.13.7, Hort (tr.) 1916), who uses *helix* for 'willow'.

23. Guthrie 1962, pp. 73–87.

24. Cook 1925, p. 459, quoting Swindler 1913.

25. According to *Greek* myth, however, this happens at Delos (Eliade 1978, p. 268).

26. Ransome 1937, p. 99.

27. Herodotus 4, 33, in Waterfield 1998; also Pausanias 1, 31, 2, in Levi (tr.) 1979.

28. Guthrie 1962, pp. 75–6, 78; Cook 1925, pp. 493–7.

29. Both Greek Odysseus and Sumerian Gilgamesh, for example, face the north wind on their way (to the vertical passage) to the underworld.

30. Pindar, *Tenth Pythian Ode*, in Bowra (tr.) 1969.

31. Other sources even ascribed the Hyperboreans to the Caucasus, the farthest east (Tibet) or the farthest west (Britain) (Guthrie 1962, pp. 79–80).

32. Four towns called Apollonia in Macedonia alone (Cook 1925, pp. 499–500).

33. *Ibid.*, p. 494.

34. From Apollo's connection with the Hyperboreans but also with swans and with amber, A.H. Krappe considered already in 1942 (*Class. Phil.* xxxvii, p. 353–4) 'that Apollo had certainly absorbed something of a god from the Frisian North Sea coast' (Guthrie 1962, p. 82). Regarding Apollo's solar connections compare the Swedish prehistorian E. Oxenstierna (1958) who interprets Ullr's name as 'the bright one' (p. 49) and links it with the sun and fertility.

35. Hageneder 2001, p. 120.

36. Diederichs 1984, p. 276.

37. An oath-ring (*at hringi Ullar*) is mentioned only once, though: *Atlakvida* 30, 8, a Norse poem in the Codex Regius (Dronke 1969, p. 65).

38. Dumitru 1992, p. 93.

39. Herodotus I, 173, in Guthrie 1962, p. 83.

40. See Guthrie 1962, pp. 83–4.

41. See full discussion in Guthrie 1962, pp. 82–7; also Dodds 1951, p. 69. That Apollo was native to Asia Minor is also illustrated by the fact that in the Trojan War he fought on the side of the Trojans, *against* the Greeks.

42. Inscriptions found on four Hittite altars, translated by Hrozny in 1936 (Guthrie 1962, p. 86).

43. KBo VI 2 II 62, in Puhvel 1984. For pillars see Chapter 41, note 9.

44. Above all, says Guthrie (1950, p. 87), Apollo is the Averter of Evil (*Apotropaios*), the god of purification (*Katharsios*), and the god of prophecy.

45. VI, 24, in Butterworth 1970, pp. 133–4.

46. About 1200 BCE, migrating Phrygians brought Dionysos from Thrace to Asia Minor (Nilsson 1950, pp. 567–9).

47. Hutton 2001, pp. 73–4.

48. Harris 1916, p. 9, after Athenaeus (flourished *c.* 200 CE) and Euripides, *Bacchae* 703. Euripides also refers to the red fruits: 'Grow green with smilax! Redden with berries!' (107–8, in Grene and Lattimore (eds) 1959).

49. Dodds 1951, p. 270.

50. Even more so as Dionysos' cult in Phrygia 'was evidently deeply influenced by native elements' (Nilsson 1950, p. 569) of the ancient Anatolian religion.

51. Cook 1925, pp. 271–7.

52. Pherekydes of Leros (Leros is a small mountainous island off the south-west coast of Turkey), in Cook 1925, p. 274.

53. Graves (1955, 27.b) places Mt Nysa in Heliconia, but the Thracian myth is not geographical but rather an allusion to the world-mountain.

54. In the later second millennium BCE, the Gilgamesh epic was also copied widely outside Mesopotamia; copies have been found at Hattusa (the ancient Hittite capital, at modern Bogazkale in Turkey), at Akhetaten (the royal city of Pharaoh Akhenaten, modern el-Amarna) in Egypt, at Ugarit (Ras Shamra) on the Syrian coast, and at Meggido in Palestine (George 1999, p. xxvi).

55. His Sumerian name is Bilgames.

56. The next one would be Jesus Christ, on Easter Saturday in about 30 CE.

57. Biblical Erech.

58. George 1999, p. xxxi.

59. George 1999, *Gilgamesh* VIII, lines 213–18.

60. Suggested by Schrott.

61. Meiggs 1982, pp. 410–20.

62. *Gilgamesh*, V, 6. George (1999, p. 39) gives 'godesses' throne', Schrott (2004, p. 213) 'the throne of Ishtar'.

63. George 1999, pp. 153, 164.

64. Version B, 'Ho, Hurrah', line 152, in George 1999, p. 166. The giant's Sumerian name is Huwawa.

65. Ishchali tablet, 38–9; in Schrott 2001, p. 218; also George 1999, p.46.

66. *Gilgamesh*, V, 154, in George 1999, p. 42; Schrott 2001, p. 216.

67. *Gilgamesh*, V, 10, in George 1999, p. 39; Schrott 2001, p. 213, translates *ballukku*-trees as styrax-trees.

68. *Gilgamesh*, V, 9; *ibid.*

69. See, for example, Groves and Rackham (2001, p. 150) on the Holocene climate of the Negev desert in Israel, which between 4500 and 1000 BCE seems to have had twice as much rainfall as today.

70. Some of the Sumerian poems mention the east, which has been understood as the Zagros Mountains towards Persia (modern Iran) – a region with continental climate that is far too hot and dry for *Taxus*. The 'east' in Sumerian poetry, however, alludes not to geography but to the symbolism of the rising sun and the direction of the otherworldy paradise (compare the 'gates of the sun', Chapter 34). The locations in the Babylonian versions are the Mts Sirion and Lebanon (tablet V, 134). The *Encyclopedia Britannica* (2004, 'Syria') interprets the 'cedar mountain' as the Amanus range.
71. The ancient city of Carchemish, for example, around the eighteenth century BCE was a trade centre for wood, 'most likely involved in shipping Anatolian timber down the Euphrates' (*Encyclopedia Britannica* 2004, 'Carchemish').
72. An interim form being Kybebe (Gimbutas 1982, p. 197).
73. However, the story is not an early version of a 'battle of the sexes'. The guardian giant and the grove were under personal protection by the very Enlil to whose temple the heroes bring the felled sacred tree. A foolish ironic insult that incurs the wrath of Enlil and the divine counsel, and in its wake Enkidu's death.
74. Genesis 6: 11–9, 19.
75. The dove occurs in the Greek Flood legend (Graves 1955, 38.c), too, but the branch does not.
76. *Ibid.*, 38.3.
77. Apollodorus, *The Library of Greek Mythology*, II, 5, 11, in Hard (tr.) 1997.
78. Graves 1955, 133.a.
79. *Ibid.*, 122 f; Pausanias II, 31, 13, in Levi (tr.) 1979.
80. Graves 1955, 145.e, after Sophocles (496–406 BCE).
81. E.g. *ibid.*, 89.7.
82. The olive (*Olea europaea*) occurs as a tree of 10 sometimes 15m in height, often with a gnarled trunk, or as a shrub up to 2(-5)m height, with dense thorny branches. The cultivar, var. *europaea*, is usually grown from sea level to about 400m, rarely up to 600–700. Its origin from the 'wild olive', var. *oleaster* or var. *sylvestris*, has long been a subject of controversy. Zohary 1975 (in Davis 1978, p. 156) made a plausible case that the cultivated crop has probably been derived from var. *sylvestris* by the artificial selection of high oil-yielding genotypes and perpetuated by clonal propagation. Seedlings of cultivars may revert to the wild form with which the cultivated one interbreeds and on which it is often grafted. The wild form is probably native to coastal regions of the Mediterranean as well as the Black Sea (Davis 1978, pp. 155–6).
83. 1925, p. 467.
84. Levi 1979, vol. 2, p. 237 n. 124.
85. Only much later, Servius in fourth-century Rome relates that Herakles wove himself a wreath from the tree that Hades had planted in the Elysian Fields as a memorial to his mistress, the beautiful nymph Leuke (in Graves 1955, 134 f). It is blatantly obvious, however, that a personalised Hades dedicating a tree to a past mistress is a late romantic addition. There has *always* been a tree at the centre of the underworld, it is the World Tree which connects all the worlds. It was not planted by any anthropomorphic deity but is an inherent part of the metaphysical structure of the universe. And in tree lore, it was not a poplar but an evergreen.
86. Levi (tr.) 1979, p. 215 n. 59.
87. Pindar, *Olympian* III, in Bowra (tr.) 1969; Pausanias, V. 7. 6–7, in Levi (tr.) 1979.
88. Guthrie 1962, p. 76.
89. Levi (tr.) 1979, p. 215 n. 59.
90. Brosse 1994, p. 231.
91. Herakles' close companion, Idas, was even named after the sacred Cretan mountain, and the statue of the serpent bore the horn (cornucopia) of Amaltheia (Willetts 1962, p. 52).
92. Pausanias, VI. 20. 1–5, in Levi (tr.) 1979, and note 168, p. 343.
93. Graves 1955, 138.i, o. Herakles was also said to have fenced off a sacred grove to Zeus, and erected six altars, one for every pair of the Olympian gods – obviously an addition of Hellenistic times which attributed the foundation of the Games to the *Dorian* Herakles.
94. Pausanias, V. 16. 2–4, in Levi (tr.) 1979.
95. Graves 1955, 53.5; Pausanias, V. 4. 6, in Levi (tr.) 1979, and note 26, p. 205.
96. Apples remained the prize at the Pythian Games (Cook 1925, p. 490 n. 5).
97. Pausanias, V. 8. 5, in Levi (tr.) 1979. Later on, the olive was replaced itself, by the laurel (bay) wreath.

Chapter 39. The Making of Kings

1. Campbell 1964, p. 6.
2. For example, eight or nine years in Minoan Crete (Campbell 1959, pp. 427–8); in other regions it could be as short as one year (Hoffner 1990, p. 11), Graves (1955) even proposes a tradition of a summer and a winter king.
3. Doht 1974, pp. 27–9, 145, 245 n. 60, 263 n. 185.
4. Haas 1977, pp. 11–14; Hoffner 1990, p. 11.
5. KUB XXIX 1 I, in Haas 1977, pp. 11–12.
6. KUB XXIX 1 IV 17–20, in Puhvel 1984.
7. KUB XXIX A 27–35, in Hoffner 1990, p. 18.
8. For example, we know about the yew-tree ritual of Nerik that the king and a priest handled cattle- and sheepskins (Haas 1977, p. 153).
9. KBo III 8 III 9, and 27, in Puhvel 1984.
10. Haas 1977, p. 67.
11. Haas 1977, pp. 117, 119.
12. KUB XXIX 1 I 50–II 8, in Haas 1977, p. 55; Wilde 1999, p. 123.
13. They might have occurred in triads, but there are many other local names for the fate goddesses (Haas 1977, p. 54).
14. *Ibid.*
15. *Ibid.*, p. 56.
16. Local or regional allusions to the marriage of Zeus (aka sky-god) with an earth-goddess are so great in number that Graves suggested that 'Zeus' in this context must be seen as a *title* for the local king rather than the name of a deity. The same would apply to the semi-divine Hellenistic hero Herakles and his many amorous 'adventures'.
17. KUB XVII 10 IV 27–8, in Puhvel 1984.
18. Compare the spouse of the goddess often attending the sacred marriage in the guise of a shepherd, e.g. Sumerian Dumuzi meeting Inanna, or Anchises meeting Aphrodite.
19. Others are the Golden Lamb of Atreus (Cook 1914, pp. 405–9), and a folktale from Epiros given by Cook (1914, p. 412). In Etruria, too, the golden fleece increased prosperity (*ibid.*, p. 403). See also *Cupid and Psyche*, p. 208.
20. Pindar, *Pythian* IV, in Bowra (tr.) 1969; Apollodorus I. 3, in Hard (tr.) 1997.
21. Cook 1914, pp. 238, 244. For her serpent chariot Graves (97.2) calls her 'a Corinthian Demeter'.

22. Graves 1955, §152.
23. *Ibid.*, 157.c, 157.1.
24. Usually, the connection between the *Argo* and the Indo-European sky-god is claimed via the oracular beam that was fitted into the prow of the ship and came from 'Zeus' oak' at Dodona. But the Dodona tree has to be questioned as well, see box p. 148.
25. Graves 1955, 149.d.
26. Eliade 1978, p. 300.
27. Graves 1955, 53.a, after Diodorus Siculus, Sophocles, Apollonius Rhodius.
28. *Ibid.*, §149, 152.j, 151.f.
29. *Ibid.*, 151.c.
30. Apollonius, IV, 1132, in Ransome 1937, p. 101; Graves 1955, 144.b.
31. Cook 1914, p. 164. Was Zeus Melosios, 'guardian of sheep', also the guardian of 'apples'?
32. Hunter 1998, pp. 102–3.
33. In Haas 1977, p. 118.
34. Apuleius, *Metamorphoses* 6, 11–13, in Cook 1914, p. 404.
35. Pausanias, I, 12, 1, in Levi (tr.) 1979.
36. Green 1995, p. 70.
37. *Ibid.*, pp. 70–88.
38. This identification is probable but not certain (*ibid.*, p. 82).
39. Dames 1996, pp. 62, 80.
40. Campbell 1964, p. 301. *Tuatha dé Danann* as a collective term for the early Irish nature gods might not go back further than the Middle Ages (Dames 1996, p. 67). The Tuatha dé Danann are said to have retired into the underworld; Irish folklore connects them with the prehistoric burial mounds and has belittled the *tuatha* into fairies – just as the modern emblem of *Ierne*, the 'yew island', is not the 'lasting' tree *(Búannan)* anymore but a tiny herb: clover (the shamrock).
41. Hageneder 2000, p. 232; Dames 1996, pp. 62, 77, 80.
42. Hageneder 2000, p. 232.
43. Le Roux and Guyonvarch (1996, pp. 500–1) suggest that the fighting might refer both to yew-wood weapons as well as to magical ogham sticks (cf. the ogham rune for yew, fig. 41.10). Dagda's epithet, transferred to sacred kingship, became a title of many Irish rulers.
44. Ellis 1987.
45. Green 1995, p. 31; Edel and Wallrath 2005, p. 122.
46. In Chetan and Brueton 1994, p. 224.
47. The *f* is silent, hence the two terms sound the same (Doht 1974, p. 16).
48. Campbell 1964, p. 31.
49. *Ibid.*, p. 74.
50. Aeneas himself was a child of a sacred marriage, conceived on sacred Mt Ida where his royal father in the guise of a shepherd met the mountain-mother, Aphrodite.
51. *Encyclopedia Britannica* 2004, 'Aeneas'.
52. Virgil, *Aeneid*, VI, 106–7, in Lewis (tr.) 1986.
53. *Ibid.*, 164, 169–70.
54. *Ibid.*, 193, 203–5, 209.
55. Virgil only *compared* it with mistletoe, a point hugely misinterpreted by Frazer, see Appendix V. Virgil's biographer, Peter Levi, shows clearly that the mistletoe in Virgil's account is only a simile (1998, p. 182).
56. The Greek colonists of Cumae came from Chalcis in Euboea in *c.* 750 BCE (J. Griffin, in Lewis 1986, p. 421 n. 2).
57. For the Latin text see Chapter 31, note 10. Is this just Statius' interpretation of Virgil or did he have other sources?
58. Cook 1925, p. 418 n. 5 (after Val. Max. I, 2, 1). *Egeria* is the Latin form of Greek *Aegeria*, the underworld tree; see discussion of Herakles' pyre, p. 202.
59. That the tree associations of deities were not set in stone is also illustrated by the fact that in the direct neighbourhood of Nemi, Diana became the goddess of beech trees (Cook 1925, p. 420 n. 1).
60. Kawase 1975. Compare the Babylonian tradition that the sceptre of the king is a branch of the World Tree (Doht 1974, p. 145).
61. Kawase 1975.
62. However, the written characters for the sceptre shaku and the unit of length are different, only their pronunciation is the same. The length of the unit varied in the past but was standardised in 1891. The royal shaku is about 1 shaku long but this could be coincidence. By the way, the shakuhachi, the traditional flute of sacred Japanese music, is 1.8 shaku in length (C. Worrall. Also cf. Chapter 38, note 1).
63. The Japanese sources consulted by Worrall refer to the 'Takeuchi Document' (*Takeuchi Bunsho*).
64. Simple wooden bows were used by the predecessors of the present Japanese and Ainu people. Since the early centuries CE, the construction of the Japanese bow has synthesised these wooden bows with the two traditions of composite Chinese bows. Japanese longbows are about 210–70cm long and exquisitely elegant (Hardy 1992, p. 183).
65. Littleton 2002, p. 46. Cf. the raven as a general symbol, in Chapter 33, note 17, the raven of Bran, in Chapter 45, note 8, and the raven of Odin.
66. Naumann 1996, pp. 82–6.
67. In Hartzell 1991, p. 130, after *Columbia University Contributions to Anthropology*, 26: 125.

Chapter 40. The Dance of the Amazons

1. According to Ukrainian archaeologists, 25 per cent of all Scythian graves containing weapons are of females (Wilde 1999, p. 57).
2. Homer, Hesiod, Plutarch, Strabo, Apollodorus, Pausanias and others.
3. A good example is the sybil Herophile, a priestess of Apollo (who 'seems to have been born before the Trojan War'). The Delphians showed Pausanias a rock where she used to stand and sing her oracles. In a trance she was possessed by the god and in her *Hymn to Apollo* 'she calls herself Artemis as well as Herophile'. Elsewhere in her oracles she describes herself as the daughter of a shepherd and one of the immortal nymphs of Ida: 'I was born between man and goddess, Slaughterer of sea-monsters and immortal nymph, Mountain-begotten by a mother from Ida, And my country is sacred to my mother, Red-earthed Marpessos, the river Aideneus' (Pausanias, X, 12, 1–4, in Levi (tr.) 1979).
4. Excavation by Renate Rolle; the grave is in Georgia, and dates from the end of the second millennium BCE; in Wilde 1999, p. 48. Amazons were said to have settled at the foot of the Caucasus (Graves 1955, 131.k, after Strabo and Servius).
5. Lazistan.
6. Graves 1955, 131.e.
7. Mainly the journey of the *Argo*, see previous chapter.
8. Graves 1955, 131.3.

9. *Karadeniz Eregli*, p. 10.
10. Persson 1950, p. 143.
11. Quote from Callimachus, *Hymn to Artemis*, 237, in Cook 1925, p. 405.
12. Persson 1950, p. 143.
13. Baumann 1982, p. 37.
14. *Ibid.*, p. 143; Graves 1955, 131.d.
15. Willetts 1962, p. 186.
16. This, however, is the *Ephesian* view; the cult of Apollo (e.g. in Lycia) is older than Apollo's appearance at Ephesus.
17. Callimachus (c. 305 BCE–c. 240 BCE) is the only classical source naming a tree, but his *phagos* (see box on p. 148) has already been disputed by Graves (1955, 131.5) on the grounds that Callimachus was an Egyptian who did not know enough about 'northern' vegetation. Anyhow, Callimachus lived centuries after the tree shrine had been converted into a stone temple. Graves himself, by the way, proposed date palm because to him all ancient goddesses were derivations of Egyptian Isis (in Graves's time, it was still early days in Anatolian research).
18. Virgil, *Georg*. 2, 112–3, in Fallon (tr.) 2004: 'Bacchus adores wide-open uplands, and yew a chilly northern aspect'.
19. Cook 1925, p. 962.
20. Willetts 1962, p. 185, after Lethaby 1917. Compare the congregation of the Dactyls (see p. 184) on nearby Mt Ida in 1460 BCE.
21. In both the older and the younger temple, an isolated column stood immediately behind the cult-image, probably as an 'architectural substitute for the sacred tree' (Cook 1925, p. 405).
22. *Ibid.*, p. 609.
23. See Hageneder 2001, pp. 29–32.
24. Cook 1925, pp. 515–16, 609.

Chapter 41. The World Tree

1. These sources are (a) the Icelandic sagas, mainly from the twelfth century; (b) short poems by Icelandic poets (Skalds), a number of which go back to the ninth century; (c) poems on mythological themes from the thirteenth-century Codex Regius, better known as the *Elder* or *Poetic Edda*; and (d) a treatise known as the *Prose Edda*, compiled by the Icelandic politician, historian and poet, Snorri Sturluson, in c. 1222–3 (Davidson 1993, pp. 7–8).
2. The apocalyptic nature of *Voluspa*, however, suggests a Christian influence. See Dronke 1997, pp. 93–101; Collins 1983, 1987; Momigliano 1988.
3. Dronke may not be the best-known scholar on the subject but I largely confine my references in this chapter to her because her *Poetic Edda* (1997) was recommended to me in 2005 by the Árni Magnússon Institute in Iceland as the best English translation available.
4. In the old Icelandic literature, *Heimdallr* occurs both with single *l* and with double *l*.
5. Dronke 1997, p. 107. 'The god Heimdallr as a hypostasis of the world axis', however, has first been fully discussed in the chapter of this name by C. Tolley in his DPhil thesis at Oxford 1993 (pp. 326–61).
6. Dronke 1997, p. 109. Also in *Gylfaginning*, in Diederichs 1984, p. 144.
7. Dronke (1997, p. 110) explains *ívidia* as 'giantess dwelling, living, in a wood, tree, tree root'.
8. The single letter I or Y for yew is quite common in medieval north-western Europe, for example as the name for the island of Iona, which in medieval Scottish was simply called I or Hi. The Nordic *i-* (*iv-*) for 'yew' might have had its influence on the development of French *if*, at least in Normandy (my guess). Scheeder (1994, p. 7) as well as Dumitru (1992, p. 93) mention **iwa-widja*, 'yew-tree', as a possible root for *ividi*.
9. Dronke 1997, p. 32. Compare Germanic *Irmin-pillars* which represented the World Tree (see Hageneder 2001, pp. 122–4) with the Greek *Hermae*, wooden posts dedicated to Hermes (*Hermes* and *Irmin* also derived from the same root word). Like Heimdallr, Hermes is the son of divine sisters: his mother is Maia, one of the *Pleiades* or dove-spirits (cf. 'Dodona', p. 148).
10. *Ibid.*, pp. 31–2. Dronke speaks of 'existence' here because *miot* is usually found in a very practical context of husbandry such as caring for the shelter and calculating fuel for the winter period.
11. Compare the Japanese shaku as a unit of length (p. 211) and the sacred flute, shakuhachi, and Chapter 39, note 62.
12. From Greek *a-mbrosia*, 'non-mortal food', and *nék-tar*, the 'death-vanquishing', or *né-kta*, the 'not-dead', all alluding to food of the gods (Cook 1940, p. 497).
13. For Celtic parallels, see Hageneder 2001, pp. 138–9.
14. While later Indian sources say that soma grows *as a plant* on the mythical world mountain (Oberlies 1998, p. 245) – the shift from tree to a substitute plant probably occurred as a result of human migration (Doht 1974, p. 271).
15. See Butterworth 1970, ch. 5.
16. See Hageneder 2001, p. 176; Dronke 1997, p. 127.
17. The other occurrence of *askr* is in association with Heimdallr's horn (cornucopia theme) and Mímir, the spirit of the well (stanza 45).

A Greek parallel for the sacredness of dew is the annual rite of the 'Dew-bearers' (Arrhephóroi). Around midsummer, possibly in the very night of the last full moon of the Attic calendar, a small procession took a sacred artefact down an underground descent and brought up another item, both of them being wrapped up and secret. At Athens, the path proceeded from the sanctuary of Aphrodite of the Gardens at the bottom of the northern side of the Acropolis rock down into an adjacent cave. The two white-clad girls (between 7 and 11 years of age) who carried the artefact given to them by the priestess of Athena (here clearly in her old function as mountain-mother) had been prepared for a year. At Eleusis and Mytilene, the ritual of the Dew-bearers was connected to Demeter and Kore, and refered to as 'the most holy mysteries'. Cook (1940, pp. 165–81) who analysed the entire context of this cult – which includes sacred serpents, Aphrodite as mountain-mother, the dove as the embodiment of the goddess, the fertility of the earth, and the birth of a divine son named Erichthónios, 'very child of the ground' – concludes that the Arrhephóroi symbolically brought the sacred seed of the sky-father [dew or pollen?] down into the womb of the earth.

18. In the Vedic hymns, soma is often called *madhu*, Sanskrit for 'mead' (Ransome 1937, p. 137).
19. Arils described as honey were first pointed out by Meredith (n.d.).
20. Pure yew aril spirit (45 per cent) is still produced by the German distillery Edelbrennerei Dirker.

21. *Cannabis,* native to central Asia, has been grown in China since *c.* 1500 BCE and had reached the Scythians in *c.* 700 BCE and central European cultures some 300 years later (Lewington 2003, p. 76); in Greece, Aphrodite was associated with the poppy (see Chapter 37, note 75) and the black mandrake (Harris, n.d.); and fly agaric is connected to soma (Eliade 1978, p. 438) and also has been a traditional 'sacred plant' in many shamanic traditions across Eurasia.
22. In Berger and Holbein 2003; in Nepal, yew is one of the *dhupi,* plants burnt in shamanic ritual. In the Pacific northwest of America, yew leaves are smoked among various native tribes (see p. 137).
23. *'dvi er oldr bazt, at aptr uf heimtir hverr sitt ged gumi',* *Havamál* 14.
24. Doht 1974, p. 244; until modern times, the Swedish great winter market has been called *disting*. Two sources, however, mention the late autumn/beginning of winter for a *disablet* (*Víga-Glúms Saga* 6 and *Egils Saga* 44, in Davidson 1993, p. 113; Doht 1974, p. 27), but this could have referred to additional ones held for special reasons.
25. Doht 1974, pp. 29, 245.
26. *Ibid.*, pp. 30, 154, referring to Adam of Bremen's late eleventh-century account.
27. Dumitru 1992, p. 93.
28. Davidson 1993, p. 107.
29. *Ibid.*, p. 113. The connection of Dísir with the rites of kingship also explains why the Old Saxon *ides* occur in the Second Merseburg Charm as female supernatural beings (Grimm 1883, in Davidson 1998, p. 23) while the composer of *Beowulf* 'consistently associates the word with queens and the power to rule' (Davidson 1998).
30. The emphasis is on *if*. I deduced *ídia* from *ividia* and *dia* (*i* = yew, *vid* = tree, *dia* = spirit, goddess).
31. *Völ.* 7, 2; 57, 2. 'Yew valley' in Scheeder (1994, p. 7). Dronke (1997, p. 23), however, translates *Idavelli* with 'Eddying plain'. Dumitru (1992, p. 93) lists further (controversial) yew translations from Norse myth: Asgard is protected by the river Ifing, 'yew river'; the names of the deities Yngwinn and Ingun might have developed from **Igwana-z*, 'yew-god', and **Ig-wano*, 'yew-goddess'.
32. Etymologists may forgive me for relating a Germanic to an Anatolian word; I believe that cultural context can override the rules of language evolution. When peoples are in contact with each other they import ideas and the related vocabulary, particularly names. Cannot names like Jesus Christ or Coca Cola be found in almost every language on the planet today?
33. Cook 1925, p. 950.
34. Most notably, the 'Neolithic revolution' (see p. 178 and Chapter 36, note 11) that spread from the Balkans westbound over Europe between *c.* 7000 and 3000 BCE.
35. And not only in Eurasia. An ethnological comparison regarding the nine worlds and nine mothers: To the Kogi Indians in the mountains of Colombia, South America, the notion of nine worlds is central to their cosmology: 'Each of the nine worlds has its mother, its sun and its moon, and there are living beings in all worlds.' (Julien 2005, p. 13).
36. Willetts 1962, p. 99 n. 211.
37. Graves 1955, 38.c.
38. Callimachus, *Hymn to Artemis*, in Graves 1955, 22.d, which is likely to have corresponded with the age choice for the girl's initiations at her temple at Brauron in Attica.
39. Graves 1955, 6.a.
40. *Vol.*, 53. Cf. *Wafthrudnismál* 43: 'I have come through nine worlds down to Niflheim [the realm of the dead].'
41. Graves 1955, 14.5, 149.e. A similar festival was held at Myrine (*ibid.*, 149.3).
42. Willetts 1962, p. 99.
43. An Anglo-Saxon harvest prayer merging Christian elements and older ritual, in Davidson 1998, p. 61.
44. Weinreb 1999, p. 67.
45. Webster 1998.
46. Personal communications with a shaman from Ulan-Ude, in Hageneder 2003.
47. *Havamál*, in Hageneder 2001, p. 121.
48. Doht 1974, p. 233.
49. *'Gunnlod mér gaf gullnom stóli á drycc ins dyra miadar'*, *Havamál* 105, in Doht 1974, pp. 40–1. The other version of the tale is in Snorri's *Skáldskaparmál* (both in Doht 1974, pp. 36–44).
50. Two familiar Skaldic kennings for gold are 'the bed ...' and 'the fire of the serpent' (Davidson 1993, p. 60).
51. The Langobards are said to have worshipped Odin in serpent form (Doht 1974, p. 51).
52. Bates 1996.
53. *Yggr*, the 'fearsome', or 'awe-inspiring one', is one of Odin's other names. In Hindu tradition, the World Tree is called *Asvattha*, 'horse abode' (Dronke 1997, p. 126). Buryat shamans from Siberia still speak of mystical horse journeys during their professional trance (see Chapter 35, note 20).
54. Coles 1998, p. 169. Compare **Ig-wano*, 'yew-goddess', note 31 above.
55. Calibrated; *ibid.*
56. The so-called 'image 3'. Its legs seem to fit best into the holes nearest the prow of the Roos Carr 'boat', which possibly indicates that this character was a kind of leader or guide to the group.
57. So does the pine figure from Dagenham, eastern England (*ibid*).
58. 'it is perhaps more likely that he [Ullr] was Odin in another guise, or rather that Odin was a later manifestation of the ancient god Ull[r]' (*ibid*, p. 168). During the first millennium CE, however, Odin's popularity was on the wane, giving way to Thor (and his oak). The rise of Thor, as Coles suggests, 'may signify a move away from shamanism, possibly but not necessarily influenced by the spread of Christianity' (p. 170).

The focal point of the inauguration site at Navan Fort (ancient Emain Macha), Co. Armagh, Ireland, comprises six concentric rings of wooden posts (nearly 40m in diameter), centred around a single pillar which might be interpreted as an image of the world pillar. The oak trunk that was set up in this position in 94 BCE (dendrochronological evidence) (Cunliffe 1997, p. 206) perhaps signals a shift towards the Irish thunder-god, similar to the rise of Thor in Germania.
59. Hageneder 2001, p. 124.
60. Wirth 1979, p. 160.
61. *Ibid.*, pp. 160, 462.
62. Cleasby and Vigfusson 1975, p. 326.
63. Chetan and Brueton 1994, p. 131; Dumitru 1992, p. 92.
64. Inscribed *edoeboda*, possibly meaning 'return, messenger' (Chetan and Brueton 1994).
65. Le Roux and Guyonvarch 1996, p. 184.
66. The same was said of Mt Dikte in Crete (scholiast on

Callimachus' *Hymn to Zeus*, 11–12, in Butterworth 1970, p. 29).

67. The disorientating effects experienced in the direct vicinity of the world axis (i.e. a sacred 'centre') also shine through in the Gilgamesh epic; in the sacred grove of the goddess, Enkidu says to Gilgamesh: 'This is a place where mysterious things happen, where we lose our hold and everything begins to slide.' (*Gilgamesh* V, in Schrott 2001, p. 214).

68. Compare the revolving 'spiral castle' of the galaxy in Welsh Bardic tradition.

69. Among other preparations, Kirke summons Boreas, the north wind (which blows from the direction of the Pole Star, see p. 182). She robes Odysseus and herself and veils her head (*Od.* 10.545); she becomes like Calypso, 'the veiled one'. Clearly, a ritual was about to begin, but Homer had no interest in preserving it and skipped the sequence (Butterworth 1970, p. 30).

70. Butterworth 1970, pp. 8, 28–30. At his return, Kirke greets him with the words 'you who die twice, when other men die once' (*Od.* 12.22). Butterworth (p. 180) points out the initiatory character of this event by adding, 'That a deep experience lies behind the story of Odysseus' and his companions' sojourn on the island of Kirke may be suspected from these words alone. Those who die twice are also twice born.' Cf. Dionysos, p. 199.

71. See also Butterworth 1970, pp. 180–1.

72. Contrary to a contemporarily widespread cliché (particularly in feminist theology) – one that sees the 'gentle' tides of the moon as an emblem of the feminine principle and the 'harsh, linear' rays of the sun as 'typically' male – many ancient goddess religions linked the permanent sun with the eternal mother archetype and the waxing and waning moon with the death and rebirth cycles of the mythical son, the vegetation-god. The 'male' moon, governing the water element, was seen as sending fertilising rain.

It was the (patriarchal) Indo-Europeans – ironically the very 'arch-enemies' for many followers of the writings of Marija Gimbutas or Robert Graves – who reversed the ancient symbology and associated the female with the moon and the night, and their own radiant heroes, kings and conquerors with the glorious sun.

73. Compare Egyptian Horus. Mithraic worshippers spoke of Helios, Kirke's father, as a hawk (Cook 1914, p. 240).

74. *Ibid.*, p. 241; see also Cook 1925, p. 252. Also cf. Gawain, the 'hawk of May', in Arthurian legend.

75. Even underground (e.g. Odin's visit at Gunnlod's underworld lair), gold suggests a bright, pleasant, generous atmosphere.

76. Campbell 1991, p. 122; Dronke 1997, p. 107.

77. 15–16, in Butterworth 1970, p. 126.

78. Naumann 1996, p. 85. The weaving hall myth: *ibid.*, pp. 74–7.

79. These options result from the fact that ancient Japanese religion was not a unified body (Naumann 1996, p. 14).

80. *Ibid.*, p. 77, quoting Vonessen (1992, p. 40f).

81. *Helgakvida Hundingsbana*, a poem in the *Poetic Edda*, quoted in Davidson 1998, pp. 119–20.

82. Dronke 1997, p. 141. The Japanese Queen of Heaven has a necklace consisting of the Pleiades (Naumann 1996, p. 84). Aphrodite has a girdle powerful enough to divert a thunderbolt thrown by Zeus (Graves 1955, 18.f).

83. *Skaldskaparmál* 1, 18.

84. Davidson 1998, pp. 104, 167.

85. Dronke 1997, p. 43. Seidr was a 'professional exercise ... designed to learn the unknown ... by communication with spirits and by the exploration of their world' (*ibid.*, p. 133).

86. Davidson 1998, p. 141. Maybe Demeter's sacrificial drink at Eleusis was fermented after all, see Chapter 35, note 20.

87. Davidson 1998, p. 176.

88. In Viking times, 'the unfailing killing power of Odin' as battle god was counterbalanced by the 'unfailing regenerative power of Freyja' (Dronke 1997, p. 44).

89. E.g. Freyja's husband Odr seems to be a doublet of Odin (Davidson 1998, p. 169). Dronke (1997, pp. 123–4) differentiates *ond* as the breathing spirit of animation that dies with an individual, and *odr* as the spirit or soul which is 'continually renewed in another life (if the proper sacrifices are performed)'. Freyja being *married* to Odr, then, speaks for itself.

90. Davidson 1998, pp. 8–10, 182.

91. Dronke 1997, p. 44.

92. Davidson 1993, p. 107.

93. *Ibid.*, p. 73. Dronke (1997, pp. 43–4) points out that the Asir needed the truce with the Vanir because they needed Freyja's powers of bringing the living into the world (see note 87 above). With her, 'death became no more than the necessary condition for the renewal of life, a sacrifice to ensure the future. So Freyja became, in social terms, the gods' ... "priestess of sacrifice" (as Snorri says).' Hence Idun seems to represent this particular aspect of Freyja. (Idun appears in Snorri's *Skáldskaparmál* and in an earlier skaldic poem, the ninth- or early tenth-century *Haustlong*.)

94. Diederichs 1984, p. 179.

95. *Ibid.*, p. 243.

96. Dronke 1997, p. 43.

97. Hageneder 2000, p. 114; cf. Hesiod's 'brazen race' (warfaring Bronze Age peoples) who had fallen from ash trees (*Works and Days*, 109–201, in Graves 1955, 5.d).

98. Dronke 1997, p. 12.

99. 'a golden image of the "bride of the Vanir", *Vanabrudr* Freyja, whose divinity characteristically manifests itself in a multiplicity of distinct figures with distinct names' (*ibid.*, p. 41).

100. *Há* is 'high, tall', *hárr* is 'hoary' in a dignified way, i.e. wise with age (*ibid.*, p. 130).

101. *Ibid.*, pp. 131–2.

102. The famous 'chorus' line from *Voluspa*.

Chapter 42. Harmonies

1. Gowen 2004; www.mglarc.com.

2. At least in Europe: there is a suggestion of an early Chinese composite instrument like pan pipes with a gourd going back to about 1500 BCE, but that is an illustration rather than the instrument (Gowen 2004).

3. O'Dwyer 2004.

4. *Odyssey*, quoted in Campbell 1964, p. 169.

5. Butterworth 1970, p. 8.

6. Plato, *Republic*, 617b, in Waterfield (tr.) 1993.

7. Hesiod was the first to mention individual names for the Muses: Clio, Euterpe, Thalia, Melpomene, Terpsichore, Erato, Polymnia (Polyhymnia), Urania and Calliope, who was their chief. Many of these names are connected with music (obviously, what the *Muses* inspire is *music*):

Melpomene is the 'Songstress', Polymnia 'She of the many hymns', Calliope 'She of the beautiful voice'; and, since dancing was a regular accompaniment of song, Terpsichore the 'Whirler of the dance'. The Muses had a festival every four years at Thespiae, near Mt Helicon, which included a contest *(Museia)*, presumably in singing and playing music. In all likelihood they were the original patron goddesses of poets and musicians and over the centuries their sphere was extended to include all liberal arts and sciences (*Encyclopedia Britannica* 2004).

8. *Fragm*. 795, Nauck, in Ransome 1937, p. 107.
9. Berendt 1983, p. 23.
10. In Hageneder 2001, p. 71; 2004, p. 119.
11. 2.2.3–4, in Butterworth 1970, pp. 133–4.
12. O'Dwyer 2004.
13. Butterworth 1970, p. 43.
14. Dronke 1997, pp. 48–9, 136, 145.
15. Artemis *Hymnia* (Pausanias, *Guide to Greece*, VIII, 5.11, in Levi (tr.) 1979).
16. Pausanias IV, 27, 6–7, in Levi (tr.) 1979.
17. Quoted in Cook 1914, p. 258. Apollo, it seems, played a pivotal role in the music of the spheres. In Aristophanes' *The Birds* (lines 219–24, in Rogers 1979), Apollo listens to the echoes of the song of the nightingale (that reaches the divine realm through the yew thicket, see Chapter 38, note 5) and draws answering strains from his lyre 'Till he stirs the dance of the heavenly choir.'
18. Doczi 1981.
19. Campbell 1964, p. 185.
20. Cf. Hageneder 2001, pp. 29–31, and figs 5, 6c, 8, 10a, 23, 25, 26, 34.
21. Local leaflet. The abbey was founded in 1098.
22. The symbolic beasts of Dionysos were the bull, the lion and the serpent (Graves 1955, 129.1).
23. Campbell 1964, pp. 265–6.
24. According to Campbell (1964, p. 259), the foundations of astrology were laid between c. 4300 and 2150 BCE.
25. *Ibid*., p. 309.
26. Luke 22: 39, John 18: 1f; see also Hageneder 2001, pp. 155–6.
27. The comprehensive knowledge about celestial movements that the ancients possessed makes it hard to believe that they should not have known that the earth was in orbit around the sun, too. The answer might be that they had *no interest* in the purely physical structure of the solar system as 'seen from outside'. We are not outside.
28. Cook 1914, p. 271.

Chapter 43. Migrations

1. The Inanna cycle 'could have been created any time between 1900 BCE and 3500, or even further back' (Kramer and Wolkstein 1983, p. 136).
2. Quoted in *ibid*., pp. 4–5.
3. In Hebrew legend, Lilith was the first bride of Adam. But she insisted on her own equality to the male gender, and 'refused to copulate with him, for she did not want to be underneath him' (*ibid*., p. 142). She preferred to leave the Garden of Eden and live somewhere else. This, obviously, was too much feminism for the Hebrew patriarchs of the first millennium BCE: Lilith was demonised and replaced with the more 'obedient' Eve (obedient, that is, until the serpent came along, again).
4. 'From the roots of the tree Inanna fashioned a *pukku* for ... [Gilgamesh]. From the crown of the tree Inanna fashioned a *mikku* for Gilgamesh, the hero of Uruk.' (quoted in *ibid*., p. 9). These items could not be identified yet. Butterworth (1970, p. 143) suggests a 'shaman' drum and drumstick because (in *Gilgamesh, Enkidu and the Underworld*) in some way unknown these objects of magical power fall into the underworld and Enkidu sets out to retrieve them.
5. Historically verified examples of tree transport include the Egyptian Queen Hatshepsut's (reigned c. 1472–1458 BCE) expedition for the myrrh trees of Punt, the arboretums of Persian and Assyrian kings, and the cutting of Buddha's 'Tree of Enlightenment' (pipal, *Ficus religiosa*) sent in the third century BCE as a gift to King Tissa of Sri Lanka by King Ashoka (Hageneder 2005, p. 101).
6. *Germania*, 40, relating the procession of the earth-goddess Nerthus who was worshipped among the north German Reudignians, Aviones, Anglians, Varians, Eudosians, Suardones and Nuitones.
7. Cook 1925, p. 403. Roman denarii in the British Museum.
8. KBo 2689 II 30, in Puhvel 1984.
9. For example, wild einkorn (*Triticum monococcum*) and emmer wheat (*T. dicoccoides*), two ancestors of domesticated wheat, can still be found growing on the rocky limestone slopes of the Taurus Mountains. Other plants of which pollen, spore or seed material have been identified in cave deposits associated with prehistoric human settlements are chamomile, rice, barley, lentil, pea, potato, sweet potato, common bean, maize, pepper and rock polypody (*Polypodium*). Other food plants that have an original habitat in hard limestone bedrock, slopes, talus slopes, mountain ravines and fissures in different parts of the world are oats, asparagus, broad bean, chickpea, cabbage, cucumber, courgette, squash, pumpkin, onion and carrot to name but a few (Larson *et al*. 2004, pp. 21–3, 36).
10. The secrets of sexual propagation have been well-known to humans since the Neolithic when interbreeding of cultivars and clonal propagation led to the species of grains, olives and dates we know today.
11. A date that coincides with archaeological evidence of a sudden change from shaft-graves to *thólos*-tombs. Homer called this people *Danaoí*, the Egyptian sources *Daanaou*; they were generally seen as masters of irrigation. Their name has been traced to Sanskrit *danu*, 'fluidity, dampness, drops', and can be found in many rivers, for example the four major streams that run into the Black Sea: the Dunarea (Danube), Dnister, Dnipro and Don rivers (Cook 1940, pp. 362–70).
12. II, 36.8–37.2, 38.4.
13. Already suggested by W. Helbig (1887, p. 440) in Cook 1940, p. 364.
14. *Ibid*., p. 180; the Hellenised names for the couple of the sacred marriage on Mt Ida are Aphrodite and Anchises.
15. Cook (*ibid*., p. 367) mentions four publications between 1884 and 1911 that discuss the possible link between the Irish and Argolian Danaeans.
16. *Dind*. III, 'Mag Mugna'.
17. Chetan and Brueton 1994, p. 231.
18. *Ibid*.
19. Pithos fragments from Datcha, Caria, in the Berlin Museum (one similar piece was found in Rhodes); described by F. Dümmler in 1896, in Cook 1925, pp. 614–17.

20. Greek *Kentavros*, from *kentima*, 'embroidery, welt', and *avra*, 'aura' (the energy field of a living being), points to the healing of the holistic human being, harmonising it with itself and possibly also with the universe (compare weaving and the cosmic loom, Chapters 37 and 41). The image of a human torso on a horse's body may symbolise the rule of human consciousness or will-power over the vital and instinctive (animalistic) forces. The god of healing, Asklepios, was said to have received his art from the Centaur Chiron.
21. The axe-bearer is thought to be either the Hittite sky-god or his Hellenised successor, Zeus *Labráyndos*. But as the images lack any sense of a ceremonial atmosphere they might actually refer to historical events rather than a mythological one.
22. In Cook 1925, pp. 680–1.

Chapter 44. Ten Hundred Angels

1. Chetan and Brueton 1994, pp. 53, 218; also Hageneder 2001, p. 140.
2. Chetan and Brueton 1994, pp. 217–18.
3. Paul Greenwood, referring to the *Inverness Transactions*.
4. Hutchison 1890.
5. I measured the two largest trees within the cathedral grounds at 408cm (at the western entrance to the building) and 362cm girth (mid-northern wall) in October 2004. There are at least two other yews of this calibre hidden in the undergrowth just outside the fence at the western end.
6. See note 9 below.
7. Based on the words of a monk, Hugh of Kirkstall, who recorded the full foundation story between 1225 and 1247, at the request of John, Abbot of Fountains Abbey (Chetan and Brueton 1994, p. 91).
8. Quoted in Bevan-Jones 2002, p. 62. For a fuller account of this site see *ibid.*, pp. 62–75.
9. Hugh de Payns, one of the founders of the Knights Templar, was a knight of Count Hugh of Champagne (Aube *département*, north-eastern France); Count Hugh was a friend of St Bernard, the founder and abbot of the Cistercian monastery called Clara Vallis, or Clairvaux, and it was he who had donated the land to the Order. The Rule of the Order of the Temple was written by Bernard, and confirmed in 1129 at the Council of Troyes, Champagne (Count Hugh's residence). Count Hugh had already joined the Knights Templar in 1125 (Nicholson 2001, pp. 22, 28).

Both Orders, the Cistercian and the Templar, followed (a strict version of) the Rule of St Benedict, and the traditional regulations and daily services of the monastic day. Both Orders shared the general principles of poverty and simplicity, and chose to build rather austere buildings which reflected this.
10. Brande 2004.
11. 'Taxus', in the University of Georgia Libraries (see Chapter 38, note 11).
12. Nine brothers: according to Archbishop William of Tyre who wrote between 1165 and 1184 (Brande 2004, p. 23).
13. Lowe (1897, p. 86), after Revd J. Kershaw, 1895.
14. In Portugal, the Order of Christ, and in Valencia, Aragon, Spain, the Order of Montesa (Nicholson 2001, p. 231).
15. As the former Templars were not moved into the Hospital it was not a formal merger but more that the Hospitallers did not drop the 'Temple' name from the Templars' properties, and lands that carried the Templars' privileges (not to pay certain taxes and dues) were still said to have 'Temple right' (personal communication from Helen Nicholson).
16. It was 'the German masons who in the 1760s introduced the idea that the Templars must have had secret wisdom and magical powers, which they had learned while they held the so-called Temple of Solomon in Jerusalem. The wisdom and power, they claimed, had been handed down a secret line of succession to the present-day masons!' (Nicholson 2001, p. 240).
17. The Friends of Rosslyn have published various books on the subject; see e.g. the bibliography in www.wikipedia.org Knights Templar Rosslyn.
18. Adkins and Adkins 1996, p. 153.

Chapter 45. 'Behold the healthful day'

1. Cimok 2000, p. 144.
2. The Alexandrian school, led by the patriarch Cyril, emphasised Christ's divine nature, the school of Antioch (in Syria) primarily saw him as the ideal human and stressed his humanity (*ibid.*, p. 195).
3. Mainly through Phoenician and Greek colonies.
4. Begg 1996, pp. 16–102. There are about 450–500 Black Madonnas in Europe, in France alone about 180.
5. Nicholson 2001, pp. 142–4.
6. Abella 2001.
7. See *Taxus* distribution map on the website of the Forestry Commission of France: http://junon.u-3mrs.fr/msc41www/pltcli/PC9049.html.
8. The town name is reminiscent of the Celtic deity, Bran, best known from the Welsh *Mabinogion*. His animal, the raven, occurs as a cast-metal statue in the centre of a fountain at the town entrance. (The Virgin at Braine, however, is not a Black Madonna.)
9. The dolmen has been removed since, but in general Le Puy is an extraordinary example of smooth and peaceful transitions from old religions to succeeding ones (Derderian 1992). Le Puy cathedral was built in the eleventh and twelfth centuries. Among the stone carvings above the healing spring at the back of the cathedral is a serpent. Mt Anis's other name is Rocher Corneille, the Rock of the Crow, a name reminiscent of the dark aspect of the (Celtic) goddess (compare Braine, note 8, and the Black Annis tales from Leicestershire, Begg 1996, pp. 56, 86); the coat of arms of Notre Dame de France includes not the dove but the crow.
10. Derderian 1992, pp. 103–5.
11. *Mala*, 'mother' but also 'black', Greek *melas*, 'black', *lucine*, 'light' (Latin *lux*) – as suggested by Derderian 1992, p. 27. On other Gaulish traditions, Melusine or Lucine was the female consort of the Celtic light-god Lugh (*ibid.*).
12. *Ibid.*, p. 24.
13. Begg 1996, p. 213.
14. The statue was inaugurated on 12 September 1860 in the presence of 120,000 pilgrims (l'Office de Tourisme de l'Agglomération du Puy-en-Velay, 03/04).
15. The building structure and masonry are typical Nabataean (Santarelli 1997, pp. 8–11). The fact that the Casa only has three walls because it was built against a grotto in a rock face suggests the possibility, though, that rather than a private Jewish dwelling (Mary's parents) it could have been a small shrine or temple only later converted into a Christian site.

16. A document dated September 1294 testifies that Niceforo Angelo, ruler of Epirus, when endowing his daughter Ithamar for her marriage to Filippo di Taranto (Philip II of Anjou), the King of Naples, included 'the holy stones carried away from the House of Our Lady, the Virgin Mother of God' (*ibid.*, p. 13).
17. Santarelli, no year, local information leaflet published by the Universal Congregation of the Holy House, Loreto.
18. As is written on the altar of Loreto: *Hic Verbum caro factum est.*
19. The pontiff quotes from the Second Vatican Council, *Lumen gentium*, 58 (in Santarelli 1997, p. 42).
20. 'that derives its inspiration and direction from the mystery that was accomplished in the Holy House'. From his Message to the Nuns of Cloistered Orders, delivered at the Holy House on 10 September 1995 (*ibid.*).
21. Referring to the Halaf period, Chapter 36.
22. Conversation of the author with ecclesiastical staff at Loreto. The following excerpts are from the full version of the Litany, published in Giorgio Basadonna: *Commento alle invocazioni delle Litanie Lauretane*.
23. See Chapter 30; also Alessio 1957.
24. See figures in Cook 1925, p. 406.
25. The robe (or rather its original design) is said to have come with the Casa from Dalmatia in 1294.
26. The original black wooden statue accidentally burned in 1921 and was replaced by one made from the wood of a cedar grown in the Vatican (Begg 1996, p. 242).
27. As a dress of seven vertical layers, the robe is also reminiscent of the sevenfold robe of the Sumerian goddess Inanna and the architecture of the Mesopotamian temples, the ziggurats.
28. Cunliffe 2002, pp. 4–5.
29. Personal communication with, and various (French) articles by Christian Vaquier, the current forestry warden of Ste Baume, in 2005.
30. Related by Cuncliffe 2002, pp. 6ff, after Strabo. No temple has been found yet.
31. Her body was officially discovered 9 September 1279 at Saint-Maximin-la-Sainte-Baume, Provence, where it is still housed inside a thirteenth-century basilica. But views on the subject have been controversial since early times: Gregory of Tours (*De miraculis*, I, xxx), for example, supports the tradition that she retired to Ephesus (like Mary the mother of Jesus) – with no mention of any connection to Gaul.
32. Cassianites 415–1079, Benedictines 1079–1295, Dominicans since 1295.
33. In the *Guide to the Sainte-Baume* of the local museum (Écomusée de la Sainte-Baume), Petit is described as a 'disciple of light' in the Association of Journeymen, Catéchisme de lumière.
34. Levi quotation from Gospel of Mary, 18: 10–15; 'what is hidden' 10: 8; see also 9: 21–4 and 17: 16–22 (in Robinson 1990, pp. 524–7).
35. For the symbolic meaning of these trees see Hageneder 2005, 'Olive' and 'Myrrh'.
36. Full account in Palmer *et al.* 1995, pp. 3–53.
37. In the Warring States period (403–221 BCE) and more so in the Han Dynasty (206 BCE–220 CE), various texts mentioning a Queen Mother of the West arose from the general shamanistic background of China. She is likely to have represented a number of different local deities which include 'a teacher, a directional deity, spirits of the holy mountains, a divine weaver, a shaman and a star goddess' (Cahill 1993, p. 13, quoted in Palmer *et al.* 1995, p. 14).
38. Quotation from Palmer *et al.* 1995, p. 14.
39. She arrived in Japan (in both male and female form) in the seventh to ninth centuries with Buddhist pilgrims returning from China (*ibid.*).
40. As protector(ess) of all life, Kuan Yin is sometimes depicted clutching and firing a crossbow or bow and arrow and carrying a fierce-looking shield (*ibid.*, pp. 42–3, 46 [fig]). Unique to Japan, where she is known as Kannon, there is a tradition that she has thirty-three manifestations (eleven for each of three distinct worlds: the world of heaven, the world of the sky, and the world of the earth). Three of them are non-human: a snake, a winged bird-like creature and a dragon. Ryushin Kannon, 'snake Kannon', links the spiritual and physical worlds and can travel across unimaginable distances and times (*ibid.*, p. 45).

APPENDICES

Appendix III. Important Occurrences of European Yew

1. Paule *et al.* 1993, after Majer 1971. Paule *et al.* continue: 'In most cases the plant associations correspond to the *Taxo-Fagetum* with certain subunits e.g., *Taxo-Fagetum bakonyicum, Taxo-Fagetum carpaticum* etc. [1, 2, 3, 4, 7, 9, 10] or *Tilieto-Taxetum* [5], *Fagetum orientalis – submontanum taxetosum* [6], *Euonymo-Taxetum* [8] and *Cephalantero-Taxetum balticum* (or *Fagetum boreo-atlanticum* according to Myczkowski (1961)) [11].'
2. Personal communication with Prof. B. Schirone.
3. Personal communication with J. Hassler.
4. Korori *et al.* 2001.
5. Eastern European data are from Paule *et al.* 1993. Other data collected from the respective site managements, unless otherwise stated.
6. Personal communication with Bosco Imbert, University of Navarra, and Ignacio Abella.
7. Tenorio *et al.* 2005, pp. 202–6.
8. Personal communication with Monsieur Christian Vacquié, forest warden of Ste Baume; also *Der Eibenfreund*, 1: 39.
9. Personal communication with Prof. B. Schirone.
10. Other important yew stands are located in Graubünden near Sagogn (Palius da Tuora; Cauma Su; Uaul da Salums); in St Gallen near Pfäfers (Gigerwald), Mosnang (Bruederwald) and Kirchberg (Iddaberg Burgwald); in Thurgau near Hüttlingen (Griesenberger Tobel); in Aargau near Baden (Unterwilerberg; Brenntrain); in Solothurn near Oftringen (Engelberg); in Neuenburg near Neuchâtel (Gorges du Seyon); in the Wallis near Saint-Maurice (Bois noir) and near Naters (Blindtal); and in Neuenburg near Boudry (Areuse gorge) (personal communication with Jürg Hassler, after A. Rudow, ETHZ/BAFU, 2009, 2013–24, and Kurt Pfieffer). More about *Taxus* in Switzerland in Hassler (1999). 'Die Eibe (*Taxus baccata* L.)', Haldenstein (Switzerland), self-published by the author. To order please write to Jürg Hassler-Schwarz, Sum Curtgins 9, CH-7013 Domat Ems, Switzerland.

11. Personal communication with Dr Berthold Heinze, Federal Research Centre for Forestry, Vienna.
12. Boratynski *et al.* 2001.
13. Kassioumis *et al.* 2004. Voliotis in 1986 also listed the following mountains: Voras; Tzena; Paikon; Kerkini, Orvilos; Falakron; Pangaeon; Athos; Vermion; Vourinos; Tymfi; Lakmos (Peristeri); Athamanika Ori; Koziakas; Agrafa; Pieria; Ossa; Pilion; Tymfristos; Oxya, Oeta; Giona; Parnassos, Kyllini; Oligyrtos, Chelmos; Maenalon; Parnon; Dirphys; Xerovouni; Skotini; Ochi Euboeae; Hypsarion in Thasos; Fengari in Samothraki; also Kryoneri; Olympias; north-eastern Chaldiki; Perivoli, Grevena; Aghia Paraskevi; Trikala; Imathia; Parnassos.
14. Pridnya 2002.
15. Davis 1978, Aksoy 1998.
16. Pridnya 2000a, 2002.
17. Pridnya 2000b.
18. Pridnya 2000a, 2002.
19. *Der Eibenfreund*, 1: 44.
20. Lickl and Heinze 2001.
21. Sagheb-Talebi and Lessani 2001. Conifers other than *Taxus* are rare, only cypress and juniper in some drier and higher places (Lickl and Heinze 2001).
22. After an evaluation from 1971; *Der Eibenfreund,* 1: 44.
23. Sagheb-Talebi and Lessani 2001.
24. Shanjani 2001.

Appendix V. A Note on Frazer's *Golden Bough*

1. A full account in Vickery 1973.
2. Hutton 1991, p. 326.
3. Ackerman 2002, p. 46.
4. Frazer was by no means alone with his preoccupation with mistletoe and oak. Since the druid revivals in the sixteenth and seventeenth centuries had picked up on Pliny's story the mistletoe 'myth' had gradually become a part of the public domain. In 1834, for example, the young William Crawford Williamson, in his first edition of the report on the Bronze-Age grave of the Gristhorpe Man (now in the Rotunda Museum, Scarborough) had no doubts that the 'berries' from inside the oak coffin were mistletoe. In 1865, the Gristhorpe mistletoe berries made it into J.B. Davis's and J. Thurnam's *Crania Britannica* (later to become the *Ecyclopedia B.*), but in 1872 were calmly withdrawn from the third edition of the report by Williamson himself, now Professor of Natural History at Owen's College, Manchester. Rightly so, as an examination at the University of Bradford, Department of Archaeological Sciences, in November 2006 revealed that the 'berries' are not even of vegetable origin.
5. Ackerman 1987, p. 108; Frazer quotation also in the abridged version, p. 703.
6. Frazer 1993 (1922), p. 704.
7. *Ibid.*, p. 163.
8. Or the one from which wreaths are made: Dionysos could still be regarded as a yew tree spirit while his followers adorn themselves with ivy and vine leaves (which are bigger and prettier, after all).
9. Another link of Nemi with the ancient Eastern religion is preserved by Ovid (*Metamorphoses*, XV, 506ff, in Melville (tr.) 1986), and by Pausanias (2, 27, 4; 2, 32, 8, in Levi (tr.) 1979). The mythical prince Hippolytus 'dies' in a horse-chariot accident caused by an *elaos* tree (see p. 202), and then is revived by Diana/Artemis and becomes the 'king' or priest of Diana's sacred grove, his new name being Virbius, from *vir bis*, 'twice a man', 'twice born' (Graves 1955, 101.1), a name reminiscent of initiation rites like those of Zeus at Mt Ida and of Dionysos. A trace of this tradition appears in the Roman calendar, where St Hippolytus is said to have been dragged to death by his horses on 13 August – Diana's own day (Campbell 1964, p. 155).
10. Out of all the Scottish clans the Fraser clan is the only one that has yew as their clan badge. (For a complete list of Scottish clans and their trees see Hageneder 2001, pp. 144–5.)
11. See Ackerman 2002.

Fortingall Yew, nineteenth-century illustration.

BIBLIOGRAPHIES

BOTANY

Ahrens, T.G. (1933). 'Schutz der britischen Eiben', *Naturschutz*, 14: 249

Akkemik, Ünal, Aytug, Burhan and Güzel, Sercay (2004). 'Archaeobotanical and dendroarchaeological studies in Ilgarini Cave (Pinarbasi, Kastamonu, Turkey)', *Turkish Journal of Agriculture and Forestry*, 28: 9–17

Aksoy, Necmi (1998). 'Monumental trees of Turkey', 16: *Koca Ardunç; The Karaca Arboretum Magazine*, IV, 4 November

—— (2000). 'Porsuk Agaci (*Taxus baccata* L.)', *Lamin' ART*, Agustos-Eylül, Sayi, 9

Alessio, G. (1957). 'Stratificazione dei nomi del tasso (*Taxus baccata* L.) in Europa', *Studi Etruschii*, 25: 219–64

Allona, I., Collada, C., Casado, R. and Aragoncillo, C. (1994). 'Electrophoretic analysis of seed storage proteins from gymnosperms', *Electrophoresis*, 15: 1062–7

Ballero, M., Loi, M.C., van Rozendaal, E.L.M., van Beek, T.A., Cees van der Haar, Poli, F. and Appendino, G. (2003). 'Analysis of pharmaceutically relevant taxoids in wild yew trees from Sardinia', *Fitoterapia*, 74: 34–9; also available at www.elsevier.com/locate/fitote

Barnea, A., Harborne, J.B. and Pannell, C. (1993). 'What parts of fleshy fruits contain secondary compounds toxic to birds and why?', *Biochemical Systematics and Ecology*, 21: 421–9

Bartkowiak, S. (1978). 'Seed dispersal by birds', in Bartkowiak *et al.* (1978), 139–46

——, Bugala, W., Czartoryski, A., Hejnowicz, A., Król, S., Rodo, A. and Szaniawski, R.K. (eds) (1978). *The Yew* –Taxus baccata L. Warsaw, Foreign Scientific Publications, Department of the National Center for Scientific and Technical, and Economic Information (for the Department of Agriculture and the National Science Foundation, Washington, DC) [Polish edn: *Cis pospolity* – Taxus baccata *L. Nasze Drzewa Lesne*, 3]

Benfield, Barbara (2006). 'A study of the lichens on some yews in eastern Devon', unpublished manuscript, available at www.ancient-yew.org/articles.shtml

Bolsinger, Charles and Jaramillo, Annabelle E. (1990). *T'axus brevifolia* Nutt. – Pacific Yew', *Silvics of Forest Trees of North America* (rev. edn). Portland, Pacific Northwest Research Station

Boratynski, A., Didukh, Y. and Lucak, M. (2001). 'The yew (*Taxus baccata* L.) population in Knyazhdvir Nature Reserve in the Carpathians (Ukraine)', *Dendrobiology* 2001, 46: 3–8

Bowman, J.E. (1837). 'On the longevity of the yew', *The Magazine of Natural History and Journal of Zoology, Botany, Mineralogy, Geology and Meteorology*, Ser. 2, 1: 28–35, 85–90

Brande, A. (2001). 'Die Eibe in Berlin einst und jetzt', *Der Eibenfreund*, 8: 24–43

—— (2003). 'Postglaziale *Taxus*-Nachweise und Waldtypen in den nördlichen Kalkalpen (Niederösterreich)', *Der Eibenfreund*, 10: 52–62

Brandis, D. (1874). *Illustrations of the Forest Flora of North-west and Central India*, London, W.H. Allen

Browicz, Kazimierz and Zielinski, Jerzy (1982). *Chorology of Trees and Shrubs in South-west Asia and Adjacent Regions*, vol. 1. Warsaw, Polish Scientific Publishers

Brzeziecki, B. & Kienast, F. (1994). 'Classifying the life-history strategies of trees on the basis of the Grimian model', *Forest Ecology and Management*, 69: 167–87

Callow, R.K., Gulland, J.M. and Virden, C.J. (1931). 'Physiologically active constituents of the yew, *Taxus baccata*. I. Taxine', *Journal of the Chemical Society* (1931) 2138–49

Cao, C. (2002). 'Untersuchungen zur genetischen Variation und zum Genfluß bei der Eibe (Taxus baccata L.)', masters degree, Georg-August-University Göttingen

Carruthers, T. (1998). *Kerry – A Natural History*, Cork, Collins Press

Christison, R. (1897). 'The exact measurement of trees. (Part 3) The Fortingall Yew', *Transactions of the Botanical Society of Edinburgh*, 13: 410–35

Conwentz, H. (1892). 'Die Eibe in Westpreussen – ein aussterbender Waldbaum', *Abhandl. z. Landeskunde der Provinz Westpreussen*, Danzig

—— (1898). 'Die Eiben in der vorgeschichtlichen Zeit', *Korrespondenzblatt für Anthropologie*, Kiel

—— (1921). 'Über zwei subfossile Eibenhorste bei Christiansholm, Kreis Rendsburg', *Berichte der deutschen Botanischen Gesellschaft*, 39: 384–90

Cooper, M.R. and Johnson, A.W. (1984). *Poisonous Plants in Britain and their Effects on Animals and Man*, Ministry of Agriculture, Fisheries and Food, Reference Book 161, London, HMSO

Cortés, Simón, Vasco, Fernando and Blanco, Emilio (2000). *El libro del Tejo (*Taxus baccata L.*) – Un proyecto para su conservación*, Madrid, Edita Arba

Coutin, Remi (2003). 'Faune entomologique de l'if, *Taxus baccata*', *Insectes*, 128 (1): 19–22

Creutz, G. (1952). 'Misteldrossel und Seidenschwanz', *Ornithologische Mitteilungen*, 4: 67

Daniewski, W.M., Gumulka, M., Anczewski, W., Masnyk, M., Bloszyk, E. and Gupta, K.K. (1998). 'Why the yew tree *(Taxus baccata)* is not attacked by insects', *Phytochemistry* 49: 1279–82

Dark, S.O.S. (1932). 'Chromosomes of *Taxus, Sequoia, Cryptomeria* and *Thuya*', *Annals of Botany*, 46: 965–77

Davis, P.H. (1965/1978). *Flora of Turkey and the East Aegean Island*, vols 1 and 6, Edinburgh, Edinburgh University Press

Dempsey, D. and Hook, I. (2000). 'Yew *(Taxus)* species – chemical and morphological variations', *Pharmaceutical Biology*, 38: 274–80

Detz, H. and Kemperman, J. (1968). 'Zaaikalender van coniferen en loufhoutgewassen', *Proefstation Boksoop Jaarboek*, 163–74

Di Sapio, O.A., Gattuso, S.J. and Gattuso, M.A. (1997). 'Morphoanatomical characters of *Taxus baccata* bark and leaves', *Fitoterapia*, 68: 252–60

DiFazio, S.P., Vance, N.C. and Wilson, M.V. (1997). 'Strobilus production and growth of Pacific yew under a range of overstorey conditions in western Oregon', *Canadian Journal of Forest Research*, 27: 986–93

Dumitru, A. (1992). 'Die Eibe (*Taxus baccata* L.) – Eine botanisch-ökologische sowie medizinische und kulturhistorische Betrachtung', diploma in Forest Science, University of Munich

Duncan, R.W., Bown, T.A., Marshall, V.G. and Mitchell, A.K. (1997). 'Yew Big Bud Mite', *Forest Pest Leaflet*, 79, Victoria BC, Canadian Forest Service, Pacific Forestry Centre

Eddelbüttel, H. (1935/1937). 'Zur Altersbestimmung von Eiben', *Mitteilungen der Deutschen Dendrologischen Gesellschaft*, 47: 147–54; 49: 47–51

Elsohly, H.N., Croom, E.M. Jr, Kopycki, W.J., Joshi, A.S., Elsohly, M.A. and McChesney, J.D. (1997). 'Taxane contents of *Taxus* cultivars grown in American nurseries', *Journal of Environmental Horticulture*, 15: 200–5

Elwes, H.J. and Henry, A. (1906). *The Trees of Great Britain and Ireland*, Edinburgh, privately printed

Emberger, Louis (1968). *Les Plantes Fossiles*, Paris, Masson & Cie

Engler, A. (ed.) (1960). *Die natürlichen Pflanzenfamilien, vol. 13. Gymnospermae*, 199–211

Erdtman, H. and Tsuno, K. (1969). '*Taxus* heartwood constituents', *Phytochemistry*, vol. 8, 931–2

Evelyn, John (1664). *Sylva – A Discourse of Forest Trees*, London, Martyn & Allestry

Ezard, John (1995). 'Relics of ancient forest found', *Guardian*, 7 February

Ferguson, D.K. (1978). 'Some current research on fossil and recent taxads', *Review of Palaeobotany and Palynology*, 26: 213–26, Amsterdam, Elsevier

Florin, R. (1931/1944). *Untersuchungen zur Stammesgeschichte der Coniferales*, Part 1

Frank, Norbert (2003). 'Eiben (*Taxus baccata* L.) im Bakony-Wald – einst und jetzt', *Der Eibenfreund*, 10: 20–5

Franklin, Jerry F., Cromack, K. Jr, Denison, W., McKee, A., Maser, C., Sedell, J., Swanson, F. and Juday, G. (1981). 'Ecological characteristics of old-growth Douglas-fir forests', General Technical Report PNW-118, Portland OR, USDA (Forest Service)

Fritts, H.C. (1971). 'Dendroclimatology and dendroecology', *Quaternary Research*, 1: 419–49

Fuller, R.J. (1982). *Bird Habitats in Britain*, Calton, Staffs, T. and A.D. Poyser

García, D., Zamora, R., Hódar, J.A., Gómez, J.M. & Castro, J. (2000). 'Yew (*Taxus baccata* L.) regeneration is facilitated by fleshy-fruited shrubs in Mediterranean environments', *Biological Conservation*, 95: 31–8

Graeter, Carlheinz (1994). 'Eibe und Knabenkraut – Baum des Jahres, Wildpflanze des Jahres', *Main-Post*, 30/31 July

Green, Ted (2003). 'The Ancient Oaks of the British Isles – The Remnants of Europe's Rainforests', Alan Mitchell Lecture 2003, London, Conservation Foundation

Gregorius, H.-R., Degen, B. (1994). 'Estimation of the extent of natural selection in seedlings from different *Fagus sylvatica* L. populations: application of new measures', *Journal of Heredity*, 85: 183–90

Groves, A.T. and Rackham, O. (2001). *The Nature of Mediterranean Europe – An Ecological History*, New Haven/London, Yale University Press

Guchelaar, H.J., ten-Napel, C.H., de Vries, E.G. and Mulder, N.M. (1994). 'Clinical, toxological and pharmaceutical aspects of the antineoplastic drug taxol: a review', *Clinical Oncology -R- Coll Radiology*, 6: 40–8

Gulland, J.M. and Virden, C.J. (1931). 'Physiologically active constituents of the yew, *Taxus baccata*. Part II. Ephedrine', *Journal of the Chemical Society*, 2148–51

Hassler, Jürg (1999). *Die Eibe* (Taxus baccata L.), Haldenstein, Switzerland, the author

—— (2003). 'Die Bedeutung der Tiere bei der Verbreitung der Eibensamen', *Der Eibenfreund*, 10: 118–20

Hassler, J., Schoch, W. and Engesser, R. (2004). 'Auffällige Stammkrebse an Eiben (*Taxus baccata* L.) im Fürstenwald bei Chur (Graubünden, Schweiz)' *Schweizer Z. Forstwesen*, 155/9, 400–3

Harris, T.M. (1961). *The Yorkshire Jurrasic Flora 1: Thallophyta–Pteridophyta*, London, The British Museum

Heath, G.W. (1961). 'An investigation into leaf deformation in *Medicago sativa* caused by the gall midge *Jaapiella medicaginis* Rübsaamen (Cecidomyiidae)', *Marcellia*, 30: 185–98

Hegi, G. (1981). 'Familie Taxaceae. 1. *Taxus*', *Illustrierte Flora von Mitteleuropa*, vol. I, part 2: 126–34

Heinze, B. (2004). 'Zur Populationsbiologie der gemeinen Eibe *(Taxus baccata)*', *Austrian Journal of Forest Science*, 121: 47–59

Heit, C.E. (1969). 'Propagation from seedpart. Part 18. Testing and growing of popular *Taxus* forms', *American Nurseryman*, 129 (2): 10–11, 118–28

Hejnowicz, A. (1978). 'The yew – Anatomy, embryology and karyology', in Bartkowiac *et al.* (1978), 33–54

Hermann, W. (2000). 'Die Stammpflanze der Säuleneibe *Taxus baccata* "fastigiata" (=*T.* "hibernica")', *Der Eibenfreund*, 7: 82

Herrera, C.M. (1987). 'Vertebrate-dispersed plants of the Iberian peninsula: a study of fruit characteristics', *Ecological Monographs*, 57: 305–31

Hertel, H., Kohlstock, N. (1996). 'Genetische Variation und geographische Struktur von Eibenvorkommen (*Taxus baccata* L.) in Mecklenburg-Vorpommern', *Silvae Genetica*, 45: 290–4

Hindson, Toby (2000). 'The growth rate of yew trees: An empirically generated growth curve', Alan Mitchell Lecture 2000, London, Conservation Foundation

Hofmann, M. (1989). 'Das Naturwaldreservat Huckstein – Baumwachstum und Flora als Ausdruck geomorphologischer Standortprägung', diploma, Georg-August-University Göttingen

Howard, P.J.A., Howard, D.M. and Lowe, L.E. (1998). 'Effects of tree species and soil physico-chemical conditions on the nature of soil organic matter', *Soil Biology and Biochemistry*, 30: 285–97

Huf, Karl (2002). 'Specht hämmert im Kronthal einzigartige Lochmuster in Eiben', *Der Eibenfreund*, 9: 174–5

Hulme, P.E. (1996). 'Natural regeneration of yew (*Taxus baccata* L.) Microsite, seed or herbivore limitation', *Journal of Ecology*, 84: 853–61

—— (1997). 'Post-dispersal seed predation and the establishment of vertebrate dispersed plants in Mediterranean scrublands', *Oecologia*, 111: 91–8

—— and Borelli, T. (1999). 'Variability in post-dispersal seed predation in deciduous woodland: relative importance of location, seed species, burial and density', *Plant Ecology*, 145: 149–56

Huntley, B. and Birks, H.J.B. (1983). *An Atlas of the Pollen Flora 13000-0BP*, Cambridge, Cambridge University Press

Jaloviar, P. (1998). 'Struktur und Naturverjüngung der Eibe in verschiedenen Waldbestandstypen der Slowakei', *Der Eibenfreund*, 5: 45–56

Kanngiesser, F., (1906). 'Über Lebensdauer und Dickenwachstum der Waldbäume, vol. 3, *Taxus baccata*', *Allgemeine Forst- und Jagd-Zeitung*, 36: 253–5

Kartusch, B. and Richter, H. (1984). 'Anatomische Reaktionen von Eibennadeln auf eine Erschwerung des Wassertransports in Pflanzenkörper' ('Anatomical reactions of yew needles to an impairment of water transport in the plant body'), *Phyton*, 24: 295–303

Kassioumis, K., Papageorgiou, K., Glezakos, T. and Vogiatzakis, I.N. (2004). 'Distribution and stand structure of *Taxus baccata* populations in Greece: Results of the first national inventory', *Ecologia Mediterranea*, 30, 2: 27–38

Kawase, M. (1975). 'Japanese yew (*Taxus cuspidata*)', *Proceedings of the International Taxus Symposium*, Horticulture Series 421, A1–A5

Kaya, Zafer (1998). 'Anit Agacin Hatira Defteri', *Kasnak Mesesi ve Türkiye Florasi Sempozyumu*, University of Istanbul (Orman Botanigi Anabilim Dali)

Kayacik, Hayrettin and Aytug, Burhan (1968). 'Gordion Kral Mezari'nin Agac Malzemesi Üzerinde Ormancilik Yönünden Arastirmalar' ('A Study of the Wooden Materials of the Gordion Royal Tomb with Special Reference to Forestry), A, 18, University of Istanbul (Orman Fakültesi Dergisi)

Keen, R.A. (1958). 'A study of the genus *Taxus*', *Dissertation Abstracts*, 18: 1196–7

Kelly, D.L. (1981). 'The native forest vegetation of Killarney, south-west Ireland – an ecological account', *Journal of Ecology*, 69: 437–72

Kirchner, O., Loew, E. and Schröter, C. (1908). *Lebensgeschichte der Blütenpflanzen Mitteleuropas*, vol. 1. 60–78. Stuttgart, Ulmer

Koch, K. (1879). *Die Bäume und Sträucher des alten Griechenlands*, 41: *Eibe*, Stuttgart, Enke

Korori, S.A.A., Matinizadeh, M. and Teimouri, M. (2001). 'Untersuchungen über Eibe (*Taxus baccata* L.) Mycorrhizen im Norden des Iran', *Der Eibenfreund*, 8: 165–7

Korpel, S. (1996). 'Das geschützte Eibenvorkommen "Pavelcovo", seine Zustandsanalyse, die naturschützerische und forstliche Bedeutung', *Der Eibenfreund*, 3: 21–32

Korpel, S. and Paule, L. (1976). 'Die Eibenvorkommen in der Umgebung von Harmanec, Slowakei', *Archiv für Naturschutz und Landschaftsforschung*, 16, 123–39

Krenzelok, E.P., Jacobsen, T.D. & Aronis, J. (1998). 'Is the yew really poisonous to you?', *Journal of Toxicology, Clinical Toxicology*, 36: 219–23

Krüssmann, G. (1985). *Manual of Cultivated Conifers*, Portland, Timber Press

Kucera, L.J. (1998). 'Das Holz der Eibe', *Schweizerische Zeitschrift für Forstwesen* 149 (5)/*Der Eibenfreund*, 4: 328–39

Kukowka, A. (1970). 'Über die Gefährlichkeit der Eibe (*T. b.*)', *Landarzt*, 46 (7)

Lange, O.L. (1961). 'Die Hitzeresistenz einheim. immer- und wintergrüner Pflanzen im Jahreslauf', *Planta*, 56 (6)

Lange, S., Rajewski, M., Leinemann, L. and Hattemer, H. (2001). 'Fremdpaarung im Wald – Das Liebesleben der Eibe', *Forschung, Magazin der Deutschen Forschungsgem.*, 4 (2001): 10-3; also in *Der Eibenfreund*, 9: 113–16

Larcher, W. (2001). *Ökophysiologie der Pflanzen. Leben, Leistung und Stressbewältigung der Pflanzen in ihrer Umwelt*, Stuttgart, Ulmer Verlag

Larson, Doug (1999). 'Ancient Stunted Trees on Cliffs', *Nature*, 398, 1 April

—— (2000). *Cliff Ecology: Pattern and Process in Cliff Ecosystems*, Cambridge, Cambridge University Press

——, Matthes, U., Kelly, P.E., Lundholm, J. and Gerrath, J. (2004). *The Urban Cliff Revolution – New Findings on the Origins and Evolution of Human Habitats*, Ontario, Fitzhenry & Whiteside

de Laubenfels, D.J. (1988). 'Coniferales', *Flora Malesiana*, 10 (3): 337–453, Leiden, Nationaal Herbarium Nederland

Leonhardt, U., Paul, M. and Wolf, H. (1998). 'Eibenwald bei Schlottwitz', *Der Eibenfreund*, 5: 65–71

Leuthold, C. (1980). 'Die ökologische und pflanzensoziologische Stellung der Eibe (*Taxus baccata*) in der Schweiz', *Veröffentlichungen des Geobotanischen Instituts der ETH, Stiftung Rübel*, Zürich, 67: 1–217

—— (1998). 'Die pflanzensoziologische und ökologische Stellung der Eibe (*Taxus baccata* L.) in der Schweiz – Ein Beitrag zur Wesenscharakterisierung des "Ur-Baumes" Europas', *Der Eibenfreund*, 4: 349–71

Lewandowski, A., Burczyk, J. and Mejmartowicz, L. (1995). 'Genetic structure of English yew (*Taxus baccata* L.) in the Wierzchlas Reserve: implications for genetic conservation', *Forest Ecology and Management*, 73: 221–7

Lewington, A. and Parker, E. (2000). *Alte Bäume. Naturdenkmäler aus aller Welt*, Augsburg, Weltbild Verlag

Lickl, E. and Heinze, B. (2001). 'Eiben im Elburs – Ein kleines Vorkommen im Wald von Kheyrudkenar', *Der Eibenfreund*, 8: 90-91

Löblein, I. (1995). *Einfluss von innerstädtischen Bodenverhältnissen auf das Durchwurzelungsverhalten von Eiben*, Prüfungsarb. Staatsexamen Univers, Münster

Lowe, John (1897). *The Yew-trees of Great Britain and Ireland*, London, Macmillan

Ludwig, A. and Bauer, M. (2000). 'Die Eibennachzucht im Bayerischen Staatswald', *Der Eibenfreund*, 7: 63–6

Majer, A. (1971). *A Bakony tiszafása [Yew forest of Bakony]*, Budapest, Akadémia Kiadó

Manandhar, Narayan P. (2002). *Plants and People of Nepal*, Portland, Timber Press

Mayer, Hannes & Aksoy, Hüseyin (1986). *Wälder der Türkei*, Stuttgart/New York, Gustav Fischer

Melzack, R.N. and Watts, D. (1982). 'Variations in seed weight, germination, and seedling vigour in the yew (*Taxus baccata* L.) in England', *Journal of Biogeography* 9: 55–63

Mitchell, A.F. (1972). *Conifers in the British Isles*, London, HMSO/Forestry Commission

Mitchell, A.K. (1998). 'Acclimation of Pacific yew (*T. brevifolia*) foliage to sun and shade', *Tree Physiology*, 18: 749–57

Mitchell, F.J.G. (1990). 'The history and vegetation dynamics of a yew wood (*Taxus baccata* L.) in S.W. Ireland', *New Phytologist*, 115: 573–7

Mehlman, P.T. (1988). 'Food resources of the wild Barbary macaque *Macaca sylvanus* in high-altitude fir forest Ghomaran Rif Morocco', *Journal of Zoology* 214: 469–90

Moir, A.K. (1999). 'The dendrochronological potential of modern yew (*Taxus baccata*) with special reference to yew from Hampton Court Palace, UK', *New Phytologist*, 144: 479–88

Moore, D.M. (1982). *Flora Europaea Check-List and Chromosome Index*, Cambridge, Cambridge University Press

Muhle, O. (1978). 'Rückgang von Eiben-Waldgesellschaften und Möglichkeiten ihrer Erhaltung', *Bericht des Symposiums des Internationalen Vereins für Vegetationskunde in Rinteln,* 483–501

Myczkowski, S. (1961). 'Zespoly lesne rezerwatu cisowego Wierzchlas' [Forest associations of the yew reserve at Wierzchlas], *Ochrona przyrody,* 27: 91–108

Mysterud, A. and Ostbye, E. (1995). 'Roe deer *Capreolus capreolus* feeding on yew *Taxus baccata* in relation to bilberry *Vaccinium myrtillus* density and snow depth', *Wildlife Biology,* 1: 249–53

Namvar, S. and Spethmann, W. (1986). 'Die Eibe', *Allgemeine Forstzeitung (AFZ),* 1986 (23): 568–71

Newbould, P.J. (1960). *The Age and Structure of the Yew Wood at Kingley Vale,* Wye, report, Wye, NCC

Núñez-Regueira, L., Rodríguez Añón, J.A. & Proupín Castiñeiras, J. (1997). 'Calorific values and flammability of forest species in Galicia. Continental high mountainous and humid Atlantic zones', *Bioresource Technology,* 61: 111–9

Parker, J. (1971). 'Unusual tonoplast in conifer leaves', *Nature,* 234: 231

Paule, L., Gömöry, D. and Longauer, R. (1993). 'Present distribution and ecological conditions of the English yew (*T. b.* L.) in Europe', unpublished report for the International Yew Resources Conference, Berkeley, CA, March 12–13, 1993. [The article was published almost identically in German as Paule, L., Radu, S., Stojko, S.M. (1996). 'Eibenvorkommen des Karpatenbogens', *Der Eibenfreund,* 3: 12-20]

Pietzarka, Ulrich (2005). 'Zur ökologischen Strategie der Eibe (*Taxus baccata* L.) – Wachstums- und Verjüngungsdynamik', doctorate at the Faculty for Forest, Geo and Hydro Sciences, Technical University Dresden

Pilcher, J.R., Baillie, M.G.L., Brown, D.M., McCormac, F.G., MacSweeney, P.B. and McLawrence, A.S. (1995). 'Dendrochronology of subfossil pine in the north of Ireland', *Journal of Ecology,* 83 (4): 665–71

Pilger (1916). 'Die Taxales', *Mitteilungen der Deutschen Dendrologischen Gesellschaft,* 25: 1–30

Pilkington, N., Proctor, J. and Reid, K.I. (1994). 'The Inchlonaig yews, their tree epiphytes, and their tree partners', *Glasgow Naturalist,* 22: 365–73

Pisek, A., Larcher, W. and Unterholzner, R. (1967). 'Kardinale Temperaturbereiche der Photosynthese und Grenztemperaturen des Lebens der Blätter verschiedener Spermatophyten. I. Temperaturminimum der Nettoassimilation, Gefrier- und Frostschadensbereiche der Blätter', *Flora,* 157: 239–64

—— *et al.* (1968). 'Kardinale Temperaturbereiche' ... part 2. Temperaturmaximum der Netto-Photosynthese und Hitzeresistenz der Blätter', *Flora,* 158: 110–28

—— *et al.* (1969). 'Kardinale Temperaturbereiche ... part 3. Temperaturabhängigkeit und optimaler Temperaturbereich der Netto-Photosynthese', *Flora,* 158: 608–30

Pridnya, Mikhail (1998). 'Pflanzensoziologische Stellung und Struktur des Khosta-Eiben-Vorkommens im Kaukasus-Biosphärenreservat', *Schweizerische Zeitschrift für Forstwesen,* 149: 5; also in *Der Eibenfreund,* 4: 387–96

—— (2000a). 'Pflanzensoziologische Stellung und Struktur des Chosta-Eibenvorkommens im West-Kaukasus Biosphärenreservat', *Der Eibenfreund,* 7: 22–7

—— (2000b). 'Eibenvorkommen im Kaukasus', *Der Eibenfreund,* 7: 28–9

—— (2001). 'Ursachen des Rückganges der Eibenvorkommen im West-Kaukasus und Massnahmen zu ihrer Erhaltung. (Forschungskonzeption)', *Der Eibenfreund,* 8: 148–52

—— (2002). '*Taxus baccata* in the Caucasus region', *Der Eibenfreund,* 9: 146–66

Prioton, J. (1976/77). 'Nouvelle contribution à l'étude de l'if (*Taxus baccata* L.) en France et dans quelques pays limitrophes. Necessité de sa protection', Castelnau-le-Lez

—— (1979). 'Étude biologique et écologique de l'if (*Taxus baccata* L.) en Europe et Occidentale', *La Forêt Privée,* 128: 19–34; 129: 19–37

Quantz, B. (1937): 'Eibenschutz in Hannover und Thüringen vor 70–75 Jahren', *Naturschutz,* 18 (4): 76–9

Rajda, Vladimír (1992). 'Electro-Diagnostics of the health of oak trees', *Ustav systematicke a ekologicke biologie CSAV,* Brno, Czech Republic

—— (1995). 'Die Elektrodiagnostik bei Bäumen als ein neues Verfahren zur Ermittlung ihrer Vitalität', *Austrian Journal of Forest Science,* 114: 348–61

—— (2004). 'Metabolische Energie und Elektrodiagnostik der Pflanzenvitalität', Talk at the 10th International Conference Elektrochemischer Qualitätstest BTQ

—— (2005). 'Die Eiben – Nadelbäume mit hoher metabolischer Energie und Vitalität' [Yew trees – Conifers with high metabolic energy and vitality]. Unpublished paper

Rajewski, M. & Lange, S. (1997). *Genetische Strukturen in verschiedenen ontogenetischen Stadien der Eibe* (Taxus baccata *L.*), diploma, Georg-August-Univ. Göttingen

Rajewski, M., Lange, S., Hattemer, H.H. (2000). 'Reproduktion bei der Generhaltung seltener Baumarten – Das Beispiel der Eibe (*Taxus baccata* L.)', *Forest Snow and Landscape Research,* 75: 251–66

Redfern, Margaret (1975). 'The life history and morphology of the early stages of the yew gall midge *Taxomyia taxi* (Inchbald) (Diptera: Cecidomyiidae)', *Journal of Natural History,* 9: 513–33

—— and Askew, R.R. (1998). *Plant Galls. Naturalist's Handbooks 17,* Slough, The Richmond Publishing Co.

—— and Hunter, Mark D. (2005). 'Time tells: long-term patterns in the population dynamics of the yew gall midge, *Taxomyia taxi* (Cecidomyiidae), over 35 years', *Ecological Entomology,* 30: 86–95

Rohde, M. (1987). 'Untersuchungen über die Pollenverteilung in einem Eibenbestand.' diploma, Georg-August-University Göttingen

Roloff, A. (1989). *Kronenentwicklung und Vitalitätsbeurteilung ausgewählter Baumarten der gemäßigten Breiten,* Frankfurt, J.D. Sauerländer

——. (1998). 'Biologie und Ökologie der Eibe (*Taxus baccata* L.)', in *Tagungsband 'Internationale Eibentagung' 1998,* Tharandt, TU Dresden; also in *Der Eibenfreund,* 5: 3–16

—— and Pietzarka, U. (2001). 'Die Gemeine Esche (*Fraxinus excelsior* L.) – Baum des Jahres 2001', *Mitteilungen der Deutschen Dendrologischen Gesellschaft,* 86: 73–84

——, —— and Schmidt, C. (2001). '*Juniperus communis* Linné', in Schütt, P., Weisgerber, H., Schuck, J., Lang, U., Roloff, A. and Stimm, B (eds), *Enzyklopädie der Holzgewächse,* 26: 1–11. Landsberg, Ecomed Verlag

Rössner, H. (1996). 'Paterzeller Eibenwald: Erinnerungen, Beobachtungen, Vermutungen', in Kölbel, M. & Schmidt, O. (eds) 'Beiträge zur Eibe', *Berichte aus der Bayrischen Landesanstalt für Wald und Forstwirtschaft*, 10: 48–55

—— (2001). 'Bemerkungen zur Diplomarbeit von Patrick Insinna (1999) "Analyse von Altbestand und Naturverjüngung der Eibe im Naturschutzgebiet von Paterzell"', *Der Eibenfreund*, 8: 157–63

Sagheb-Talebi, K. and Lessani, M.-R. (2001). 'Das Eibenvorkommen im Iran', *Der Eibenfreund,* 8: 85-89

Sainz, M.J., Iglesias, I., Vilariño, A., Pintos, C. and Mansilla, J.P. (2000). 'Improved production of nursery stock of *Taxus baccata* L. through management of the arbuscular mycorrhizal symbiosis', *Acta Horticulturae*, 536: 379–84

Salisbury, E.J. (1927). 'On the causes and ecological significance of stomatal frequency, with special reference to the woodland flora', *Philosophical Transactions of the Royal Society of London*, B, 216, 1–65

Saniga, M. (1996). 'Zustand, Struktur und Regenerationsprozesse im Eibenreservat "Harmanecka tisina"', *Der Eibenfreund*, 3: 33–7

Sax, K. and Sax, H.J. (1933). 'Chromosome number and morphology in the conifers', *Journal of the Arnold Arboretum*, 14: 356—74 (and two end plates)

Schaede, R. and Meyer, F.H. (1962). *Die pflanzlichen Symbiosen,* 3rd edn, Stuttgart, Fischer

Scheeder, Thomas (1994). *Die Eibe (Taxus baccata L.) – Hoffnung für ein fast verschwundenes Waldvolk*, Eching, IHW-Verlag

—— (1996). 'Ursachen des Rückganges der Eibenvorkommen und die Möglichkeit des Schutzes durch forstlich integrierten Anbau', in Kölbel, M. and Schmidt, O. (eds), 'Beiträge zur Eibe', *Berichte aus der Bayrischen Landesanstalt für Wald und Forstwirtschaft* 10: 9–16

Scher, S. and Schwarzschild, B.S. (1998). 'The role of non-governmental organizations in protecting the threatened Pacific Yew – a case history', *Der Eibenfreund*, 5: 57–62

—— (1998). 'Do browsing ungulates diminish avian foraging? – Studies of woodpeckers in forest understorey communities of central Europe and western North America show cause for concern', *Der Eibenfreund*, 4: 411–19

—— (2000). 'Weltweite Eibenvorkommen (*Taxus*) neu betrachtet', *Der Eibenfreund*, 6: 109–18

—— (2005a). 'YewCon2005 Meeting Report', *Der Eibenfreund*, 12: 117–23

—— (2005b). 'Genetic diversity of yew trees in China: Questions raised ...', *Der Eibenfreund*, 12: 124–7

—— (2005c). 'Gene flow in yew (*Taxus*) *(hongdoushan)*. A geographic information system (GIS) approach to identify populations at risk and estimate gene transport from introduced to native *Taxus* populations in China', *Der Eibenfreund*, 12: 128–30

Schirone, B., Bellarosa, R. and Piovesan, G. (eds) (2003). *Il tasso – Un albero da conoscere e conservare*, Penne (PE), Cogecstre Edizioni

Schönichen, W. (1933). *Dt. Waldbäume und Waldtypen*, Jena, Gustav Fischer

Shanjani, P. (2001). 'Quantitative und qualitative Untersuchung von Eiben-Peroxidasen in den Wäldern Arasbaran und Gorgan, Iran', *Der Eibenfreund,* 8: 164

Sharp, A.J., Crum, H. and Eckel P. (eds) (1994). *The Moss Flora of Mexico*, vol. 2, New York, The New York Botanical Garden Press

Shemluck, M.J., Estrada, E., Nicholson, R. and Brobst, S.W. (2003). 'A preliminary study of the taxane chemistry and natural history of the Mexican yew, *Taxus globosa* Schltdl.', *Boletín de la Sociedad Botánica de México*, 72: 119–27

Simms, Eric (1971). *Woodland Birds*, New Naturalist Series 52, London, Collins

Sitte, P., Weiler, E.W., Kadereit, J.W., Bresinsky, A. and Körner, C. (2002). *Lehrbuch der Botanik für Hochschulen*, Begr. von E. Strassburger. Heidelberg/Berlin, Spektrum Akad. Verlag

Siwecki, R. (1978). 'Diseases and parasitic insects of the yew', in Bartkowiak *et al.* (1978), 103–9

—— (2002). 'Krankheiten und parasitäre Insekten bei der Eibe', *Der Eibenfreund*, 9: 120–6

Skorupski, M. & Luxton, M. (1998). 'Mesostigmatid mites (Acari: Parasitiformes) associated with yew *(Taxus baccata)* in England and Wales', *Journal of Natural History*, 32, 419–39

Smal, C.M. & Fairley, J.S. (1980a). 'Food of wood mice and bank voles in oak and yew woods in Killarney, Ireland', *Journal of Zoology*, 191: 413–18

—— & —— (1980b). 'The fruits available as food to small rodents in two woodland ecosystems', *Holarctic Ecology*, 3: 10–18

Snow, B. & Snow, D. (1988). *Birds and Berries: A Study of an Ecological Interaction*, Calton, Staffs, T. and A.D. Poyser

Spjut, R. (1996). '*Niebla* and *Vermilacinia* (Ramalinaceae) from California and Baja California', *Sida Botanica Miscellany*, 14: 27

Stahr, R. (1982). 'Untersuchungen zum Vorkommen der Eibe (*Taxus baccata* L.) im Tharandter Gebiet', diploma, Tharandt, TU Dresden

Stern, Horst (1979). *Rettet den Wald*, Munich, Kindler

Stewart, W.N. (1983). *Palaeobotany and the Evolution of Plants*, Cambridge, Cambridge University Press

Strauss-Debenedetti, S. and Bazzaz, F.A. (1991). 'Plasticity and acclimation to light in tropical Moraceae of different successional positions', *Oecologia*, 87: 377-87

Strouts, R.G. (1993). 'Phytophthora root disease', *Arboriculture Research Note* 58/93/PATH, Farnham, Surrey, Arboricultural Advisory and Information Service

—— and Winter, T.G. (1994). 'Diagnosis of Ill-Health in Trees', *Research for Amenity Trees*, 2, London, HMSO/ Forestry Commission

Suszka, B. (1978). 'Generative and vegetative reproduction', in Bartkowiak *et al.* (1978), 87–102

Svenning, J.-C. and Magård, E. (1999). 'Population ecology and conservation status of the last natural population of English yew *Taxus baccata*' [in Denmark], *Biological Conservation*, 88: 173–82

Swift, M.J., Healy, I.N., Hibberd, J.K., Sykes, J.M., Bampoe, V. and Nesbitt, M.E. (1976). 'The decomposition of branchwood in the canopy and floor of a mixed deciduous woodland', *Oecologia* 26: 138–49

Szaniawski, R.K. (1978). 'An outline of yew physiology', in Bartkowiac *et al.* (1978), 55–63

Tabbush, P. (1997). 'Estimating the age of churchyard yews', *Proceedings from Veteran Trees: Habitat, Hazard or Heritage?*, Royal Agricultural Society of England and the Royal Forestry Society, March 1997

Tabbush, P. and White, J. (1996). 'Estimation of tree age in ancient yew woodland at Kingley Vale', *Quarterly Journal of Forestry*, 90: 197–206

Tansley, A.G. and Rankin, W.M. (1911). 'The plant formation of calcareous soils. B. The sub-formation of the Chalk', in Tansley, A.G. (ed.), *Types of British Vegetation*, Cambridge, Cambridge University Press, pp. 161–86.

Tenorio, M.C., Juaristi, C.M. and Ollero, H.S. (eds) (2005). *Los Bosques Ibéricos – Una Interpretación Geobotánica*, Barcelona, Editorial Planeta

Thoma, S. (1992). 'Genetische Variation an Enzymgenloci in Reliktbeständen der Eibe (*Taxus baccata* L.)', diploma, Georg-August-University, Göttingen

—— (1995). 'Genetische Unterschiede zwischen vier Reliktbeständen der Eibe (*Taxus baccata* L.)', *Forst und Holz*, 50: 19–24

Thomas, P.A. and Polwart, A. (2003). 'Biological Flora of the British Isles. *Taxus baccata* L.', *Journal of Ecology*, 91: 489–524

Tittensor, R.M. (1980). 'Ecological history of yew (*Taxus baccata* L.) in southern England', *Biological Conservation*, 17: 243–65

United States Dept. of Agriculture (1948). *Woody-Plant Seed Manual*, Miscellaneous Publication no. 654. Washington DC, USDA, Forest Service

—— (1974). *Seeds of Woody Plants in the United States*. Agricultural Handbook 450. Washington DC, United States Department of Agriculture, Forest Service

Van Ingen, G., Visser, R., Peltenburg, H., van der Ark, A.M. and Voortman, M. (1992). 'Sudden unexpected death due to *Taxus* poisoning. A report of five cases, with review of the literature', *Forensic Science International*, 56: 81–7

Vidensek, N., Lim, P., Campbell, A. and Carlson, C. (1990). 'Taxol content in bark, wood, root, leaf, twig, and seedling from several *Taxus* species', *Journal of Natural Products* 53 (6): 1609–10

Vogler, P. (1904). 'Die Eibe (*Taxus baccata* L.) in der Schweiz', *Jahrbuch der St. Gallischen Naturwissenschaftlichen Gesellschaft für das Vereinsjahr 1903*, 436–91

Voliotis, D. (1986). 'Historical and environmental significance of the yew (*T. b.* L.)', *Israel Journal of Botany*, 35: S.47–52

Vor, T. and Lüpke, B. v. (2004). 'Das Wachstum von Roteiche, Traubeneiche und Rotbuche unter verschiedenen Lichtbedingungen in den ersten beiden Jahren nach der Pflanzung', *Forstarchiv*, 75: 13–19

de Vries, B.W.L. and Kuyper, T.W. (1990). 'Holzbewohnende Pilze auf Eibe (*Taxus baccata*)', *Zeitschrift für Mykologie* 56 (1): 87–94

Walter, K.S. and Gillitt, H.J. (eds) (1997). *IUCN Red List of Threatened Plants*, Gland, Switzerland, World Conservation Union

Watt, A.S. (1926). 'Yew communities of the South Downs', *Journal of Ecology*, 14: 282–316

White, James W. (1912). *Flora of Bristol*, Bristol, Wright & Sons

White, John (1994). *Estimating the Age of Large Trees in Britain*, Information note 250, Farnham, Surrey, Forestry Commission

—— (1998). *Estimating the Age of Large and Veteran Trees in Britain*, Information Note FCIN12, November 1998. Edinburgh, Forestry Commission; also available at www.forestry.gov.uk

Willerding, W. (1968). 'Beiträge zur Geschichte der Eibe (*Taxus baccata*) – Untersuchungen über das Eibenvorkommen im Plesswalde bei Göttingen', *Plesse-Archiv*, 3: 97–155

Wilks, J.H. (1972). *Trees of the British Isles in History and Legend*, London, Muller

Wilson, E.H. (1929). *China – Mother of Gardens*, Boston, The Stratford Co.

—— (1916). *The Conifers and Taxads of Japan*, Cambridge, MA, Publications of the Arnold Arboretum, 8.

Wolf, Christian (2002). 'Anmerkungen zu den Spechteinschlägen in der Eibe', *Der Eibenfreund*, 9: 169–74

Wolff, R.L., Deluc, L.G. and Marpeau, A.M. (1996). 'Conifer seeds: oil content and fatty acid composition', *Journal of the American Oil Chemists Society*, 73: 765–71

Worbes, M., Hofmann, M., and Roloff, A. (1992). 'Wuchsdynamik der Baumschicht in einem Seggen-Kalkbuchenwald in Nordwestdeutschland (Huckstein)', *Dendrochronologia*, 10: 91–106

Yadav, Ram R. and Singh, Jayandra (2002). 'Tree-ring analysis of *Taxus baccata* from the Western Himalaya, India, and its dendroclimatic potential', *Tree-Ring Research*, 58 (1/2): 23–9

Yaltirik, Faik and Efe, Asuman (1994). 'Dendroloji ders Kitabi', *Orman Endüstri Mühendisligi Bölümü Ögrencileri icin*. Yayin University no. 3836, Yayin Faculty no. 431

CULTURE

Abella, Ignacio (2001). 'La magia de los Árboles', tr. H. Rössner, in *Der Eibenfreund*, 8: 104–21

Ackerman, Robert (1987). *J.G. Frazer: His Life and Work*, Cambridge, Cambridge University Press

—— (2002). *The Myth and Ritual School: J.G. Frazer and the Cambridge Ritualists*, New York, Routledge

Ackroyd, Peter (2006). 'The poets who built the modern world', *The Times*, 14 January 2006, 12–13

Adkins, Lesley and Adkins, Roy A. (1996). *Dictionary of Roman Religion*, Oxford, Oxford University Press

Adler, B. (1915). 'Die Bogen der schweizer Pfahlbauer', *Anzeiger für Schweizer Altertumskunde*, vol. XVII

Albright, William. F. (1943). 'The excavations of Tell Beit Mirsim', 3, *Annual of the American Schools of Oriental Research*, 21–32

Alessio, G. (1957). 'Stratificazione dei nomi del tasso (*Taxus baccata* L.) in Europa', *Studi Etruschii*, 25: 219–64

d'Alviella, Count Eugene Goblet (1894). *The Migration of Symbols*, Westminster, Archibald Constable & Co.

Anderson, Fiona (2005). 'Yews under Threat', *Tree News*, Autumn/Winter 2005

Bach, Axel *et al.* (2004). 'Lebenskünstler Baum', script for *Quarks & Co.*, Cologne, Westdeutscher Rundfunk

Bates, Brian (1996). *The Wisdom of the Wyrd – Teachings for Today from Our Ancient Past*, London, Rider

Baumann, H. (1999). *Die griechische Pflanzenwelt in Mythos, Kunst und Literatur*, Hirmer, Munich

Beauvisage, G. (1895). 'Cercueils pharaoniques en bois d'if',

Extrait des Annales de la Societé Botanique de Lyon, 20: 33–8

—— (1896). 'Recherches sur quelques bois pharaoniques. I. Le bois d'if', *Recueil de Traveaux Relatifs à la Philologie et à l'Archéologie Egyptiennes et Assyriennes*, 23: 78–90

Beckhoff, K. (1963). 'Die Eibenholz-Bogen vom Ochsenmoor am Dümmer', *Die Kunde*, 1963, 63–81

Begg, Ean (1996). *The Cult of the Black Virgin*, London, Penguin Arkana

Berendt, J.-E. (1983). *Nada Brahma – Die Welt ist Klang*, Reinbek, Rowohlt

Berger, M. and Holbein, U. (2003). 'Eibe: *Taxus* spp. – Eine psychoaktive Gattung?', *Entheogene Blätter* 10: 108–15

Bertoldi, V. (1928). 'Sprachliches und kulturhistorisches über die Eibe und den Faulbaum', *Wörter und Sachen*, 11: 145–61

Bevan-Jones, R. (2002). *The Ancient Yew*, Bollington, Windgather Press

Billington, Sandra and Green, Miranda (eds) (1996). *The Concept of the Goddess*, London, Routledge

Blacker, Carmen (1996). 'The mistress of the animals in Japan: Yamanokami', in Billington and Green (eds) (1996)

Boardman, J. (1961). *The Cretan Collection in Oxford: The Dictaean Cave and Iron Age Crete*, Oxford, Clarendon Press

Bowra, C.M. (tr.) (1969). *Pindar: The Odes*, London, Penguin Classics

Brande, A. (2001). 'Die Eibe in Berlin einst und jetzt', *Der Eibenfreund*, 8: 24–43

—— (2004). 'Hugo Conwentz 1855-1922', *Der Eibenfreund*, 11: 168–72

Briehn, Georg (2001). 'Die Eiben im Kronberger Burggelände', in *Kronberger Burgbote* 2001: 42–3, reprinted in *Der Eibenfreund,* 9: 132–3

Brosse, Jacques (1994). *Mythologie der Bäume*, Düsseldorf, Walter-Verlag

Butterworth, E.A.S. (1970). *The Tree at the Navel of the Earth*, Berlin, Walter de Gruyter

Cahill, Suzanne (1993). *Transcendence and Divine Passion – The Queen Mother of the West in Medieval China*, Stanford CA, Stanford University Press

Cameron, Dorothy O. (1981). *Symbols of Birth and Death in the Neolithic Era*, London, Kenyon-Deane

Campbell, Joseph (1959). *The Masks of God: Primitive Mythology*, New York, Penguin Compass

—— (1964). *The Masks of God: Occidental Mythology*, New York, Penguin Compass

—— (1991). 'The mystery number of the Goddess', in Campbell and Musès (eds) (1991), 55–130

—— and Musès, Charles (eds) (1991). *In All Her Names – Explorations of the Feminine in Divinity,* San Francisco, Harper Collins

Chapman, Geoff and Young, Bob (1979). *Box Hill*, Lyme Regis, Serendip

Chetan, Anand and Brueton, Diana (1994). *The Sacred Yew – Rediscovering the Ancient Tree of Life through the Work of Allen Meredith*, London, Penguin Arkana

Cimok, Fatih (2000). *Biblical Anatolia – From Genesis to the Councils*, Istanbul, A Turizm Yayinlari

Cleasby, Richard and Vigfusson, Gudbrand (1975). *An Icelandic–English Dictionary*, Oxford, Oxford University Press

Coles, Bryony (1998). 'Wood species for wooden figures: a glimpse of a pattern', in Gibson *et al.* (1998), 163–73

Collins J.J. (1983). 'Sybilline oracles (2nd cent. BC–7th cent. AD)', in Charlesworth, J.H. (ed.), *The Old Testament pseudoepigrapha I: Apocalyptic Literature and Testaments*, 317–472, London, Bantam Doubleday Dell

—— (1987). 'The development of the Sybilline tradition', in Haase, W. (ed.) *Aufstieg und Niedergang der römischen Welt*, pt. 2, vol. 20, I, 421–59, Berlin, De Gruyter

Conwentz, H. (1898). 'Die Eibe in der vorgeschichtlichen Zeit', talk 08 12 1897, *Correspondenz, Blatt der deutschen Gesellschaft für Anthropologie, Ethnologie und Urgeschichte*, 29: 13–14

—— (1921). 'Über zwei subfossile Eibenhorste bei Christiansholm, Kreis Rendsburg', *Berichte der Deutschen Botanischen Gesellschaft*, 39: 384–90

Cooper, J.C. (1978). *An Illustrated Encyclopedia of Traditional Symbols*, London, Thames & Hudson

Cook, Arthur Bernard (1914). *Zeus – A Study in Ancient Religion*, vol. 1, Cambridge, Cambridge University Press

—— (1925). *Zeus – A Study in Ancient Religion,* vol. 2, Cambridge, Cambridge University Press

—— (1940). *Zeus – A Study in Ancient Religion,* vol. 3, Cambridge, Cambridge University Press

Cook, Roger (1992). *The Tree of Life – Image of the Cosmos*, London, Thames & Hudson

Coote, H.C. (1878). *The Romans in Britain*, London, F. Norgate

Cornish, Vaughn (1946). *The Churchyard Yew and Immortality*, London, Frederick Muller Ltd

Croft, L.R. (1989). *The Life and Death of Charles Darwin*, Elmwood, Chorley

Cunliffe, Barry (1997). *The Ancient Celts*, Oxford, Oxford University Press

—— (2002). *The Extraordinary Voyage of Pytheas the Greek*, New York, Walker Books

Curry, Anne (2005). *Agincourt – A New History*, Stroud, Tempus

Dakyns, H.G. (tr.) (n.d.). *The Sportsman by Xenophon*

Dallimore, W. (1908). *Holly, Yew and Box*, London, John Lane

Dames, Michael (1996). *Mythic Ireland*, London, Thames & Hudson

Davidson, Hilda E. (1993). *The Lost Beliefs of Northern Europe*, London/New York, Routledge

—— (1998). *Roles of the Northern Goddess*, London/New York, Routledge

Davies, Jonathan C. (1911). *Folk-Lore of West and Mid-Wales*, Aberystwyth, *Welsh Gazette* Offices

Deissmann, Marieluise (ed., tr.) (1980). Caesar, *De bello Gallico/Der Gallische Krieg*, Stuttgart, Reclam

Delahunty, J.L. (2002). 'Religion, war, and changing landscapes: An historical and ecological account of the yew tree (*Taxus baccata* L.) in Ireland', dissertation, University of Florida

Demandt, A. (2002). *Über allen Wipfeln – Der Baum in der Kulturgeschichte*, Cologne/Weimar/Vienna, Böhlau

Derderian, Jacques (1992). *Le Puy: Haut lieu ésotérique. Capitale des enfers ou Jérusalem céleste?* Paris, Éditions Dervy

Desmond, A. and Moore, J. (1991). *Darwin*, London, Michael Joseph

Didron, A.N. (1907). *Christian Iconography, or The History of Christian Art in the Middle Ages*, vol. 2, London, George Bell & Sons

Diederichs, Ulf (ed.) (1984). *Germanische Götterlehre: Nach den Quellen der Lieder und der Prosa-Edda* (tr. F. Genzmer & G. Neckel), Cologne, Diederichs Gelbe Reihe

Dieterle, Martina (1999). 'Dodona – Religionsgeschichtliche und historische Untersuchungen zu Entstehung und Entwicklung des Zeus-Heiligtums', doctorate, University of Hamburg

Doczi, György (1981). *The Power of Limits*, Boulder, CO, Shambala

Dodds, E.R. (1951). *The Greeks and the Irrational*, Berkeley and Los Angeles, University of California Press

Doht, R. (1974). *Der Rauschtrank im germanischen Mythos*, Wiener Arbeiten z. germanischen Altertumskunde und Philologie 3, Vienna, Halosar

Dronke, Ursula (1969). *The Poetic Edda*, vol. 1: *Heroic Poems*, Oxford, Oxford University Press

—— (1997). *The Poetic Edda*, vol. 2: *Mythological Poems*, Oxford, Oxford University Press

Dunbar, Janet (1970). *J.M. Barrie – The Man Behind the Image*, London, Collins

Dyggve, Ejnar (1948). *Das Laphrion – Der Tempelbezirk von Kalydon*, Kopenhagen, Ejnar Munksgaard

Earwood, Caroline (1993). *Domestic Wooden Artefacts in Britain and Ireland from Neolithic to Viking Times*, Exeter, University of Exeter Press

Edel, M. and Wallrath, B. (2005). *Die Kelten – Europas spirituelle Kindheit*, Saarbrücken, Neue Erde

Eliade, Mircea (1978). *A History of Religious Ideas – From the Stone Age to the Eleusinian Mysteries*, vol. 1, Chicago IL, University of Chicago Press

—— (1996). *Patterns in Comparative Religion*, Lincoln NE/London, University of Nebraska Press

Ellis, P.B. (1987). *A Dictionary of Irish Mythology*, London, Constable

Evans, Sir Arthur J. (1901a). *The Mycenaean Tree and Pillar Cult and its Mediterranean Relations*, London, Macmillan

—— (1901b). 'Mycenaean tree and pillar cult', *Journal of Hellenic Studies*, 21, 99–204

—— (1925). 'A signet ring from Nestor's Pylos and a royal hoard from Thisbe in Boeotia', *Journal of Hellenic Studies*, 45: 17–24

—— (1921–35). *The Palace of Minos – A Comparative Account of the Successive Stages of the Early Cretan Civilization as Illustrated by the Discoveries at Knossos*, 4 vols, London, Macmillan

Fallon, Peter (tr.) (2004). Virgil, *Georgics*, Oxford, Oxford University Press

Fitzgerald, R. (tr.) (1974/1999). Homer, *Iliad*, Oxford, Oxford University Press

Fontane, Theodor (1873). *Wanderungen durch die Mark Brandenburg*, vol. 3, Berlin, Havelland (excerpt in *Der Eibenfreund*, 8: 36–41)

Frazer, Sir James G. (1906). *Adonis, Attis, Osiris – Studies in the History of Oriental Religion*, London, Macmillan

—— (1993). *The Golden Bough – A Study in Magic and Religion* (reprint of 1922 abridged version, Ware, Herts, Wordsworth Editions)

Fuhrmann, Manfred (tr.) (1995). Tacitus, *Germania*, Stuttgart, Reclam

George, Andrew (tr.) (1999). *The Epic of Gilgamesh*, London, Penguin Classics

Gibson, Alex and Simpson, Derek (eds) (1998). *Prehistoric Ritual and Religion*, Stroud, Sutton

Gimbutas, Marija (1982). *The Goddesses and Gods of Old Europe – 6500–3500 BC – Myths and Cult Images*, London, Thames & Hudson

—— (1991). 'The "monstrous Venus" of prehistory – Divine creatrix', in Campbell and Musès (eds) (1991), pp. 25–54

—— (1995). *Die Sprache der Göttin – Das verschüttete Symbolsystem der westlichen Zivilisation*, Frankfurt, Zweitausendeins [originally published in English as *The Language of the Goddess – Unearthing the Hidden Symbols of Western Civilization*, 1989, San Francisco, Harper & Row]

Goethe, Johann W. von (1819). *West-oestlicher Divan* [reference to John Whalley's 1974 translation, *West-eastern Divan*, London, Oswald Wolff]

Goff, Beatrice L. (1963). *Symbols of Prehistoric Mesopotamia*, New Haven/London, Yale University Press

Goodman, J. and Walsh, V. (2001). *The Story of Taxol – Nature and Politics in the Pursuit of an Anti-cancer Drug*, Cambridge, Cambridge University Press

Gowen, Margaret (2004). *4000-year-old music? Unique Prehistoric Musical Instrument Discovered in Co. Wicklow*, Margaret Gowen & Co., Dublin 17 May 2004

Gradishar, W.J. *et al.* (2005). 'Phase III trial of nanoparticle albumin-bound paclitaxel compared with polyethylated castor oil-based paclitaxel in women with breast cancer', *Journal of Clinical Oncology*, 23/31: 7794–803

Graves, Robert (1955). *The Greek Myths*, vols 1 & 2, Harmondsworth, Penguin

Green, Miranda (1995). *Celtic Goddesses – Warriors, Virgins and Mothers*, London, British Museum Press

—— (1998). *Exploring the World of the Druids*, London, Thames & Hudson, 1997

Grene, D. and Lattimore, R. (eds) (1959). *The Complete Greek Tragedies: Euripides V*, Chicago, University of Chicago Press

Grimm, Jakob (1882). *Teutonic Mythology*, London, George Bell

Grimm, Jacob and Grimm, Wilhelm (2004). *Deutsches Wörterbuch – Elektronische Ausgabe der Erstbearbeitung*, Frankfurt a.M., Zweitausendeins

Gupta, Sankar Sen (ed.) (1965). *Tree Symbol Worship in India*, Calcutta, Indian Publications

Gurney, O.R. (1952). *The Hittites*, London, Penguin/Pelican

Guthrie, W.K.C. (1950/1962). *The Greeks and their Gods*, London, Methuen

Gwynn, Edward (ed.) (1903–35). *The Metrical Dindsenchas*, vols 1–5, Dublin, Hodges, Figgis

Haas, Volkert (1977). *Magie und Mythen im Reich der Hethiter*, vol. 1, Hamburg, Merlin

Hageneder, Fred (2000). *The Spirit of Trees – Science, Synthesis and Inspiration*, Edinburgh, Floris; New York, Continuum (2001)

—— (2001). *The Heritage of Trees – History, Culture and Symbolism*, Floris, Edinburgh

—— (2003). 'Sacred trees in Siberian shamanism (Buryat tradition)', *Friends of the Trees Research Paper* 001, April 2003. Unpublished essay (available online at www.friendsofthetrees.org.uk)

—— (2004). *Geist der Bäume – Eine ganzheitliche Sicht*

ihres unerkannten Wesens, 3rd edn, Saarbrücken, Neue Erde
—— (2005). *The Living Wisdom of Trees – Natural History, Folklore, Symbolism, Healing*, London, Duncan Baird. [Published in the US as *The Meanings of Trees*, San Francisco, Spectacle]
—— and Singh, Satya (2007). *Tree Yoga*, Baden-Baden, Earthdancer
Hammond, N.G.L. (1967). *Epirus – The Geography, the Ancient Remains, the History and the Topography of Epirus and Adjacent Areas*, Oxford, Oxford University Press
Hansard, G.A. (1841). *The Book of Archery*, Dallington, The Naval and Military Press
Hard, Robin (tr.) (1997). Apollodorus, *The Library of Greek Mythology*, Oxford, Oxford University Press
Hardy, Robert (1992). *Longbow – A Social and Military History*, Sparkford, Patrick Stephens
—— (2003). 'Longbow', *Living History*, August 2004, 14–19
Harris, J. Rendel (n.d.). *The Origin of the Cult of Aphrodite*, repr. 1999 by Holmes Publishing, Edmonds, WA
—— (1916). *The Ascent of Olympus*, Manchester, Manchester University Press
—— (1919).*Origin and Meaning of Apple Cults*, Manchester, Manchester University Press
Hartzell, Hal Jr. (1991). *The Yew Tree – A Thousand Whispers*, Eugene, OR, Hulogosi
Helbig, W. (1887). *Das homerische Epos aus den Denkmälern erläutert*, Leipzig, Teubner
Hellmund, Monika (2005). 'Geböttcherte Eibenholzeimer aus der römischen Kaiserzeit – Funde von Gommern, Ldkr. Jerichower Land, Sachsen-Anhalt', *Der Eibenfreund*, 12, 157–64
Henslow, George (1906). *Plants of the Bible*, London, Samuel Bagster
Herzhoff, B. (1990). 'FHGOS – Zur Identifikation eines umstrittenen Baumnamens', *Hermes*, 118: 257–72, 385–404
Hilf, R.B. (1926). 'Die Eibenholzmonopole des 16. Jahrhunderts', *Vierteljahrsschrift für Sozial- und Wirtschaftgeschichte* 18: 183–91
Hoenn, K. (1946). *Artemis – Gestaltwandel einer Göttin*, Zürich, Artemis
Hoernle, August F.R. (ed.) (1893–1912). *The Bower Manuscript; Facsimile Leaves, Nagari Transcript, Romanised Transliteration and English Translation with Notes*, Archaeological Survey of India. [Reports]: New Imperial Series, 22
Hoffner, H.A. Jr (1998). *Hittite Myths*, Society of Biblical Literature, Writings from the Ancient World Series, Atlanta GA, Scholars Press
Holmes, Richard (1982). *Coleridge*, Oxford, Oxford University Press
Holt, J.C. (1982). *Robin Hood*, London, Thames & Hudson
Hooker, Sir Joseph (1854). *Himalayan Journals; or Notes of a Naturalist*, 8 vols, London, John Murray
Hort, Arthur (tr.) (1916). Theophrastus, *Enquiry into Plants*, Loeb Classical Library, Cambridge, MA/London, Harvard University Press
Hrozny, B. (1917). *Die Sprache der Hethiter*, Leipzig, J.C. Hinrichs
—— (1924). *Das hethitische Ritual des Papanikri von Kowana*
Hunter, R. (tr.) (1998). Apollonius of Rhodes, *Jason and the Golden Fleece (The Argonautica)*, Oxford, Oxford University Press
Hutchison, Robert (1890). 'On the old and remarkable yew-trees of Scotland', *Proceedings of the Antiquaries of Scotland*
Hutton, Ronald (1991). *The Pagan Religions of the Ancient British Isles – Their Nature and Legacy*, Oxford, Blackwell
—— (2001). *Shamans – Siberian Spirituality and the Western Imagination*, London/New York, Hambledon & London
Jablonski, Eike (2001). 'Die Bedeutung der Eibe im Gartenbau', *Der Eibenfreund*, 8: 60–9
Johnson, W. (1908). *Byways in British Archaeology*, Cambridge, Cambridge University Press
Jones, W.H.S. (tr.) (1960). Pliny, *Natural History*, London, Heinemann
Julien, Eric (2005). *Le chemin des neuf mondes – Les indiens kogis de Colombie peuvent nous enseigner les mystères de la vie*, Paris, Editions Albin Michel
Karadeniz Eregli '99, 11-12-13 Haziran, local town magazine published by the official Festival Committee
Kayacik, H. and Aytug, B. (1968). *Gordion Kral Mezari'nin Agac Malzemesi Üzerinde Ormancilik Yönünden Arastirmalar* ['A study of the wooden materials of the Gordion Royal Tomb with special reference to forestry'], University of Istanbul (Orman Fakültesi Dergisi)
Keel, Othmar and Uehlinger, Christoph (1998). *Gods, Goddesses, and Images of God in Ancient Israel*, Edinburgh, T. & T. Clark
Koch, K. (1879). *Die Bäume und Sträucher des alten Griechenlands*, Stuttgart, Enke
Kramer, Samuel Noah (1956). *From the Tablets of Sumer*, Indian Hills, CO, Falcon's Wing Press
—— (1961). *Sumerian Mythology – A Study of Spiritual and Literary Achievement in the Third Millennium* B.C. (rev. edn 1972), Philadelphia, PA, University of Pennsylvania Press
—— (1963). *The Sumerians – Their History, Culture, and Character*, Chicago/London, University of Chicago Press
—— and Wolkstein, D. (1983). *Inanna – Queen of Heaven and Earth, Her Stories and Hymns from Sumer*, New York, Harper & Row
Küchli, Christian (1987). *Auf den Eichen wachsen die besten Schinken – Zehn intime Baumporträts*, Zürich, Im Waldgut
Lethaby, W.R. (1917). 'The earlier temple of Artemis at Ephesus', *Journal of Hellenic Studies*, 37: 1–16
Levi, Peter (tr.) (1971). Pausanias, *Guide to Greece*, vols 1 & 2 (rev. edn 1979), London, Penguin Classics
—— (1998). *Virgil – His Life and Times*, London, Duckworth
Lewington, Anna (2003). *Plants for People*, London, Eden Project Books, Transworld Publishing
Lewis, C.D. (tr.) (1983). Virgil, *Eclogues*, Oxford, Oxford University Press
Lewis, C.D. (tr.) (1986). Virgil, *The Aeneid*, Oxford, Oxford University Press
Littleton, C. Scott (2002). *Understanding Shinto – Origins, Beliefs, Practices, Festivals, Spirits, Sacred Places*, London, Duncan Baird
Llewellyn, Roddy (1997). 'Genius with a wild streak – William Robinson "invented" modern gardening', *Mail on Sunday*, 30 March
Loudon, J.C. (1841–4). *Arboretum et Fruticetum Brittanicum*, 8 vols, London, Longman *et al.*

Mackenzie, D.A. (1917). *Myths of Crete and Pre-Hellenic Europe*, London, Gresham

—— (1922). *Ancient Man in Britain*, London, Blackie & Son; repr. 1996, London, Senate

Mannhardt, Wilhelm (1858). *Germanische Mythen*, Berlin, F. Schneider

March, Jenny (1998). *Dictionary of Classical Mythology*, London, Cassell

Markale, Jean (1989). *Die Druiden – Gesellschaft und Götter der Kelten*, Munich, Goldmann

Matthews, Caitlín & John (2003). *Encyclopaedia of Celtic Wisdom – A Celtic Shaman's Sourcebook*, London, Rider

Matthews, John (1996). *The Druid Source Book*, London, Blandford

Mawer (1920). *Place-names of Northumberland and Durham*, Cambridge, Cambridge University Press

Meiggs, Russell (1982). *Trees and Timber in the Ancient Mediterranean World*, Oxford, Oxford University Press

Melville, A.D. (tr.) (1986). Ovid, *Metamorphoses*, Oxford, Oxford University Press

Menzel, Wolfgang (1870). *Die vorchristliche Unsterblichkeit*, Leipzig

Meredith, Allen (n.d.). 'The Secret Seed', unpublished manuscript

Meyer, R.M. (1910). *Altgermanische Religionsgeschichte*, Leipzig, Quelle & Meyer

Mills, A.D. (1998). *Oxford Dictionary of English Place-names*, Oxford, Oxford University Press

Milner, J.E. (1992). *The Tree Book – The Indispensible Guide to Tree Facts, Crafts and Lore*, London, Collins & Brown

Moerman, Daniel E. (1998). *Native American Ethnobotany*, Portland OR, Timber Press

Momigliano, A. (1988). 'From the pagan to the Christian sybil: prophecy as history of religion', in Dionisotti, A.C., Grafton, A. & Kraye, J. (eds), *The Uses of Greek and Latin: Historical Essays*, 3–18. Warburg Institute Surveys and Texts, 16, London, Warburg Institute

Morris, Richard (1989). *Churches in the Landscape*, London, Dent

Mozley, J.H. (tr.) (1967). *Statius I: Thebaid I–IV*, Loeb Classical Library, Cambridge, MA/London, Harvard University Press

Mozley, J.H. (tr.) (1969). *Statius II: Thebaid V–XII*, Loeb Classical Library, Cambridge, MA/London, Harvard University Press

Mozley, J.H. (tr.) (1998). *Valerius Flaccus – Argonautica*, Loeb Classical Library, Cambridge, MA/London, Harvard University Press

Müller-Beck, H. (1991). 'Die Holzartefakte', in Waterbolk and van Zeist (eds) (1991), 13–233

Munch, P.A. (1926). *Norse Mythology* (tr. S.B. Hustvedt), New York, American-Scandinavian Foundation

Museo Archeologico Nazionale dell'Umbria (2005). 'Perugian urns from the late third century to the first century BC', information leaflet

Musès, Charles (1991). 'The ageless way of Goddess – divine pregnancy and higher birth in ancient Egypt and China', in Campbell and Musès (eds) (1991), 131–64

Mutschlechner, G. and Kostenzer, O. (1973). 'Zur Natur- und Kulturgeschichte der Eibe in Nordtirol', *Veröffentlichungen des Tiroler Landesmuseums Ferdinandeum* 53: 247–87

Mylonas, George E. (1961). *Eleusis and the Eleusinian Mysteries*, Princetown, New Jersey, Princetown University Press

Naumann, Nelly (1996). *Die Mythen des alten Japan*, Munich, Beck

Nicholson, Helen (2001). *The Knights Templar – A New History*, Stroud, Sutton

Nilsson, Martin P. (1950). *The Minoan-Mycenaean Religion and its Survival in Greek Religion*, 2nd rev. edn, Lund, Biblio & Tannen

Oberlies, Thomas (1998). *Die Religion des Rigveda*, Vienna, Nobili Research Library

O'Dwyer, Simon (2004). *Prehistoric Music of Ireland*, Stroud, Tempus

Osthoff, H. (2001). 'Medizin aus der Eibe', *Der Eibenfreund*, 8: 70–5

Oxenstierna, Eric Graf (1958). *Die Nordgermanen*, repr. (n. d.) Essen, Phaidon

Palmer, Martin, Ramsay, Jay and Kwok, Man-Ho (1995). *Kuan Yin – Myths and Prophecies of the Chinese Goddess of Compassion*, London/San Francisco, Thorsons

Pande, Trilochan (1965). 'Tree-worship in ancient India', in Gupta (1965), 35–40

Patai, Raphael (1990). *The Hebrew Goddess*, Detroit MI, Wayne State University Press

Persson, A.W. (1950). *The Religion of Greece in Prehistoric Times*, Sather Classical Lectures, vol. 17, Berkeley/Los Angeles, University of California Press

Petersmann, H. (1986). 'Der homerische Demeterhymnus, Dodona und südslawisches Brauchtum', *Wiener Studien*, 99, 1986, 69–85

Puhvel, Jaan (1984). *Hittite Etymological Dictionary*, vols 1–2, Berlin/New York/Amsterdam, Mouton

Rashid, Subhi A. (1984). *Musikgeschichte in Bildern, vol. 2: Musik des Altertums/Lieferung 2: Mesopotamien*, Leipzig, VEB

Ransome, Hilda M. (1937). *The Sacred Bee in Ancient Times and Folklore*, London, George Allen & Unwin

Reaney, P.H. (1964). *The Origin of English Place-names*, London, Routledge & Kegan Paul

Reinerth, H. (1926). *Die jüngere Steinzeit der Schweiz*, Augsburg, Filser

Robinson, J.M. (ed.) (1990). *The Nag Hammadi Library in English*, San Francisco, Harper

Rodger, D., Stokes, J. and Ogilvie, J. (2003). *Heritage Trees of Scotland*, London, The Tree Council

Rogers, B.B. (tr.) (1979). *Aristophanes II: The Peace, The Birds, The Frogs*, Loeb Classical Library, Cambridge, MA/ London, Harvard University Press

Rolleston, T.W. (1993). *The Illustrated Guide to Celtic Mythology*, London, Studio Editions

Rössner, H. (2004). 'Was wir noch nicht über die Eibe wissen', *Der Eibenfreund*, 11: 60–70

Roux, Françoise le, and Guyonvarch, Christian-J. (1996). *Die Druiden*, Engerda, Arun

Santarelli, G. (1997). *Loreto in Art and History*, Ancona, Edizioni Aniballi

Sayce, A.H. (1893). *Assyria: Its Princes, Priests and People*, London, The Religious Tract Society

Scheeder, Thomas (2000). 'Zur anthropogenen Nutzung der Eibe (*Taxus baccata* L.)', *Der Eibenfreund*, 7: 67–81

—— and Brande A. (1997). 'Die Bedeutung der Eiben-forschung von Hugo Conwentz für die Geschichte des Naturschutzes', *Arch. Für Nat.- Lands* 36: 295–304

Schefold, K. and Jung, F. (1989). *Die Sagen von den Argonauten, von Theben und Troja in der klassischen und hellenistischen Kunst*, Munich, Hirmer

Schliemann, Heinrich (1880). *Ilios – The City and Country of the Trojans*, John Murray, London

Schrott, Raoul (2004). *Gilgamesh*, Frankfurt am Main, Fischer

Shéaghdha, Nessa ní (ed.) (1967). *Tóruigheacht Dhiarmada agus Ghráinne* ('The pursuit of Diarmuid and Grainne'), Dublin, Irish Text Society

Sheridan, Alison (n.d.). 'The Rotten Bottom Bow: The Story of Britain's Oldest Bow', unpublished paper (Dr A Sheridan, Archaeology Department, National Museums of Scotland, Chambers Street, Edinburgh EH1 1JF)

Sherr, J. and Dynamis School (2002). *Dynamic Provings – Volume II*, Dynamis Books, Malvern

Simek, R. (1993). *Dictionary of Northern Mythology*, Cambridge, D.S. Brewer

Simmonds, Norman (1979). 'Warblington Church Guide'

Sommer, Siegfried (1998). 'Die Eibe in der Lanschaftsarchitektur – früher und heute', *Der Eibenfreund*, 5: 17–22

Spindler, Konrad (2004). 'Der Eibenholzbogen des Mannes im Eis/The Yew Bow of the Man in the Ice', *Austrian Journal of Forest Science*, 121/1: 1-24

Sprengel, K. (ed.) (1971). *Theophrasts Naturgeschichte der Gewächse*, Darmstadt, WBG

Stadler, J. (1981). 'Eigenschaften und Verwendung von Eibenholz (T. b. *L.)*', diploma in Forest Science, Munich, Inst. For Wood Research

Stehli, Ulrich (2004). 'Der englische Langbogen', in *Das Bogenbauer-Buch – Europäischer Bogenbau von der Steinzeit bis heute*, Ludwigshafen, Angelika Hörnig, 132–66

Strickland, M. and Hardy, R. (2005). *The Great Warbow – From Hastings to the Mary Rose*, Stroud, Sutton

Suggs, M.J., Sakenfield, K.D. and Mueller, J.R. (eds) (1992). *The Oxford Study Bible – Revised English Bible with the Apocrypha*, New York, Oxford University Press

Swindler, M.H. (1913). *Cretan Elements in the Cult and Ritual of Apollo*, Bryn Mawr, Bryn Mawr College

Tolley, C. (1993). 'A comparative study of some Germanic and Finnic myths', D.Phil. thesis, Oxford

Tylor, E.B. (1891). *Primitive Culture*, London, John Murray

Underwood, G. (1969). *The Pattern of the Past*, Old Woking, Pitman

Vickery, J. (1973). *The Literary Impact of* The Golden Bough, Princeton, Princeton University Press

Vonessen, F. (1992). *Signaturen des Kosmos – Welterfahrung in Mythen, Märchen und Träumen. Gesammelte Aufsätze*, Witzenhausen, Die Graue Edition

de Vries, J. (1977). *La Religion des Celtes*, Paris, Payot

Waddington, C. (1997). *Land of Legend*, Wooler, Northumbria, County Store

Walchensteiner, K.R. (2006). *Die Kathedrale von Chartres – Ein Tempel der Einweihung*, Saarbrücken, Neue Erde

Walker, Barbara G. (1988). *The Women's Dictionary of Symbols and Sacred Objects*, San Francisco, HarperCollins

Warneck, I. (2000). '*Die Eibe* – Taxus baccata', *Der Eibenfreund*, 7: 55–6

Waterbolk, H.T. and van Zeist, W. (eds) (1991). *Niederwil: Eine Siedlung der Pfyner Kultur*, vol. 4: *Holzartefakte und Textilien*, Bern, Academica Helvetica

Waterfield, R. (tr.) (1993). Plato, *Republic*, Oxford, Oxford University Press

—— (tr.) (1998). Herodotus, *The Histories*, Oxford, Oxford University Press

Webster, R. (1998). *Chinese Numerology – The Way to Prosperity and Fulfillment*, St Paul, MN, Llewellyn

Weinreb, Friedrich (1999). *Zahl – Zeichen – Wort: Das symbolische Universum der Bibelsprache*, Weiler/Allgäu, Thauros

Wetzel, G. (1966). 'Ein Eibenholzbogen von Barleben, Kr. Wolmirstedt', *Ausgrabungen und Funde* 11: 9–10

Wilde, Lyn W. (1999). *On the Trail of the Women Warriors*, London, Constable

Wilks, J.H. (1972). *Trees of the British Isles in History and Legend*, London, Frederick Muller

Willetts, R.F. (1962). *Cretan Cults and Festivals*, London, Routledge & Kegan Paul

Williamson, R. (1978). *The Great Yew Forest – The Natural History of Kingley Vale*, London, Macmillan

Williamson, W.C. (1834). *Description of the Tumulus lately opened at Gristhorpe, near Scarborough*, Scarborough, C.R. Todd (1st edn)

—— (1872). *Description of the Tumulus opened at Gristhorpe, near Scarborough*, Scarborough, S.W. Theakston (3rd edn)

Wilson, E.H. (1929). *China – Mother of Gardens*, Boston, MA, Stratford Co.

Wirth, Herman (1979). *Die heilige Urschrift der Menschheit*, vols I–XII, Frauenberg, Mutter Erde Verlag

Wordsworth, W. (1803). 'Yew Trees', *Poems by William Wordsworth, Including Lyrical Ballads, and the Miscellaneous Pieces of the Author*, vol. 1, 303–4. London, 1815. (Repr. 1989, Woodstock Books, Oxford.) In *Der Eibenfreund*, 12: 189–91

Wright, E.V. and Churchill, D.M. (1965). 'The boats from north Ferriby, Yorkshire, England, with a review of the origins of the sewn boats of the Bronze Age', *Proceedings of the Prehistoric Society*, 31: 1–24

Wujastyk, Dominik (2003). *The Roots of Ayurveda – Selections from Sanskrit Medical Writings*, London, Penguin Classics

Zohary, Michael (1982). *Plants of the Bible*, Cambridge, Cambridge University Press

INDEX

Numbers in italics refer to illustrations, those in bold refer to tables and diagrams.